Corrosion Resistance of Nickel and Nickel Alloys Against Acids and Lyes

Edited by
Michael Schütze, Raul B. Rebak
and Roman Bender

Corrosion Resistance of Nickel and Nickel Alloys Against Acids and Lyes

DECHEMA

WILEY-VCH
Verlag GmbH & Co. KGaA

Editors

Prof. Dr.-Ing. Michael Schütze
DECHEMA-Forschungsinstitut
Chairman of the Executive Board
Theodor-Heuss-Allee 25
60486 Frankfurt am Main
Germany

Dr. Raul B. Rebak
General Electric Global Research Center
Schenectady, NY USA

Dr. rer. nat. Roman Bender
Chief Executive of GfKORR e. V.
Society for Corrosion Protection
Theodor-Heuss-Allee 25
60486 Frankfurt am Main
Germany

Cover Illustration
Source: DECHEMA-Forschungsinstitut,
Frankfurt (Main), Germany

Warranty Disclaimer

This book has been compiled from literature data with the greatest possible care and attention. The statements made only provide general descriptions and information.

Even for the correct selection of materials and correct processing, corrosive attack cannot be excluded in a corrosion system as it may be caused by previously unknown critical conditions and influencing factors or subsequently modified operating conditions.

No guarantee can be given for the chemical stability of the plant or equipment. Therefore, the given information and recommendations do not include any statements, from which warranty claims can be derived with respect to DECHEMA e. V. or its employees or the authors.

The DECHEMA e. V. is liable to the customer, irrespective of the legal grounds, for intentional or grossly negligent damage caused by their legal representatives or vicarious agents.

For a case of slight negligence, liability is limited to the infringement of essential contractual obligations (cardinal obligations). DECHEMA e. V. is not liable in the case of slight negligence for collateral damage or consequential damage as well as for damage that results from interruptions in the operations or delays which may arise from the deployment of this book.

This book was carefully produced. Nevertheless, editors, authors and publisher do not warrant the information contained therein to be free of errors. Readers are advised to keep in mind that statements, data, illustrations, procedural details or other items may inadvertently be inaccurate.

Library of Congress Card No.: Applied for.

British Library Cataloguing-in-Publication Data:
A catalogue record for this book is available from the British Library.

Bibliographic information published by Die Deutsche Bibliothek
Die Deutsche Bibliothek lists this publication in the Deutsche Nationalbibliografie; detailed bibliographic data is available in the Internet at <http://dnb.ddb.de>.

Printed in the Federal Republic of Germany
Printed on acid-free paper

Typesetting Kühn & Weyh, Satz und Medien, Freiburg
Printing Strauss GmbH, Mörlenbach
Binding Strauss GmbH, Mörlenbach
Cover Design Graphik-Design Schulz, Fußgönheim

ISBN: 978-3-527-33849-8

Contents

Preface

Practically all industries face the problem of corrosion – from the micro-scale of components for the electronics industries to the macro-scale of those for the chemical and construction industries. This explains why the overall costs of corrosion still amount to about 2 to 4% of the gross national product of industrialized countries despite the fact that billions of dollars have been spent on corrosion research during the last few decades.

Much of this research was necessary due to the development of new technologies, materials and products, but it is no secret that a considerable number of failures in technology nowadays could, to a significant extent, be avoided if existing knowledge were used properly. This fact is particularly true in the field of corrosion and corrosion protection. Here, a wealth of information exists, but unfortunately in most cases it is scattered over many different information sources. However, as far back as 1953, an initiative was launched in Germany to compile an information system from the existing knowledge of corrosion and to complement this information with commentaries and interpretations by corrosion experts. The information system, entitled "DECHEMA-WERKSTOFF-TABELLE" (DECHEMA Corrosion Data Sheets), grew rapidly in size and content during the following years and soon became an indispensable tool for all engineers and scientists dealing with corrosion problems. This tool is still a living system today: it is continuously revised and updated by corrosion experts and thus represents a unique source of information. Currently, it comprises more than 12,000 pages with approximately 110,000 corrosion systems (i.e., all relevant commercial materials and media), based on the evaluation of over 100,000 scientific and technical articles which are referenced in the database.

Last century, an increasing demand for an English version of the DECHEMA-WERKSTOFF-TABELLE arose in the 80's; accordingly the first volume of the DECHEMA Corrosion Handbook was published in 1987. This was a slightly condensed version of the German edition and comprised 12 volumes. Before long, this handbook had spread all over the world and become a standard tool in countless laboratories outside Germany. The second edition of the DECHEMA Corrosion Handbook was published in 2004. Together the two editions covered 24 volumes.

The present handbook compiles new and updated information on the corrosion behavior of nickel and nickel alloys in contact with the following aggressive media: acetic acid, alkanecarboxylic acids, carbonic acid, fluorine and hydrofluoric acid, formic acid, hydrochloric acid, mixed acids, nitric acid, phosphoric acid, sulfonic acids, sulfuric acid, alkaline earth hydroxides, lithium hydroxide, potassium hydroxide and sodium hydroxide.

Nickel and nickel-base alloys are vitally important to modern industry due to their ability to withstand a wide variety of severe operating conditions involving corrosive environments, high temperatures, high stresses, and combinations of these factors.

Nickel and its alloys range in composition from commercially pure nickel to complex alloys containing many alloying elements, and like the stainless steels, offer a wide range of corrosion resistance. However, nickel can accommodate larger amounts of alloying elements – mainly chromium, molybdenum, and tungsten – in solid solution than iron. Therefore, nickel-base alloys in general can be used in more severe environments than the stainless steels.

Nickel-base alloys are used for corrosion resistance or for combined corrosion resistance and high temperature strength in a wide range of commercial applications. These various applications may demand resistance to aqueous corrosion mechanisms, such as general corrosion, localized attack, and stress corrosion cracking.

Understanding how to improve the corrosion resistance of nickel and nickel alloys used as reaction, transport and storage vessels against these omnipresent chemicals is crucial for all industries involved. This book is therefore an indispensable tool for all mechanical, civil and chemical engineers, material scientists and chemists working with nickel and nickel alloys in alkaline or acidic media.

This handbook highlights the limitations of nickel and nickel alloys in acids and lyes and provides vital information on corrosion protection measures. The chapters are arranged by the media leading to individual corrosion reactions, and a vast number of alloys are presented in terms of their behavior in these media. The key information consists of quantitative data on corrosion rates coupled with commentaries on the background and mechanisms of corrosion behind these data, together with the dependencies on secondary parameters, such as flow-rate, pH, temperature, etc. This information is complemented by more detailed annotations where necessary, and by an immense number of references listed at the end of each chapter.

An important feature of this handbook is that the data was compiled for industrial use. Therefore, particularly for those working in industrial laboratories or for industrial clients, the book will be an invaluable source of rapid information for day-today problem solving. The handbook will have fulfilled its task if it helps to avoid the failures and problems caused by corrosion simply by providing a comprehensive source of information summarizing the present state-of-the-art. Last but not least, in cases where this knowledge is applied, there is a good chance of decreasing the costs of corrosion significantly.

Finally the editors would like to express their appreciation to Dr. Rick Durham and Dr. Horst Massong for their admirable commitment and meticulous editing of a work that is encyclopedic in scope.

Michael Schütze, Raul B. Rebak and Roman Bender

How to use the Handbook

The Handbook provides information on the chemical resistance and the corrosion behavior of nickel and nickel alloys in acids and lyes.

The user is given information on the range of applications and corrosion protection measures.

Research results and operating experience reported by experts allow recommendations to be made for the selection of materials and to provide assistance in the assessment of damage.

The objective is to offer a comprehensive and concise description of the behavior of nickel in contact with a particular medium.

The information on resistance is given as text, tables, and figures. The literature used by the authors is cited at the corresponding point. There is an index of materials as well as a subject index at the end of the book so that the user can quickly find the information given for a particular keyword.

The Handbook is thus a guide that leads the reader to nickel and nickel alloys that have already been used in certain cases, that can be used or that are not suitable owing to their lack of resistance.

The resistance is labeled with three evaluation symbols in view of concise presentation. Uniform corrosion is evaluated according to the following criteria:

Symbol	*Meaning*	*Area-related mass loss rate x*		*Corrosion rate y*
		g/m^2 h	g/m^2 d	mm/a
+	resistant	≤ 0.1	≤ 2.4	≤ 0.1
⊕	moderately resistant	> 0.1 to ≤ 1.0	> 2.4 to ≤ 24.0	> 0.1 to ≤ 1.0
–	not resistant	> 1.0	> 24.0	> 1.0

The evaluation of the corrosion resistance of metallic materials is given

- for uniform corrosion or local penetration rate, in: mm/a and mpy
- or if the density of the material is not known, in: g/m^2 h or g/m^2 d.

Pitting corrosion, crevice corrosion, and stress corrosion cracking or non-uniform attack are particularly highlighted.

The following equations are used to convert mass loss rates, x, into the corrosion rate, y:

from x_1 in g/m^2 h

$$\frac{x_1 \cdot 365 \cdot 24}{\rho \cdot 1{,}000} = y\ (mm/a)$$

from x_2 in g/m^2 d

$$\frac{x_2 \cdot 365}{\rho \cdot 1{,}000} = y\ (mm/a)$$

where

x_1: value in g/m^2 h
y: value in mm/a
x_2: value in g/m^2 d
d: days
ρ: density of material in g/cm^3
h: hours

In those media in which uniform corrosion can be expected, if possible, isocorrosion curves (corrosion rate y = 0.1 mm/a) are given.

Unless stated otherwise, the data was measured at atmospheric pressure and room temperature.

The resistance data should not be accepted by the user without question, and the materials for a particular purpose should not be regarded as the only ones that are suitable. To avoid wrong conclusions being drawn, it must be always taken into account that the expected material behavior depends on a variety of factors that are often difficult to recognize individually and which may not have been taken deliberately into account in the investigations upon which the data is based. Under certain circumstances, even slight deviations in the chemical composition of the medium, in the pressure, in the temperature or, for example, in the flow rate are sufficient to have a significant effect on the behavior of the materials. Furthermore, impurities in the medium or mixed media can result in a considerable increase in corrosion.

The composition or the pretreatment of the material itself can also be of decisive importance for its behavior. In this respect, welding should be mentioned. The suitability of the component's design with respect to corrosion is a further point which must be taken into account. In case of doubt, the corrosion resistance should be investigated under operating conditions to decide on the suitability of the selected materials.

Warranty disclaimer

This book has been compiled from literature data with the greatest possible care and attention. The statements made in this book only provide general descriptions and information.

Even for the correct selection of materials and correct processing, corrosive attack cannot be excluded in a corrosion system as it may be caused by previously unknown critical conditions and influencing factors or subsequently modified operating conditions.

No guarantee can be given for the chemical stability of the plant or equipment. Therefore, the given information and recommendations do not include any statements, from which warranty claims can be derived with respect to DECHEMA e.V. or its employees or the authors.

The DECHEMA e.V. is liable to the customer, irrespective of the legal grounds, for intentional or grossly negligent damage caused by their legal representatives or vicarious agents.

For a case of slight negligence, liability is limited to the infringement of essential contractual obligations (cardinal obligations). DECHEMA e.V. is not liable in the case of slight negligence for collateral damage or consequential damage as well as for damage that results from interruptions in the operations or delays which may arise from the deployment of this book.

Acetic Acid

Author: *G. Elsner* / Editor: *R. Bender, R. Durham*

Nickel

– Acetic acid –

Boilers made from nickel or NiCu-alloys have proven to be suitable for the hydrogenation (even under pressure) of substances dissolved in glacial acetic acid [1, 2].

The electrochemical behavior of nickel in acetic acid solutions was investigated in [3].

However, the corrosion resistance of nickel in acetic acid depends greatly on the presence of air, oxygen and other oxidizing agents; see Table 1.

Metal	Acetic acid %	Temperature °C	Corrosion rate mm/a	Remarks	Literature
Nickel	50	20	< 0.1		[4]
	50	100	< 3.0		
	conc.*	20	< 0.1		
	conc.	100	< 10.0		
	1–100	20	0.04–0.202	air free	[2]
	1–100	boiling	0.202–0.607	slightly aerated	
	1–100	20	0.202–0.607	moderately aerated	
	1–100	20	1.01–1.62	vigorously aerated, stirred	
	6	30	0.001–0.01	not aerated	[5]
	10	boiling	1.01–3.03		
	50	boiling	0.101–1.01		
	6	30	0.101–1.01	aerated	[5]
	10	30	1.01–3.03	aerated	
	95	115	> 10.1	aerated, technical grade acetic acid, 95%	
	5	boiling	0.267	liquid phase	[1, 2, 4]
	5	boiling	0.186	vapor phase	

* concentrated

Table 1: Corrosion behavior of nickel in acetic acid of various concentrations and temperatures

Table 1: Continued

Metal	Acetic acid %	Temperature °C	Corrosion rate mm/a	Remarks	Literature
Nickel	50	boiling	0.485	liquid phase	
	50	boiling	0.364	vapor phase	
	98	boiling	0.299	liquid phase	
	98	boiling	0.097	vapor phase	
	99.9	boiling	0.344	liquid phase	
	99.9	boiling	0.057	vapor phase	
	6	20	0.05	6% acetic acid + hydrogen	[4]
	50	50	0.25	50% acetic acid + hydrogen	
	100	20	0.10	100% acetic acid + hydrogen	
	99.5	112.2	0.1325	99.5% acetic acid + 0.05% salicylic acid + 0.45% water, vapor/ 1,022 h, operational test	[1, 6]
Nickel, 99%	10–100	24	> 1.25	aerated	[7]
	10–20	24	< 0.05	air-free	[7, 8]
	30	24	0.5–1.25		
	40–100	24	< 0.5		
	50	52	0.5–1.25		
	90–100	100	< 0.5		
	100	100	< 0.5	acetic acid vapor, air-free	[7]
	100	100	0.5–1.25	acetic acid + acetic anhydride + peracetic acid	[7, 8]
	100	24	> 1.25	acetic acid + hydrobromic acid	[7, 8]

* concentrated

Table 1: Corrosion behavior of nickel in acetic acid of various concentrations and temperatures

Table 1: Continued

Metal	*Acetic acid %*	*Temperature °C*	*Corrosion rate mm/a*	*Remarks*	*Literature*
Nickel, 99%	100	24	> 1.25	acetic acid + hydrochloric acid	[7, 8]
	100	100	< 0.5	acetic acid + salicylic acid vapors	[7, 8]
	100	100	0.5–1.25	acetic acid + cupric acetate + chlorides	[8]

* concentrated

Table 1: Corrosion behavior of nickel in acetic acid of various concentrations and temperatures

Anhydrous acetic acid dissolved in organic solvents, for example benzene, is virtually non-aggressive towards nickel [1, 9].

In contrast to the stainless steels, nickel and Monel® are more resistant to acetic acid/formic acid mixtures than to glacial acetic acid, in particular in the presence of more than 2% formic acid at temperatures of more than 120°C [10].

NiP-coatings deposited at zero current containing 14% phosphorus and having a layer thickness of 0.025 to 0.035 mm are attacked by glacial acetic acid even at room temperature. Heat treatment (1 h at 400°C) slightly improves the corrosion resistance [11].

Ni-coatings deposited at zero current are severely attacked, amongst others, by 5% acetic acid [12].

Ni-coatings (about 0.076 mm thick) produced at zero current by the Kanigen process have corrosion rates listed in Table 2 after a heat treatment at 760°C.

Acetic acid concentration, %	*Corrosion rate, mm/a*
5	0.0033 (0.13)*
10	0.0017
50	0.0020
Glacial acetic acid	0.00025

* value for the non-heat treated material

Table 2: Corrosion rates of Ni-coatings (heat-treated) in acetic acid of various concentrations [13]

The self-fluxing Ni-hard alloy PG-Sr3 deposited on the steel 12Ch18N10T was investigated for its corrosion and wear resistance in acetic acid. The material consumption rate of the coating was determined to be 0.003 to 0.005 g/m^2 h. A coating formed in 10 s at 917°C at 52 MPa under friction in acetic acid showed the best wear resistance [14].

According to a patent, Ni-coatings on stainless hardened nitrated or galvanized steels can be removed by anodization in solutions with ammonium nitrate, ammonium chloride and acetic acid without corrosion of the steels [15]. Cast nickel ((%) Ni-1.6Si-0.5C) and Duranickel® 301 ((%) Ni-0.05Cu-0.15Fe-0.55Si-0.25Mn-0.15C-4.5Al-0.5Ti) [1, 16] show the same corrosion behavior in acetic acid as pure nickel (Nickel® 200 (%) Ni-0.05Cu-0.15Fe-0.05Si-0.25Mn-0.06C and Nickel® 201 (%) Ni-0.05Cu-0.15Fe-0.05Si-0.20Mn-0.01C) [16, 17].

The acetic salt spray, the Corrodcote or the Kesternich test are used as accelerated test methods for the evaluation of the usability of NiP-coatings for the oil and gas industry [5].

According to numerous tables, the addition of mercury sulfate is said to decrease the corrosion of nickel; however, these data should be examined [2].

– Acetic acid vapor –

Above 300°C, acetic acid vapors are more or less extensively catalytically decomposed at the surface of nickel and its alloys [1, 2].

– Acetic anhydride –

If air is excluded, nickel and NiCu-alloys are resistant to acetic anhydride containing up to 5% acetic acid up to the boiling point [18].

As for the corrosion behavior of nickel in acetic anhydride see Table 3.

– Acetic anhydride vapor –

At temperatures above 400°C, acetic anhydride vapors are decomposed catalytically on nickel and its alloys [18].

– Vinegar –

Food products do not or only slightly attack nickel. Traces of nickel which are absorbed by certain food products are not harmful to health [19].

Nickel-chromium alloys

– Acetic acid –

For the corrosion behavior of NiCr-alloys [20] in acetic acid, see Table 4.

– Acetic anhydride –

For the corrosion behavior of NiCr-alloys in acetic anhydride, see Table 5.

Nickel-chromium-iron alloys

– Acetic acid –

High-nickel alloys such as Incoloy® 825 are resistant to organic acids over very long periods of time and have proven to be suitable for plants in which acetic acid and acetic anhydride are prepared, in which stainless steels have a tendency for pitting corrosion, stress corrosion cracking and severe uniform corrosion [17]. The alloy Incoloy® 825 ((%) Ni-21.5Cr-30Fe-3.0Mo-2.2Cu-1Ti-0.04C) is used instead of conventional stainless steel for heat exchangers, in boilers and condensers, for false bottoms and the like. After continuous operation for 3 years, no pitting corrosion had occurred, so that the inspection period could be extended from 3 to 18 months [17, 21, 22]. Operational tests in the reactor part of the same plant have proven the superiority of this material even compared with titanium, which was confirmed by tests, using sieve bottoms in boiling 83% acetic acid. The same result was obtained in tests using sieve bottoms in mixtures of 80% acetic acid and 20% acetic anhydride, in which stainless steels barely reached a longer service life than 1 year [17].

Acetic anhydride, %	Temperature °C	Corrosion rate mm/a	Remarks	Literature
100	up to 79	< 0.5		[7]
100	at 100°C	0.5–1.25		
98	boiling	0.11–0.14		[4]
50	125	0.5–1.25	acetic anhydride in acetic acid, air-free	[8]
90	125	< 0.05		
100	125	< 0.05		
70	boiling	0.58	acetic anhydride + glacial acetic acid, liquid phase	[23]
98	boiling	0.11		
70	boiling	0.86	acetic anhydride + glacial acetic acid, gas phase	[23]
98	boiling	0.14		

Table 3: Corrosion behavior of nickel in acetic anhydride of various concentrations and temperatures

Metal	Acetic acid %	Temperature °C	Corrosion rate mm/a	Remarks	Literature
NiCr-alloy (no further details given)	10	20	0.02		[4]
	30	20	0.26		
	glacial acetic acid	20	1.27		
	10	20	< 0.1		[4]
	10	boiling	> 1.0		
	100	20	< 0.1		
	100	boiling	> 1.0		
	10–100	20	< 0.1		
	10–100	boiling	< 1.0		
Inconel® Ni15Cr10Fe1Mn	10	boiling	0.19	reflux	[4]
	80–100	boiling	2.89	reflux	
	80–100	boiling	0.19	vapor	
	0.5	20	0.05	not aerated	
	0.5	70	0.10		
	2	20	0.05		
	2	70	0.11		
	10	30	0.02	air-saturated	
	80	20	0.004	technical grade acetic acid, 80%	
Inconel® 600*	10	100	< 0.5	aerated	[7, 8]
	50	100	> 1.25		
	80	24	0.5–1.25		
	80	52	> 1.25		
	100	100	< 0.5		
	100	204	> 1.25		

* Inconel® 600 (%) 76 Ni, 16 Cr, 7 Fe
Inconel® 601 (%) 58–63 Ni, 21–25 Cr, 0.10 C, 1.0–1.7 Al, 1.0 Mn, 0.015 S, 0.5 Si, 1.0 Cu, balance Fe
Incoloy® 825 (%) 38–46 Ni, 19.5–23.5 Cr, max. 0.05 C, max. 0.5 Si, max. 1.0 Mn, max. 0.20 Al, max. 0.03 S, 0.6–1.2 Ti, 1.5–3.0 Cu, 2.5–3.5 Mo, balance Fe

Table 4: Corrosion behavior of NiCr-alloys and NiCrFe-alloys in acetic acid of various concentrations and temperatures

Table 4: Continued

Metal	Acetic acid %	Temperature °C	Corrosion rate mm/a	Remarks	Literature
Inconel® 600*	10	24	< 0.5	air-free	[7, 8]
	10	52	> 1.25		
	50	52	0.5–1.25		
	50	100	> 1.25		
	80	100	> 1.25		
	90	24	< 0.05		
	90	100	> 1.25		
	100	52	< 0.5		
	100	100	> 1.25		
	100	100	< 0.5	acetic acid vapor	[7, 8]
	30	52	< 0.05	acetic acid + acetaldehyde	[7, 8]
	100	100	0.5–1.25	acetic acid + ethyl alcohol	[7, 8]
	100	24	> 1.25	acetic acid + hydrobromic acid	[7, 8]
	100	24	> 1.25	acetic acid + hydrochloric acid	[7, 8]
	100	100	> 0.5	acetic acid + salicylic acid vapors	[7, 8]
Inconel®	99.5	112.2	0.1725	99.5% acetic acid + 0.05% salicylic acid + 0.45% water, vapor/1,022 h, operational test	[1, 6]

* Inconel® 600 (%) 76 Ni, 16 Cr, 7 Fe
Inconel® 601 (%) 58–63 Ni, 21–25 Cr, 0.10 C, 1.0–1.7 Al, 1.0 Mn, 0.015 S, 0.5 Si, 1.0 Cu, balance Fe
Incoloy® 825 (%) 38–46 Ni, 19.5–23.5 Cr, max. 0.05 C, max. 0.5 Si, max. 1.0 Mn, max. 0.20 Al, max. 0.03 S, 0.6–1.2 Ti, 1.5–3.0 Cu, 2.5–3.5 Mo, balance Fe

Table 4: Corrosion behavior of NiCr-alloys and NiCrFe-alloys in acetic acid of various concentrations and temperatures

Table 4: Continued

Metal	*Acetic acid %*	*Temperature °C*	*Corrosion rate mm/a*	*Remarks*	*Literature*
Inconel®	56	100	0.2	6 parts of 56% acetic acid + 1 part of 90% formic acid, operational test/150 h	[1, 6]
	56	104.6	0.375	as above: in vapors operational test/ 150 h	[1, 6]
Inconel® 601*	10	no temperature given	0.002	10% acetic acid/7 d	[24]
	10	no temperature given	0.554	10% acetic acid + 0.5% sodium chloride/30 d	
	10	no temperature given	1.161	10% acetic acid + 0.5% sulfuric acid/7 d	
Incoloy® 825*	20	boiling	0.003	duration 700 h	[25]
	50	boiling	0.059		
	75	boiling	0.047		
	100	boiling	0.033		
	20	60	0.0006		
	50	60	0.0005		
	75	60	0.0003		
	100	60	0.0003		

* Inconel® 600 (%) 76 Ni, 16 Cr, 7 Fe
Inconel® 601 (%) 58–63 Ni, 21–25 Cr, 0.10 C, 1.0–1.7 Al, 1.0 Mn, 0.015 S, 0.5 Si, 1.0 Cu, balance Fe
Incoloy® 825 (%) 38–46 Ni, 19.5–23.5 Cr, max. 0.05 C, max. 0.5 Si, max. 1.0 Mn, max. 0.20 Al, max. 0.03 S, 0.6–1.2 Ti, 1.5–3.0 Cu, 2.5–3.5 Mo, balance Fe

Table 4: Corrosion behavior of NiCr-alloys and NiCrFe-alloys in acetic acid of various concentrations and temperatures

Metal	Acetic anhydride, %	Temperature °C	Corrosion rate mm/a	Remarks	Literature
Illium® G*	all conc.	up to the boiling point	< 0.13	with water or acetic acid	[18, 23]
Inconel®	100	79	0.5		[7]
	100	100	0.5–1.25		
Nickel-chromium iron (%) 76/16/7	50	125	0.5	acetic anhydride in acetic acid, air-free	[8]
	90	125	< 0.05		
	100	150	< 0.05		
Nickel-chromium iron (%) 32/47/20	100	150	< 0.05	acetic anhydride in acetic acid, air-free	[8]

* Illium G (%) 60 Ni, 21 Cr, 8 Fe, 5 Mo, 3 Cu, 1 W, 1 Si, 1 Mn

Table 5: Corrosion behavior of nickel-chromium alloys and nickel-chromium-iron alloys in acetic anhydride of various concentrations and temperatures

NiCrFe-alloys of the type Inconel® [20] are resistant to dilute acetic acid and are better than Monel® in the case of vigorously aerated acetic acid solutions [10]. However, in general, the NiCrFe-alloys ((%) Ni-13 to 15 Cr – 6 to 20 Fe) as well as the NiCr-alloys with 14 to 20% chromium do not offer any advantages in contact with acetic acid compared to nickel or NiCu-alloys, since they are often slightly less resistant than the latter with aeration and are not recommended for hot concentrated acetic acid solutions [1, 10].

In pure 80% acetic acid, Inconel® ((%) Ni-15.5Cr 7.5Fe) loses 1 mm/a at room temperature [1, 2].

The corrosion loss of the molybdenum-containing alloy Inconel® 625 ((%) Ni – 20 to 23 Cr – max. 5.0 Fe – 8.0 to 10.0 Mo – 3.15 to 4.15 Nb + Ta – max. 0.1 C – max. 0.50 Mn – max. 0.50 Si – max. 0.015 P – max. 0.015 S – max. 0.40 Al – max. 0.40 Ti – max. 1.0 Co) during esterification (acetic acid, at times acetic anhydride, ester, alcohols, not aerated, duration 6 d) corresponds to a corrosion rate of 0.0125 mm/a [26].

For the corrosion behavior of NiCrFe-alloys in acetic acid, see Table 4.

– Acetic anhydride –

For the corrosion behavior of NiCrFe-alloys in acetic anhydride, see Table 5.

– Vinegar –

NiCr-alloys such as Inconel® are resistant to boiling vinegar and other types of fermentation vinegar [1, 2].

Nickel-chromium-molybdenum alloys

– Acetic acid –

In the BASF high-pressure process for the preparation of acetic acid from methanol with carbon monoxide, Hastelloy® B ((%) 61 Ni – 26 to 30 Mo – 4 to 6 Fe – max. 0.12 C) and in particular Hastelloy® C ((%) Ni– 16 Cr – 16 Mo – 5 Fe) are used as lining materials for the hot parts of the apparatus; see [27].

Recommendations are made for the use of Hastelloy® C as a material for heat exchangers for the production of acetic acid by the liquid-phase oxidation of benzene, butane or other hydrocarbons [28].

In a column in which volatile components were drawn off from an acetic acid solution, the lower section of the column and the pipelines were made from Hastelloy®, the upper sections from stainless steel SAE 316 and the heating lines from Karbate® [29].

The Russian alloy ChN65MV (NiCrMoW) is resistant, according to laboratory and operational tests, in boiling acetic acid (99.4%), in boiling acetic acid (98%) with 0.06 mol/l bromide additive, in boiling acetic acid with 0.5% formic acid and in the media present in the preparation of terephthalic acid (88% acetic acid, 1.3 g/l cobalt ions, 1.42 g/l bromide), for example in the recovery of acetic acid and the recrystallization of terephthalic acid. The alloy is distinctly superior to the materials which were investigated for comparison (12Ch18N10T (c.f. 1.4541), 10Ch17N13M2T (c.f. 1.4571), 03Ch21N21M4N6) [30–32].

NiCrMo-alloys, such as Hastelloy® C, Hastelloy® N ((%) Ni-7Cr-16 Mo-5 Fe), Hastelloy® F ((%) Ni-22Cr, 6 Mo, 22 Fe, 2 Nb/Ta) [20, 33], are among the most resistant materials to acetic acid of all concentrations and all temperatures [1, 34, 35]. Due to their high cost, they are mainly used for pumps, valves, nozzles and the like [1, 36], and for stems of gate valves [37]. Hastelloy® C and F have the best corrosion resistance to acetic acid; their corrosion rates over the entire concentration range up to the boiling point are less than 0.005 mm/a [38]. Hastelloy® F also has good resistance to mixtures of acetic acid and formic acid [33].

However, in halide-containing acid, Hastelloy® C is subject to intercrystalline corrosion; in particular ferric chloride must be excluded under all circumstances [1, 34].

As for the corrosion behavior of the various Hastelloy® types, see Tables 6–8.

No intergranular corrosion was observed in the weld seams of the Russian alloy EP-567, which is also resistant in acetic acid vapor (90 and 98% acid). The corrosion rates are 0.54 and 0.3 mm/a and in glacial acetic acid 0.3 mm/a [39]. This alloy is highly resistant to aerated anhydrous glacial acetic acid which contains 2.1% lithium chloride [40].

Material	Test duration, h		
	450	1,769	1,384
	Corrosion rate, mm/a		
Hastelloy® C	0.0025 (increase)	0.0025	
SAE 317	0.01	0.01	0.03
SAE 316	0.01	0.015	0.0425
SAE 316 ELC		0.015	
Illium®	0.005		
Aloyco® 20		0.0025	
SAE 309		0.01	
Hastelloy® B	0.0075	0.0175	
Hastelloy® A	0.0825		
Inconel®	3.25	2.75	
Monel®	destroyed		
Nickel	destroyed		
SAE 410			0.0375
SAE 430			0.03
SAE 446			0.005

Table 6: Corrosion behavior in glacial acetic acid vapors at 677°C. Vapor velocity 53 m/s. The specimens were located at the point of entry of the vapor into the reaction tube (operational test) [1, 6]

Material	Corrosion rate, mm/a		
	a	b	c
SAE 316	destroyed	1.375	0.0425
SAE 317	0.325	0.3	0.025
Copper	0.1	0.1	0.01
Aluminum bronze, 10% Al	0.075	0.1	0.0175
Silicon bronze	0.075	0.1	0.0125

a) Operational test in the middle of the fractionating column. 216 d in 75% acetic acid, 20% high-boiling substances, 5% water, temperature 125.6°C.
b) Operational test in the same column and the same duration, but in the top section of the column 95% acetic acid, 5% high-boiling substances and water, temperature 215.6°C.
c) Operational test on the uppermost plate of the column. Duration 241 d in 99.7% acetic acid at 117.8°C.

Table 7: Operational tests with different alloys under various conditions [1, 6]

Table 7: Continued

Material	*Corrosion rate, mm/a*		
	a	*b*	*c*
Copper-nickel 70/30	0.1	0.125	0.015
SAE 347	destroyed	destroyed	0.35
Monel®	0.075	0.225	0.0375
Nickel	0.375	destroyed	0.1625
Inconel®	0.525	destroyed	0.1
Hastelloy® C	0.0075	0.005	0,25

a) Operational test in the middle of the fractionating column. 216 d in 75% acetic acid, 20% high-boiling substances, 5% water, temperature 125.6°C.
b) Operational test in the same column and the same duration, but in the top section of the column 95% acetic acid, 5% high-boiling substances and water, temperature 215.6°C.
c) Operational test on the uppermost plate of the column. Duration 241 d in 99.7% acetic acid at 117.8°C.

Table 7: Operational tests with different alloys under various conditions [1, 6]

– Acetic acid vapor –

The Russian alloy 0Ch15N65M16V (EP-567 (%) Ni-14.7 Cr-15.8Mo-4W-1.3Fe) is more resistant in 90 to 99.8% acetic acid and also in the vapor phase at 200°C than the stainless steels Ch18N10T, Ch17N13M3T, 000Ch21N21M4B on test durations up to 80 h in air or nitrogen and is therefore recommended for these applications [34, 39, 42, 42].

– Acetic anhydride –

NiCrMo-alloys are among the most resistant materials to acetic anhydride; see Table 9.

Metal	*Acetic acid %*	*Temperature °C*	*Corrosion rate mm/a*	*Remarks*	*Literature*
Hastelloy® B*	10	20	0.00008		[2]
	10	boiling point	0.00016		
	99	20	0.000016		
	99	boiling point	0.0099		

* Hastelloy® B (%) 26 to 30 Mo, 4 to 6 Fe, max. 0.12 C, balance Ni
** Hastelloy® C (%) 16 Cr, 16 Mo, 5 Fe, balance Ni

Table 8: Corrosion behavior of Hastelloy® types in acetic acid of various concentrations and temperatures

Table 8: Continued

Metal	*Acetic acid %*	*Temperature °C*	*Corrosion rate mm/a*	*Remarks*	*Literature*
Hastelloy® C**	10	20	0		[2, 4]
	10	boiling point	0		
	99	20	0		
	99	boiling point	0		
Hastelloy®	10–100	100	< 0.05	Hastelloy® C, F; aerated, $FeCl_3$-free	[7, 8]
	100	204	> 1.25		
	10–100	100	< 0.05	Hastelloy® B, C, F; air-free	
	100	204	> 1.25	$FeCl_3$-free	
	100	204	< 0.05	Hastelloy® C; acetic acid vapor	
	100	100	< 0.05	Hastelloy® C; acetic acid + formic acid	[7, 8, 43]
	10	24	< 0.05	Hastelloy® B; acetic acid + hydrochloric acid	[7, 8]
	30	52	0.5–1.25	Hastelloy® B; 90% acetic acid + acetaldehyde + 7% water	
	100	100	< 0.5	Hastelloy® C; acetic acid + salicylic acid vapors	
	50	24	< 0.05	Hastelloy® B; acetic acid + sulfuric acid	
Hastelloy® C**	100	127	0.02	86–91% acetic acid + 8.4% salicylic acid vapors	[1, 6]

* Hastelloy® B (%) 26 to 30 Mo, 4 to 6 Fe, max. 0.12 C, balance Ni
** Hastelloy® C (%) 16 Cr, 16 Mo, 5 Fe, balance Ni

Table 8: Corrosion behavior of Hastelloy® types in acetic acid of various concentrations and temperatures

Table 8: Continued

Metal	*Acetic acid %*	*Temperature °C*	*Corrosion rate mm/a*	*Remarks*	*Literature*
Hastelloy® C**	25	104	0.015	25% acetic acid + 1.5% formic acid operational test/ 129 d	
	25	104	0.01	25% acetic acid + 4% formic acid, as above	
	30–50	104	0.175	30–50% acetic acid + 2–10% formic acid, as above	

* Hastelloy® B (%) 26 to 30 Mo, 4 to 6 Fe, max. 0.12 C, balance Ni
** Hastelloy® C (%) 16 Cr, 16 Mo, 5 Fe, balance Ni

Table 8: Corrosion behavior of Hastelloy® types in acetic acid of various concentrations and temperatures

Alloy	*Acetic anhydride, %*	*Temperature °C*	*Corrosion rate mm/a*	*Remarks*	*Literature*
Nickel-molybdenum (%) 62/28/Fe, V	10–100	100	< 0.05	acetic anhydride in acetic acid	[8]
	100	150	< 0.05		
Nickel-chromium-molybdenum (%) 54/15/16/ Fe, W	10–100	100	< 0.05		[8]
	10	150	< 0.05	acetic anhydride in acetic acid	
	50	125	< 0.5		
	90	125	< 0.05		
	100	150	< 0.05		
Hastelloy® B	10–100	100	< 0.05		[7]
	99	boiling point	up to 0.12	resistant	[44]

* Hastelloy® C (%) 12–21 Cr, 15–16 Mo, 1.5 Fe, 0.4–0.8 Mn, balance Ni
** Chlorimet® 3 (%) 18 Mo, 18 Cr, 3 Fe, 1 Si, 1 Mn, 0.07 C, balance Ni
*** Chlorimet® 2 (2) 32 Mo, 3 Fe, 1 Si, 1 Mn, 0.1 C, balance Ni

Table 9: Corrosion behavior of nickel-molybdenum alloys and nickel-chromium-molybdenum alloys in acetic anhydride of various concentrations and temperatures

Table 9: Continued

Alloy	Acetic anhydride, %	Temperature °C	Corrosion rate mm/a	Remarks	Literature
Hastelloy® B	all conc.	boiling point	< 0.5		[45]
	no data given	room temp.	< 0.13		[46]
	no data given	boiling point	0.13–1.3		
Hastelloy® C	10–100	100	< 0.05		[7]
	all conc.	94	< 0.05		[47]
Chlorimet® 3**	all conc.	125	< 0.5	including mixtures with acetic acid	[48]
Chlorimet® 2***	all conc.	125	< 0.5	including mixtures with acetic acid	[49]

* Hastelloy® C (%) 12–21 Cr, 15–16 Mo, 1.5 Fe, 0.4–0.8 Mn, balance Ni
** Chlorimet® 3 (%) 18 Mo, 18 Cr, 3 Fe, 1 Si, 1 Mn, 0.07 C, balance Ni
*** Chlorimet® 2 (2) 32 Mo, 3 Fe, 1 Si, 1 Mn, 0.1 C, balance Ni

Table 9: Corrosion behavior of nickel-molybdenum alloys and nickel-chromium-molybdenum alloys in acetic anhydride of various concentrations and temperatures

Nickel-copper alloys

– *Acetic acid* –

NiCu-alloys such as Monel® 400 ((%) 65.97Ni-31.59Cu-1.27Fe-0.87Mn-0.14Si-0.13C) are resistant at room temperature to air-free acetic acid solutions of all concentrations. However, the corrosion rate increases rapidly in hot aerated acetic acid solutions, so that these materials cannot be recommended under these conditions [1, 10, 20, 38].

Monel® is more resistant to acetic acid/formic acid mixtures than to glacial acetic acid and is better than CrNiMo-steel in hot mixtures of acetic acid of higher concentration in the presence of more than 2% formic acid (> 121°C), whereas the latter is preferred for dilute acetic acid/formic acid solutions [10].

Boilers made from NiCu-alloys have proven to be suitable in the hydrogenation of substances dissolved in glacial acetic acid even under pressure. Monel® valves are suitable for handling acetic acid at higher temperatures, and, amongst others, in the food industry [1, 50].

In the preparation of acetic acid from calcium carbide, Monel® is used for the crude acetic acid pump [1].

Reactors and condensers made of Monel® can also be used favorably in the preparation of amyl acetate from amyl alcohol and glacial acetic acid in the presence of a small amount of sulfuric acid as catalyst [1].

Likewise, Monel® is used as a lining material of recovery columns in the processing of furfural [51].

Monel® is therefore extensively used for reactors, heating coils, pipelines, stirrers and pumps for air-free acetic acid solutions even in the presence of sulfuric acid [10]. However, it must be kept in mind that under heat transfer conditions Monel® 400, annealed at 927°C for 1 h and quenched with water, loses 0.36 mm/a in 56% acetic acid + 1% sulfuric acid and 1.41 mm/a in 56% acetic acid + 5% sulfuric acid and 2.02 mm/a in 56% acetic acid + 4% formic acid in a test (96 h) with a specimen temperature of 110°C and a solution temperature of 100°C [43, 52].

Acetic acid vapors are decomposed catalytically by Monel® above 300°C [2].

For the corrosion behavior of NiCu-alloys in acetic acid, see Table 10.

– Acetic anhydride –

Centrifuges and fittings made from Monel® are used in the acetylation of cellulose with acetic anhydride to give acetyl cellulose [18, 23].

NiCu-alloys are resistant to acetic anhydride which contains less than 5% acetic acid in the absence of air up to the boiling point [18].

For the corrosion behavior of NiCu-alloys in acetic anhydride, see Table 11.

Metal	Acetic acid %	Temperature °C	Corrosion rate mm/a	Remarks	Literature
Monel® ((%) 23 Cu, 3.5 Fe, 3.5 Mn)	10–30	24	< 0.5	aerated	[7, 8]
	40	24	> 1.25	aerated	[2]
	50–60	24	0.5–1.25	aerated	
	80	24	< 0.05	aerated	
	90	24	< 0.5	aerated	
	100	24	< 0.05	aerated	
	10–40	100	< 0.5	air-free	[7]
	50	100	< 0.05	air-free	
	60	100	< 0.5	air-free	
	70	24	< 0.05	air-free	
	70	100	< 0.5	air-free	

* Monel® (%) 23 Cu, 3.5 Fe, max. 3.5 Mn, 0.3 C, 0.5 Al, 0.5 Si, balance Ni
** Silverin® (%) 28–34 Cu, 1–2.5 Fe, 1.25 Mn, 0.5 Si, 0.5 Al, 0.15 C, balance Ni

Table 10: Corrosion behavior of nickel-copper alloys in acetic acid of various concentrations and temperatures

Table 10: Continued

Metal	*Acetic acid %*	*Temperature °C*	*Corrosion rate mm/a*	*Remarks*	*Literature*
Monel® ((%) 23 Cu, 3.5 Fe, 3.5 Mn)	80–100	100	< 0.5	air-free	
	10	100	< 0.5	acetic acid vapor, air-free	[7]
	50	100	< 0.5		
	100	100	< 0.5		
	5	boiling	0.02	reflux	[1, 2, 4]
	5	boiling	0.03	vapor	
	50	boiling	0.05	reflux	
	50	boiling	0.11	vapor	
	98	boiling	0.04	reflux	
	98	boiling	0.04	vapor	
Monel®*	99.9	boiling	0.15	reflux	[1, 2, 4]
	99.9	boiling	0.05	vapor	
	20	52	0.5–1.25	acetic acid + acetaldehyde	[7]
	10	boiling	0.35	technical grade acetic acid, 10%	[1, 2]
	20	boiling	0.44	technical grade acetic acid, 20%	
	56	boiling	0.57	technical grade acetic acid, 56%	
	90	boiling	0.61	technical grade acetic acid, 90%	
	100	100	> 1.25	acetic acid + acetic anhydride + peracetic acid	[7, 8]
	100	100	< 0.5	acetic acid + ethyl alcohol	[7]
	100	79	< 0.05	acetic acid + formic acid	[7]

* Monel® (%) 23 Cu, 3.5 Fe, max. 3.5 Mn, 0.3 C, 0.5 Al, 0.5 Si, balance Ni
** Silverin® (%) 28–34 Cu, 1–2.5 Fe, 1.25 Mn, 0.5 Si, 0.5 Al, 0.15 C, balance Ni

Table 10: Corrosion behavior of nickel-copper alloys in acetic acid of various concentrations and temperatures

Table 10: Continued

Metal	*Acetic acid %*	*Temperature °C*	*Corrosion rate mm/a*	*Remarks*	*Literature*
Monel®*	100	100	0.5–1.25		
	100	24	> 1.25	acetic acid + hydrobromic acid	[7, 8]
	100	24	> 1.25	acetic acid + hydrochloric acid	[7, 8]
	100	100	< 0.5	acetic acid + salicylic acid vapors	[7]
	100	24	< 0.5	acetic acid + sulfuric acid	[7, 8]
	100	100	< 0.5		
	25	104	0.03	25% acetic acid + 1.5% formic acid, operational test/129 d	[1, 6]
	25	104	0.015	25% acetic acid + 4% formic acid, operational test/129 d	
	30–50	104	0.5	30–50% acetic acid + 2–10% formic acid, operational test/129 d	
	56	100	0.2	6 parts of 56% acetic acid + 1 part of 90% formic acid, operational test/150 h	[1, 6]
	56	105	0.02	as above: in vapors operational test/150 h	[1, 6]

* Monel® (%) 23 Cu, 3.5 Fe, max. 3.5 Mn, 0.3 C, 0.5 Al, 0.5 Si, balance Ni
** Silverin® (%) 28–34 Cu, 1–2.5 Fe, 1.25 Mn, 0.5 Si, 0.5 Al, 0.15 C, balance Ni

Table 10: Corrosion behavior of nickel-copper alloys in acetic acid of various concentrations and temperatures

Table 10: Continued

Metal	*Acetic acid %*	*Temperature °C*	*Corrosion rate mm/a*	*Remarks*	*Literature*
Monel®*	99.5	112	0.275	99.5% acetic acid + 0.05% salicylic acid + 0.45% water vapor/1,022 h, operational test	[1, 6]
NiCu-Alloy ((%) 66/32)	10–100	125	< 0.05	air-free	[8]
	10–100	300	> 1.25	air-free	
	10	25	0.5–1.25	90% acetic acid + acetaldehyde + 7% water	[8]
	100	100	< 0.5	acetic acid + ethyl alcohol	[8]
	100	75	< 0.05	acetic acid + formic acid	[8]
	100	100	0.5–1.25		
	10	100	< 0.5	acetic acid vapors	[8]
	50	100	< 0.5	acetic acid vapors	
	100	100	< 0.5	acetic acid vapors	
Silverin®**	5	boiling	0.01–0.10	reflux, liquid	[53]
	5	boiling	0.01–0.10	reflux, vapor	
	50	boiling	0.01–0.10	reflux, liquid	
	50	boiling	0.10–1.01	reflux, vapor	
	98	boiling	0.01–0.10	reflux, liquid	
	98	boiling	0.01–0.10	reflux, vapor	
	6	30	0.01–0.10	air-free	
	6	30	0.10–1.01	aerated	

* Monel® (%) 23 Cu, 3.5 Fe, max. 3.5 Mn, 0.3 C, 0.5 Al, 0.5 Si, balance Ni
** Silverin® (%) 28–34 Cu, 1–2.5 Fe, 1.25 Mn, 0.5 Si, 0.5 Al, 0.15 C, balance Ni

Table 10: Corrosion behavior of nickel-copper alloys in acetic acid of various concentrations and temperatures

Metal	Acetic anhydride, %	Temperature °C	Corrosion rate mm/a	Remarks	Literature
Monel®	100	79	< 0.5		[7]
	100	100	0.5–1.25		
Monel®	99	boiling	0.12–0.9	limited resistance	[44]
	99.5	100	0.12–0.9	not aerated	
NiCu-Alloy ((%) 66Ni/32Cu)	40	150	< 0.05	acetic anhydride in acetic acid, little air	[8]
	90	125	< 0.05		
	100	150	< 0.05	acetic anhydride + about 0.1% acetic acid	

Table 11: Corrosion behavior of nickel-copper alloys in acetic anhydride of various concentrations and temperatures

Nickel-molybdenum alloys

– *Acetic acid* –

NiMo-alloys [20] along with NiCrMo-alloys are among the most resistant materials to acetic acid of all concentrations at temperatures up to the boiling point [1, 10, 38]. The amounts of metal removed from Hastelloy® B in contact with the reaction mixture present in the preparation of amyl acetate (44% acetic acid, 34% amyl alcohol, 2% sulfuric acid, balance water) at 150°C over a period of 76 d corresponds to a corrosion rate of 7.75 mm/a [16]. However, this value appears to need re-examining, because according to a different source, this alloy loses only 0.15 mm/a in 10% acetic acid at 70°C, and other sources gave even lower values [33] (see also Table 8).

Hastelloy® B is hardly attacked by pure acetic acid [54, 55].

Hastelloy® B-2 ((%) Ni- 26 to 30 Mo – max. 2 Fe – max. 1 Co – max. 0.5 W – max. 0.10 Si – max. 0.015 C) has similar corrosion resistance to the alloy Hastelloy® B and is resistant to acetic acid, even in the presence of formic acid [56, 57]. Hastelloy® B 282 ((%) Ni-28Mo-5Fe-2V-0.6Cr-0.35Si-0.6Mn-0.02C) (solution-annealed) loses 0.1 mm/a in 10% acetic acid at 65.5°C [58].

The Russian NiMo-alloy EP-496 ((%) Ni-28.9Mo-3.86Fe-1.64V) is highly resistant in anhydrous solutions of acetic acid which contain 0.5 mol/l lithium chloride and 5% acetic anhydride: the corrosion rates are about 0.015 mm/a at 90°C [40].

However, NiMo-alloys can be sensitive to intercrystalline corrosion, for example in the proximity of welds; as for measures to prevent intercrystalline corrosion, examples include alloys which are very low in carbon and iron or the addition of carbon-binding elements such as vanadium, see [33].

For the corrosion behavior of Hastelloy® B in acetic acid, see Table 8.

In an acetylation mixture consisting of 25% acetic acid, 20% acetic anhydride and 55% carbon tetrachloride, a NiMo-alloy Ni28Mo is as resistant as titanium at 60°C [59].

– *Acetic acid vapor* –

Hastelloy® B is also resistant to 50 and 99.6% acetic acid under heat transfer conditions: at metal temperatures of 110, 125 and 140°C, corrosion rates of 0.125, 0.050 and 0.050 mm/a in 50% acid and 0.175, 0.175 and 0.075 mm/a in 99.6% acid are given.

– *Acetic anhydride* –

For the corrosion behavior of NiMo-alloys towards acetic anhydride, see Table 9.

Other nickel alloys

– *Acetic acid* –

The NiSi-alloy Hastelloy® D ((%) Ni-10Si-3Cu-max. 1Cr-max.2Fe-0.9Mn-max. 1.5Co-max.0.12C) loses 0.025 mm/a in boiling acetic acid [38].

The alloy Illium® 98 ((%) Ni-28Cr-8.5Mo-5.5Cu-1.25Mn-1Fe-0.7Si-0.05C) must be considered completely resistant for the preparation of ethyl and butyl acetate at 100°C having a corrosion rate which is hardly measurable [60].

For the chemical polishing of the surface of the NiFe-alloy Permalloy® 52 N (Ni-48Fe) a mixture of acetic acid with nitric and hydrochloric acid (70:30:1 (percent by volume)) is used. The layer which was removed by polishing (at 70°C for 6 to 9 s) was determined to be 3.39×10^{-4} cm [61].

– *Acetic anhydride* –

In boiling acetic anhydride Hastelloy® D loses less than 0.5 mm/a [62].

Alkanecarboxylic Acids

Author: *L. Hasenberg* / Editor: *R. Bender, R. Durham*

Nickel

Nickel is resistant at 20°C to propionic acid in all concentrations up to 99% acid when air is excluded. The corrosion rates under these conditions are stated to be 0.2 mm/a. They increase to 1.2 mm/a with vigorous aeration [63]. The corrosion rate of nickel 200 (99.5% Ni, 0.08 C, 0.18 Mn, 0.2 Fe, 0.005 S, 0.18 Si, 0.13 Cu) in boiling 100% unaerated propionic acid is 0.12 mm/a [64]. The corrosion rate in the same acid contaminated with sulfates increases to 0.9 mm/a at 121°C. A corrosion rate of 0.23 mm/a was determined in the vapor above this liquid [64]. In boiling propionic acid, the corrosion rate for all concentrations and in 99% acid is between 0.28 and 0.49 mm/a [63]. A corrosion rate of 0.6 mm/a is to be expected in a solution of 66% propionic acid, 17% butyric acid and 17% isobutyric acid at 149°C [63, 64].

Nickel is resistant to butyric acid at 20°C. In 99% butyric acid at 110°C the corrosion rate increases to 0.9 mm/a [63]. Small percentages of sulfates have no influence on corrosion. The corrosion rate in the vapor space is 0.23 mm/a. The corrosion resistance of nickel increases with increasing chain length of the alkanecarboxylic acids. Oxygen and water increase corrosion [65].

Nickel-chromium alloys

Nickel-chromium-iron alloys

The corrosion behavior of nickel-chromium alloys is roughly comparable to that of nickel and nickel-copper alloys. Nickel and nickel-copper alloys are particularly susceptible to corrosion when aerated [66].

Tables 12 and 13 show the corrosion resistance of the NiCrFe-alloys, while Table 14 shows the compositions of these materials. It can be seen that the alloy Inconel® 600 (75% Ni) which contains more nickel is attacked more by acetic acid at the boiling point than the Incoloy® 800 which contains less nickel (32% Ni) (see Table 12). The corrosion rates of the alloy containing less nickel are significantly less than 0.1 mm/a (except for glacial acetic acid), those of the alloy which contains more nickel significantly above that value. The corrosion rates in butyric acid (see Table 12 and Table 13) are lower than those in acetic acid (see [64, 67, 68]). In comparison, the exception is the corrosion rate in 100% butyric acid because of the higher boiling point. In propionic acid a corrosion rate between those of the two acids just mentioned is to be expected or, at the most, of the same order of magnitude of acetic acid. The corrosion data show an expected continuous decrease with increasing chain length of the alkanecarboxylic acids.

Concentration %	Temperature, °C	Corrosion rate, mm/a							
		Nickel 200	Monel® 400	Inconel® 600	Incoloy® 800	Incoloy® 825	Inconel® 625	SAE 304	SAE 316
20	boiling	0.48	1.18	0.398	0.095	0.003	0.014	0.089	0.004
50	boiling	0.12	0.05	0.145	0.032	0.059	0.007	0.028	0.004
75	boiling	0.26	0.26	0.274	0.002	0.047	0.003	0.002	0.001
100	boiling	0.16	1.83	0.153	0.132	0.033	0.002	0.016	0.003
20	60	0.21	0.29	0.003	0.0004	0.0006	0.0004	0.0004	0.0005
50	60	0.29	0.48	0.004	0.0005	0.005	0.0007	0.0003	0.0004
75	60	0.37	0.27	0.044	0.0006	0.0003	0.0023	0.0003	0.0003
100	60	0.33	0.18	0.052	0.099	0.0003	0.0003	0.0001	0.0002

Table 12: Results of corrosion investigations in the laboratory on nickel, nickel alloys and chromium-nickel steels in acetic acid [66] (Chemical analysis of the nickel materials in Table 14)

Concentration %	Temperature, °C	Corrosion rate, mm/a							
		Nickel 200	Monel® 400	Inconel® 600	Incoloy® 800	Incoloy® 825	Inconel® 625	SAE 304	SAE 316
20	boiling	0.079	0.013	0.121	1)	0.005	0.011	1)	1)
50	boiling	0.041	0.012	0.114	0.066	0.001	0.021	1)	0.001
75	boiling	0.218	0.053	0.142	0.218	0.030	0.001	0.090	0.012
concentrated	boiling	0.198	0.221	0.212	0.172	0.013	0.025	0.129	0.023

1) no measurable corrosion

Table 13: Results of corrosion investigations in the laboratory on nickel, nickel alloys and chromium-nickel steels in butyric acid [66] (Chemical compositions of the nickel materials in Table 14)

Alloy	Alloying elements, %									
	Ni	C	Mn	Fe	Cu	Cr	Mo	Al	Ti	Others
Nickel 200	99.2	0.08	0.18	0.2	0.13	–	–	–	–	
Nickel 201	99.2	0.01	0.18	0.2	0.13	–	–	–	–	–
Monel® 400	65.0	0.12	1.0	1.5	32.0	–	–	–	–	–
Monel® K 500	64.0	0.13	0.8	1.0	30.0	–	–	2.8	0.6	–
Inconel® 600	75.0	0.05	–	8.0	–	15.5	–	–	–	–
Inconel® 625	61.0	0.05	0.25	2.5	–	21.5	9.0	–	0.25	Nb 3.65
Incoloy® 800	32.0	0.04	–	45.0	–	21.0	–	0.3	0.4	–
Incoloy® 800H	32.0	0.08	–	45.0	–	21.0	–	0.3	0.4	–
Incoloy® 801	32.0	0.05	–	45.0	–	20.5	–	–	1.1	–
Incoloy® 802	32.0	0.35	–	45.0	–	21.0	–	0.6	0.7	–
Incoloy® 825	42.0	0.04	–	30.0	2.2	21.5	3.0	–	1.0	–

Table 14: Chemical compositions of nickel and nickel alloys [66]

Better corrosion behavior against alkanecarboxylic acids is exhibited by the high-iron Incoloy® 800 alloy than by a low-iron nickel-chromium alloy, Inconel® 600. According to [66], the alloys containing molybdenum are recommended. When the corrosion rates are compared with those of the chromium-nickel steels, in particular with those of the chromium-nickel steels containing molybdenum, they are the better material to use in alkanecarboxylic acids.

Nickel-chromium-molybdenum alloys

These materials are mainly to be used in alkanecarboxylic acids at the boiling point. The corrosion rates of the alloys Inconel® 625 and Incoloy® 825 are roughly comparable with those of the chromium-nickel steels containing molybdenum. In 50% butyric acid at the boiling point the Incoloy® 625 alloy which contains more nickel and molybdenum is corroded at the rate of 0.021 mm/a, and the Incoloy® 825 alloy which contains more iron at the rate of only 0.001 mm/a (see Table 13). In acetic acid of the same concentration but with a lower boiling point, the corrosion rates are higher (0.007 mm/a for Inconel® 625 and 0.059 mm/a for Incoloy® 825). In this case the alloy which contains more iron is corroded more severely by acetic acid [66].

The alloy NiMo16Cr16Ti (Hastelloy® C) exhibits very good resistance in carboxylic acids (formic to butyric acids). The corrosion rate in these acids does not exceed 0.05 mm/a at the boiling point [68, 69]. In boiling propionic acid a corrosion rate of 0.005 mm/a is stated, while for the chromium-free alloy NiMo 28 (Hastelloy® B) 0.12 mm/a is given [64].

The corrosion rate of Hastelloy® C in alkanecarboxylic acid mixtures is stated to be 0.02 mm/a, i.e. in a mixture consisting of 80% propionic acid, 0.1% acetic acid, 2.5% butyric acid, 15.4% higher acids and 2% nitric acid at 135°C. The corrosion rate of NiMo 28 in concentrated butyric acid containing small quantities of sulfates is relatively high at 121°C, namely 0.16 mm/a. From the tabulated data and the literature it may be concluded that, when handling alkanecarboxylic acids, the use of nickel-chromium-molybdenum alloys is to be preferred to that of CrNi-steels, since the latter are no longer resistant in the simultaneous presence of propionic anhydride. The corrosion rate in a mixture consisting of 64% propionic acid, 35% propionic anhydride and 1% nickel acetate at 260°C is stated to be 0.5 to 1.27 mm/a for NiMo 28 and 0.25 to 0.51 mm/a for NiMo16Cr16Ti [70].

With the Russian alloy ChN65MV (15–17% Mo, 14.5–16.5% Cr, 3.0–4.5% W, balance Ni), the corrosion rates in alkanecarboxylic acids C_5–C_{10} at 180 to 300°C and 0.3 to 4 kPa (test duration 120 hours) were 0.071 to 0.39 mm/a in the liquid and 0.021 to 0.082 mm/a in the vapor. The corrosion rate on heating surfaces was 0.27 to 0.78 mm/a [71]. Titanium should be used under these conditions.

Nickel-copper alloys

The corrosion rates of the nickel-copper alloys are dependent on aeration since oxygen promotes corrosion.

Propionic acid solutions in all concentrations attack nickel-copper alloys at a maximum rate of 0.2 mm/a at 20°C if they are not aerated. When they are vigorously aerated, the corrosion rate increases to 0.6 mm/a [63]. The corrosion rates are applicable to alloys containing 32% Cu. The corrosion rates of these alloys by acetic acid are shown in Table 12. Corrosion rates of 0.5 mm/a are exceeded only in boiling 20% and boiling concentrated acetic acid, where corrosion rates of 1.18 and 1.83 mm/a are measured.

In butyric acid the corrosion rates, as expected, are lower. The alloy containing 32% Cu, balance Ni, is corroded at the rate of only 0.221 mm/a in boiling, concentrated butyric acid (see Table 13) [66]. The corrosion rate in propionic acid is expected to be of the order of the corrosion in acetic acid. In boiling, concentrated propionic acid a very high corrosion rate of 9.35 mm/a (unaerated) in comparison with acetic acid has been determined [64].

The corrosion rate increases to 1.27 mm/a in boiling butyric acid contaminated by sulfates. The same corrosion rate was determined in butyric acid containing small quantities of water.

Nickel-copper alloys have a wide range of applications in the storage and processing of carboxylic acids with 6 or more carbon atoms. There are limits, however, to their use in boiling acids having 3 carbon atoms or more, mainly due to oxygen and water [66, 72–77].

Nickel-molybdenum alloys

Nickel-molybdenum alloys are among the most chemically resistant materials. They are resistant to propionic acid in all concentrations up to the boiling point [63]. The alloy NiMo 28 (Hastelloy® B-2) containing 26 to 30% Mo and maximum contents of 1% Co, 1% Cr, 2% Fe, 0.1% Si, 1% Mn, 0.01% C and 0.025 P, together with 0.010 S loses less than 0.02 mm/a in formic and acetic acids at the boiling point [78]. This corrosion resistance is also applicable against formic acid in concentrations of 10, 20, 30, 40, 60 and 89% and in acetic acid in concentrations of 10, 30, 50, 70 and 99%. This alloy is also resistant in these concentrations to propionic and butyric acids, since the corrosivity of the carboxylic acids decreases with increasing chain length [64, 67, 68].

The corrosion rate in unaerated, concentrated propionic acid at the boiling point is 0.12 mm/a [64].

The corrosion rate in concentrated butyric acid at 121°C contaminated by small quantities of sulfates is 0.16 mm/a in the liquid and 0.06 mm/a in the vapor above the liquid [64].

In a mixture consisting of 64% propionic acid, 35% propionic acid anhydride and 1% nickel acetate at 260°C NiMo 28 loses between 0.5 and 1.27 mm/a, whereas NiMo16Cr16Ti loses only between 0.25 and 0.51 mm/a [79]. The chromium-free alloy containing molybdenum is therefore less resistant in boiling propionic and acetic acids than the alloy NiMo16Cr16Ti.

These alloys are of interest for practical use at temperatures only in the region of the boiling point, since less expensive materials are sufficiently resistant at lower temperatures (see Table 12 and Table 13).

In propionic acid (99%) and 1% sulfuric acid the alloy containing 44–47% Ni, 21–23% Cr, 5.5–7.5 Mo, a maximum of 2.5% Co and 1.75–2.5% Ta is corroded at the rate of 0.09 to 0.23 mm/a. In a mixture of propionic acid (94%), acetic acid (5%), butyric acid (0.5%) and 0.5% solids the corrosion rate at 150°C is 0.003 mm/a [64]. Solutions of propionic and butyric acid at 138 and 157°C attack these alloys with corrosion rates of 0.03 mm/a.

Other nickel alloys

Silicon in nickel alloys increases their resistance to acids [80].

Nickel-silicon alloys containing 10% Si, 3% Cu, 1% Cr, a maximum of 2% Fe, a maximum of 1.5% Co, 0.9% Mn and a maximum of 0.12% C (Hastelloy® D) are corroded by acetic acid up to 65°C with a surface corrosion rate of less than 0.025 mm/a. However, corrosion increases above this temperature. These alloys are no longer used in the region of the boiling point at concentrations of 80% and higher [81].

These corrosion rates favor the use of these alloys in propionic, butyric and valeric acids and quite definitely justify their use in the higher alkanecarboxylic acids [64, 67, 68].

The corrosion rates of the alloy mentioned (Hastelloy® D) are below 0.5 mm/a in butyric acid above 110°C in all concentrations. It is not necessary to exclude air when these alloys are used [72]. The good resistance in capric acid is surpassed only by NiMo16Cr16Ti (no further data given) [75].

Carbonic Acid

Author: *P. Drodten* / Editor: *R. Bender, R. Durham*

Nickel

Minimal consumption corresponding to a corrosion rate of 0.002 mm/a was measured in carbon dioxide-containing waters with a partial pressure of CO_2 of 13.8 bar and a temperature of 55°C after an exposure duration of 7 days. After longer test durations of 28 and 70 days the values were even lower, implying the formation of a passivating cover layer [82].

According to older examinations nickel with a corrosion rate of 0.05 mm/a is considered resistant in pure water at a carbon dioxide pressure of 2.5 bar and a temperature of 70°C. Specifications of manufacturers also confirm resistance in carbon dioxide-containing waters even at temperatures up to 450°C and in the concurrent presence of sulfur dioxide.

Investigations into corrosion of nickel in water containing carbon dioxide with a CO_2 partial pressure of 1.38 MPa at 54°C showed corrosion rates after 7 days of 0.002 mm/a and after 28 and 70 days test duration these rates were even lower. The corrosion rate of a nickel cladding after 28 and 70 days test duration, 0.01 mm/a, was significantly greater than the initial corrosion rate after a test duration of 7 days with the nickel metal [83, 84].

The results of earlier investigations [85] have already illustrated the marked differences between the corrosion of sheet nickel and nickel coating, on exposure to CO_2 and O_2. In water saturated with 30% CO_2 and 70% O_2 sheet nickel exhibited virtually no corrosive loss. The corrosion rate of the cladding, however, was very much dependent on the rate of the supply of gas.

In distilled water at 70°C, in which carbon dioxide is dissolved at a pressure of 0.25 MPa, the corrosion rate for nickel is stated as 0.05 mm/a. With these corrosion rates, nickel is corrosion-resistant under the conditions mentioned [86]. In water containing CO_2, nickel is corrosion-resistant up to 464°C [87]. Nickel is especially resistant to water that contains sulfur dioxide or free CO_2 [88].

The water analysis of a nickel-plated saturator for carbon dioxide operating at a pressure of 1.4 MPa exhibited a corrosion rate with a test duration 10 days of 0.005 mm/a at 20°C. Nevertheless, pitting corrosion may occur even in the presence of traces of chlorides in water containing CO_2 [86, 88] International Nickel has obtained the following results in tests carried out on nickel:

In mains water saturated with CO_2 and oxygen the corrosion rate of nickel after a test duration of 20 hours was 0.11 mm/a, after 80 hours 0.16 mm/a, and after 120 hours 0.38 mm/a. In distilled water saturated with CO_2 and oxygen the corrosion rates after 20 hours were 0.08, after 60 hours 0.01 mm/a and fell after 120 hours to approximately zero.

Nickel-chromium alloys

Nickel-chromium-iron alloys

Nickel-chromium-molybdenum alloys

In carburizing gases containing carbon dioxide, carbon monoxide or hydrocarbons, the behavior of Ni-Cr or NiCrFe-alloys differs. In CO_2, for instance, not only damage to the material by carburization but also by oxidation must be expected. Concurrent carburization and oxidation impede the formation of a protective oxide layer such that Ni-Cr and Ni-Cr-Fe alloys should only be used in this atmosphere up to 1,000°C.

In carbon monoxide atmospheres, the so-called green rot of NiCr-alloys can be observed in a temperature range from 815°C to 980°C caused by the selective inner oxidation of chromium. The partial oxygen pressure of the atmosphere is so low that, in particular, nickel oxide playing a special role in the initial oxidation phase cannot be formed.

Therefore, NiCrFe-alloys are used in such media since the iron oxide is resistant also under the low partial oxygen pressures occurring here and hence can contribute to the protective effect of the material [89]. Alloying additions of silicon lead to a clear reduction of the corrosive attack in carbon monoxide atmospheres.

Generally, NiCrFe-alloys with 14 to 16% chromium and about 6% iron permit operating temperatures up to 1,000°C without any significant internal carburization or oxidation occurring. Chromium contents of 20 up to 22% lead to a clear increase in the corrosive attack in the temperature range of chromium carbide precipitation (about 800°C). Silicon or aluminum-alloyed NiCrFe-materials can be used even at temperatures above 1,000°C. Both alloying elements form protective oxide layers reducing the carbon diffusion.

The nickel alloy Alloy 825 (Incoloy® 825), i.e. a titanium-stabilized alloy, does not only exhibit good corrosion resistance both under oxidizing and reducing conditions, but also has good resistance to hot gases, e.g. combustion gases with sulfur dioxide and steam. The work described in [90] examined the behavior of 1 mm thick sheet specimens of Alloy 825 in water vapor, synthetic air (20/21% oxygen in nitrogen) and pure carbon dioxide as well as in a carbon dioxide atmosphere diluted with argon by a factor of 100. The composition of the specimens is specified in Table 15. The tests were performed at 800°C, 1,000°C and 1,200°C. The test duration was between 15 minutes and 24 hours.

C	*Si*	*Cu*	*Fe*	*Mn*	*Cr*	*Ti*	*Al*	*Co*	*Mo*	*Ni*
0.01	0.24	2.00	31.10	0.11	22.00	0.98	0.12	0.67	3.29	39.58

Table 15: Chemical composition of the examined Alloy 825 [90]

Figure 1 summarizes the results of the mass increase in a double logarithmic representation.

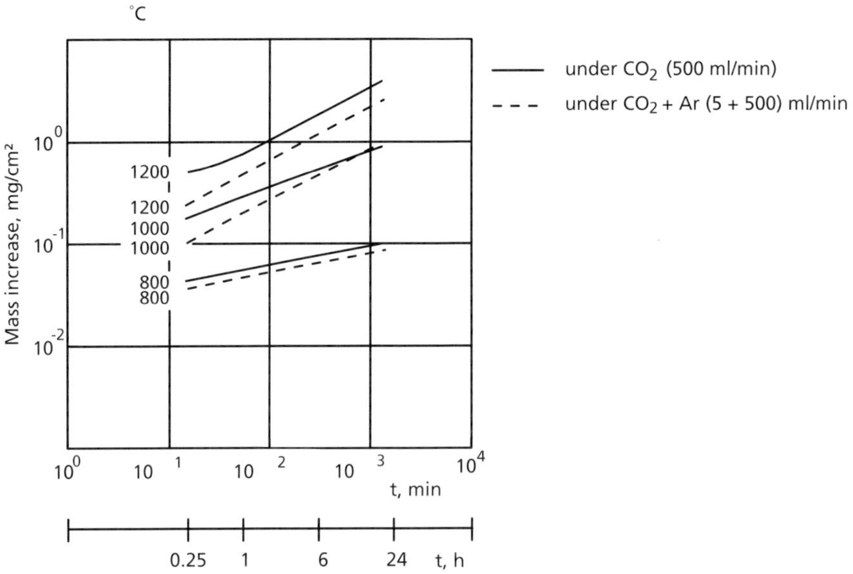

Figure 1: Mass increase of Alloy 825 specimens in pure carbon dioxide and carbon dioxide diluted with argon (1:100) as a function of temperature and exposure duration [90]

Dilution with inert gas noticeably reduces the oxidation of the nickel alloy compared with the reaction in pure carbon dioxide. In pure carbon dioxide the reaction rate at 800°C is lower than corresponding to a quadratic time dependence, follows a quadratic time law at 100°C, whereas the reaction kinetics are determined by a parabolic time law at 1,200°C.

As the parallel curve path shows, the reaction kinetics in dilute carbon dioxide at 800°C correspond to that in pure carbon dioxide. If the temperature is increased, the time dependence of oxidation is shifted into the direction of a parabolic curve path.

At 800°C and 1,000°C firmly adherent cover layers are formed, whereas some of the layers spall off at 1,200°C. Spalling is higher after short testing times than after longer exposure durations since the increasing inner oxidation and the formation of voids result in the formation of thicker and more uniform layers. The improvement in adhesion and thermal shock resistance of the layer observed with increasing exposure duration prevents or impedes spalling. It is assumed that the voids provide enough space to absorb the increase in volume of the growing layer.

In aqueous carbon dioxide-containing solutions the passivating NiCr, NiCrFe and NiCrMo-alloys are practically resistant. Stress corrosion cracking of these materials is not even found in CO_2/H_2S-containing solutions [91].

An alloy containing (%) 70 Ni, 14 to 16 Cr, 2.25 to 2.75 Ti, 0.7 to 1.2 Nb, 0.4 to 1.0 Al, 5 to 9 Fe, 0.3 to 1.0 Mn, < 0.5 Si, < 0.2 Cu, < 0.08 C and < 0.01 S (Inconel® X c.f. 2.4669) is resistant to fresh and distilled water as well as to most natural waters containing free carbonic acid, iron compounds, chlorides or dissolved air [92].

The corrosion rate of this alloy is stated to be 0.005 mm/a when it is exposed at 70°C and 3.53 MPa to an aqueous solution saturated with 30 or 70% carbon dioxide, with the balance air [93].

Nickel-chromium alloys containing about 15% Cr are not attacked in gas condensates from geothermal steam. They exhibit no susceptibility to pitting corrosion and stress corrosion cracking. In aerated steam, however, they lose about 2.0 mm/a.

The corrosion tests in the condensate were carried out in the gas-free condensate at 100°C (total CO_2 and H_2S together less than 10^{-4}%), in the aerated condensate at 60°C (0.002% CO_2 and 0.003% H_2S), and in the gas-rich condensate at 71°C (0.04% CO_2 and 0.003% H_2S). The test durations were 145 days in the gas-free condensate, 96 days in the aerated condensate and 122 days in the gas-rich condensate. Nickel-chromium alloys are recommended for use in above-ground acid gas plants. Precipitation-hardening nickel-chromium alloys should be used only with a Rockwell hardness of less than 35 [84, 94]. These alloys are also employed in the mineral oil industry and have been tried and proven in this field. Cold-worked nickel-chromium alloys in condensates containing carbon dioxide/carbonic acid, hydrogen sulfide and chlorides are susceptible to stress corrosion cracking [95].

Corrosion investigations under operating conditions showed corrosion rates below 0.05 mm/a in monoethanolamine solutions containing dissolved carbon dioxide and hydrogen sulfide at 110°C [96]. Other literature confirms the very good corrosion resistance under these conditions. The temperatures are stated to be 43 and 127°C with test durations of 5246 hours [97].

Even nickel-chromium alloys with iron contents are virtually not attacked at 20°C in aqueous solutions containing carbon dioxide.

Specimens of nickel-chromium-molybdenum alloys which were not cold worked exhibited no stress corrosion cracking after a test duration of 1,000 hours. Testing was carried out in a 5% sodium chloride solution at 90°C saturated with gas ((percent by volume) 99 CO_2 + 1 H_2S).

Even cold working up to 50% did not lead to stress corrosion cracking under the test conditions even at elevated temperatures [98].

The corrosion rate of a nickel-chromium-molybdenum alloy containing (%) 41.9 Ni, 20.1 Cr and 8.1 Mo in an aqueous solution of carbon dioxide (CO_2 partial pressure 1.01 MPa) and hydrogen sulfide (H_2S partial pressure 0.1 MPa) at temperatures up to 300°C is stated to be a maximum of 0.06 mm/a [99].

Nickel-copper alloys

The nickel-copper alloys behave rather like nickel in aqueous solutions containing carbonic acid [86].

It may be seen from the DECHEMA-Werkstoff-Tabelle, that nickel-copper alloys containing 70% nickel and 30% copper were corroded at the rate of 1.5 mm/a in distilled water at 70°C into which a mixture of carbon dioxide and air ((percent by volume) 70% CO_2 and 30% air) was introduced up to a pressure of 0.25 MPa in the test rig. The composition of the gas produces the maximum corrosion under the test conditions.

However, under identical test conditions with reduced carbon dioxide content ((percent by volume) 30% CO_2 and 70% air) the corrosion rate is decreased to 0.02 mm/a. The percentage of free carbonic acid is definitive for the usability of these alloys in solutions containing carbon dioxide/carbonic acid. Under the stated conditions they are not resistant to corrosion by excessive free carbonic acid.

At a pressure of 3.53 MPa under the test conditions described corrosion rates of 0.22 mm/a result for a gas mixture consisting of 30% carbon dioxide and 70% air and 1.5 mm/a with 70% carbon dioxide.

The corrosion rate of an alloy containing 45% Ni and 55% Cu is stated to be 0.4 mm/a with 30% and 70% carbon dioxide in the gas mixture [100]. Nickel-copper alloys with not less than 63% nickel are resistant to corrosion in water at 20°C and in distilled water with varying carbon dioxide/air contents. The maximum corrosion rates are about 0.1 mm/a. They may increase to up to 1 mm/a in the condensate of distilled water at 70°C [101].

Nickel-molybdenum alloys

According to the DECHEMA-Werkstoff-Tabelle, aqueous solutions of carbon dioxide virtually do not attack nickel-molybdenum alloys even in the presence of air.

This is also true of alloys containing chromium. Nickel-molybdenum alloys are recommended for use in acidic gas plants , and the condition of the material does not have to meet any special requirements [102].

Other nickel alloys

According to the experience of the author of the DECHEMA-Werkstoff-Tabelle, alloys containing 10% Si, 3% Cu, balance Ni are virtually not attacked in aqueous solutions of carbon dioxide even in the presence of air. Alloys containing 30% Fe, balance Ni in an aqueous solution of carbon dioxide containing a high percentage of free carbonic acid at 80°C lose 0.15 mm/a. These alloys are therefore corrosion resistant.

Fluorine and Hydrofluoric Acid

Author: *K. Hauffe* / Editor: *R. Bender, R. Durham*

Nickel

Nickel and its alloys, in particular the NiCu-alloy Monel®, are relatively resistant not only to fluorine but also to hydrogen fluoride up to 527°C. They are therefore used for containers in which fluorine and hydrogen fluoride are stored and in plant and equipment in which fluorination reactions are carried out [103].

Nickel reacts with fluorine between 1,077 and 1,197°C causing fire. Between 297 and 697°C, the fluorination of nickel in fluorine of 0.1 MPa initially follows a parabolic and later a cubic rate law. The growth of the NiF_2 film within the parabolic rate law is determined by the diffusion of fluorine ions to the phase boundary Ni/NiF_2 below 697°C [104, 105].

The temperature dependence of the kinetics of the fluorination (Figure 2) and the scattering of the values are predominantly caused by the different impurities of nickel, such as Mn, Si, and Fe and of fluorine, such as HF, oxygen, and water vapor, which also explain the pressure dependence. It should be noted that the corrosion data calculated from short term tests cannot be extrapolated without reservation, since the corrosion for the above-mentioned reasons can initially proceed at an accelerated pace.

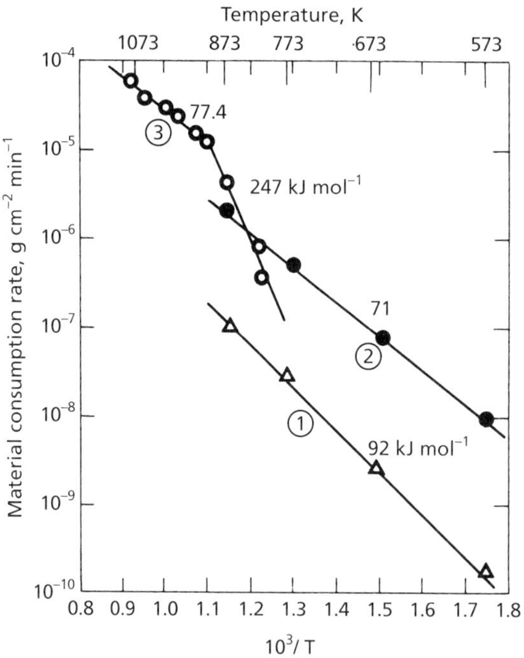

Figure 2: Temperature dependence of the corrosion of nickel in fluorine [105, 106]
① after extended fluorination at 9.3×10^{-2} MPa of fluorine, non-parabolic curve
② at the beginning of fluorination at 9.3×10^{-2} MPa of fluorine, parabolic curve
③ 0.1 MPa of fluorine, parabolic curve [106]

Despite the difficulties mentioned in extrapolating the accelerated tests with nickel for the duration of one year, the essential experimental data for comparative orientation are summarized in Table 16.

Temperature, °C	Fluorine pressure MPa	Test duration, h	Fluorination rate mm/a	Literature
–196	liquid	8,760	0.00025	[107]
–80	0.026–0.092	4	0.00017	[108]
27	0.1	4	0.0000012	[109]
30	0.026–0.092	4	0.00000055	[108]
84	0.026–0.092	4	0.00021	[108]
300	0.092	8	0.005	[104]
300	0.092	32	0.00073	[104]
400	0.092	0.5	0.015	[104]
400	0.092	28	0.0012	[104]
400	0.1	–	0.21	[110]
500	0.092	0.25	0.018	[104]
500	0.092	30	0.003	[104]
500	0.1	–	1.5	[110]
550	0.1	2	0.036	[106]
600	0.092	32	0.017	[104]
660	0.1	1.5	0.150	[106]
720	0.1	2	0.240	[106]
810	0.1	1.7	0.310	[106]

Table 16: Fluorination of nickel at different temperatures and fluorine pressures

The considerable change in the energy of activation of the fluorination of nickel above 602°C is said to be caused by the transition from anion to nickel ion diffusion [106]. This experimentally unproven assumption is probably correct, given the pore formation at the phase boundary Ni/NiF_2, since pore formation in the proximity of this phase boundary could be caused by the preferred diffusion of metal ions to the outer phase boundary NiF_2/F_2(gas). Above 797°C, nickel fluoride evaporates as NiF_2 which contains a steadily increasing proportion of NiF, as the temperature increases.

The sublimation rate of NiF_2, for example at 800°C and a fluorine pressure of 1.33×10^{-8} MPa, is 4.4×10^{-5} g/cm^2 h. The proportion of NiF can be disregarded. At 1100°C, the sublimation rate of NiF_2 or NiF is 1.1×10^{-4} and 2.9×10^{-4} g/cm^2 h [111].

The good resistance of nickel and its alloys follows from the following data: fluorine from 0.05 to 0.1 MPa, 57 to 77°C, corrosion rate less than 0.002 mm/a [112].

The addition of 2% by volume ThO_2 in order to improve the adhesion of the growing oxide film in cyclic oxidation, which is so favorable in the oxidation of nickel and NiCr-alloys, has an adverse effect when exposed to hot fluorine: the corrosion rate of nickel in fluorine at 9.2×10^{-2} MPa and 600°C is increased, for example, by a factor of 8 to 12 [113].

Nickel-containing materials have proven to be suitable as bearing shells, for example in pumps, for liquid fluorine and/or fluorine-oxygen mixtures (FLOX), since they form NiF_2 films which have good lubricating properties. Because of their limited life, these films must be renewed frequently [114].

Nickel coatings deposited at zero current and containing phosphorus or boron additives, for example on base flanges of turbomolecular pumps, valve units, pressure chambers, and vacuum gauge casings, which are also required to be resistant to fluorine, are comparable to compact nickel [115].

When exposed to hydrogen fluoride, minimum foreign-metal contents in the nickel can also significantly change the rate of fluorination (Figure 3 and 4). The composition of the two types of nickel is shown in Table 17.

Chemical elements	Nickel alloy	
	ZLP	NP-2A
	Chemical composition, wt%	
C	0.015	0.050
O	0.008	0.017
H	0.0001	0.0005
Cu	0.0026	0.030
Fe	0.002	0.230
Co	0.0002	0.270
Mn	0.0003	0.002
Mg	0.0003	0.060
Zn	0.0002	0.0014
Pb	0.0002	0.0010
Si	0.0010	0.0024
Ni	balance	balance

Table 17: Chemical compositions of the nickel types ZLP and NP-2A in wt% [116]

Thus, the pure material is passive after as little as 10 h, the technical grade nickel only after 40 h, and in addition to that the corrosion loss is 5 times as high [116].

Oxide films produced by initial oxidation only slightly affect the resistance of nickel in hydrogen fluoride. In particular above 350°C, a steady weight decrease takes place, which indicates a volatile corrosion product (which, however, has not yet been identified) [117].

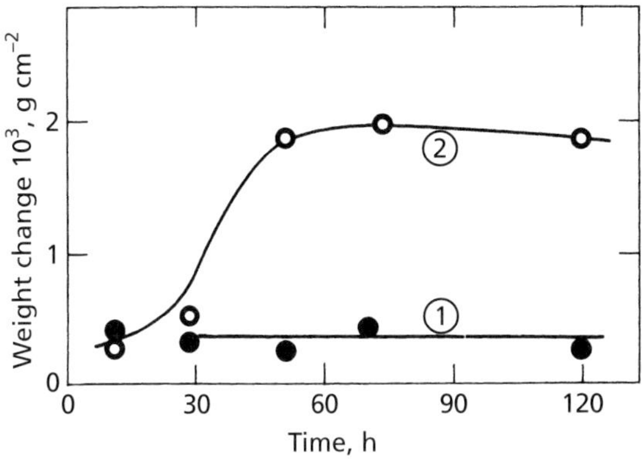

Figure 3: Time dependence of the corrosion of nickel in hydrogen fluoride, 2.0 MPa, 500°C [116]
① ZLP
② NP-2A

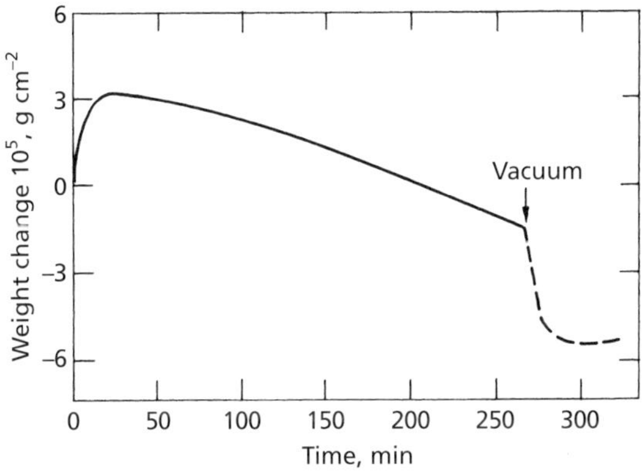

Figure 4: Time dependence of the fluorination of initially oxidized nickel (NP-2) in hydrogen fluoride at 1.33×10^{-2} MPa and 400°C [117]

In liquid hydrogen fluoride containing less than 0.005 mol/l water, films consisting of nickel oxyfluorides are said to form on anodically polarized nickel when treated one hour at 0.5 V [118].

Anodic passivation of nickel in hydrofluoric acid becomes more difficult compared to that in sulfuric acid. While nickel is passivated in 0.5 mol/l sulfuric acid between 400 and 1,300 mV at currents of about 0.1 mA/cm^2, as little as 0.01 mol/l hydrofluoric acid in the sulfuric acid causes significantly higher currents of 10 to 100 mA/cm^2 [119, 120]. The same is true for 0.1 mol/l hydrofluoric acid containing 0.9 mol/l potassium fluoride at 25°C, in which case a distinct current density maximum of about 200 mA/cm^2 occurs.

Figure 5 shows the possible fields of application of pure nickel and some nickel alloys in hydrogen fluoride [121, 122]. Thus, pure nickel is less resistant to hydrogen fluoride than Monel® (Ni30CuFe). At room temperature, especially Ni-21CrMo is better than pure nickel. It is important that the hydrofluoric acid is not aerated [122].

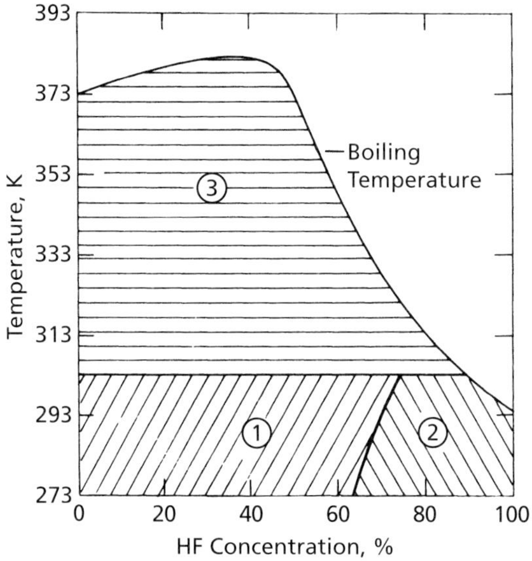

Figure 5: Areas of application of pure nickel and nickel alloys in hydrofluoric acid of different concentrations and temperatures [121]
① Ni99.2 (up to 50%, not aerated), Ni-20Cr-15Mo, Ni-30Mo, Ni-30CuFe (not aerated),
② Ni99.2 (above 90%), Ni-30Mo, Ni-30CuFe (not aerated),
③ Ni99.2 (up to 50% not aerated), Ni-15CrFe (not aerated).

As can be seen from Figure 6, nickel is not passivated in pure hydrofluoric acid until the acid exceeds 40% and in the presence of 5% HCl not until 70% HF is reached, although the corrosion-promoting effect of hydrofluoric acid is largely removed as low as about 50% HF and completely removed at 70% [111].

Nickel and Monel® are recommended as materials for the transport and storage of hydrofluoric acid, since the losses at 25°C remain under 0.1 mm/a even at higher hydrofluoric acid concentrations [123].

The alloy Ni-0.1 to 0.3Ti-0.1 to 0.2C-0.1 to 0.3Si is proposed as a welding material for nickel which is resistant to hydrogen fluoride. The alloy additions are introduced by way of the molten material consisting of CaO-CaF_2-Al_2O_3 during arc welding of 10 mm thick nickel NP2 (Russian type). Owing to the pore-free NiF_2 film, the weld seam of the nickel alloy is resistant at 552°C in hydrogen fluoride at 2.0 MPa [124].

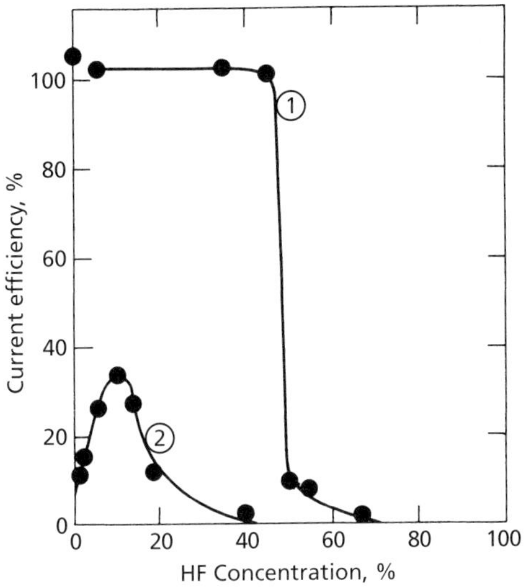

Figure 6: Dependence of the current efficiency of the nickel dissolution in HF + 5% HCl solution [111]
① on the concentration of hydrofluoric acid in hydrochloric acid
② on the concentration of hydrofluoric acid without hydrochloric acid

Nickel-chromium alloys

Nickel-chromium-iron alloys

Nickel-chromium-molybdenum alloys

NiCr-alloys with and without additional alloying additions are generally less resistant to fluorine, hydrogen fluoride and hydrofluoric acid than nickel. Between 347 and 747°C, for example, Inconel® types corrode in fluorine about 10 to 140 times faster than nickel [110, 125, 126].

In liquid fluorine, the Inconel® alloy Ni-15.2Cr-7Fe is comparable to nickel between 25 and 200°C [125, 127]. The corrosion behavior of the alloy Illium® (Ni-22Cr-5Mo) in gaseous and liquid fluorine is comparable to the Inconel® alloy between –3 and 597°C [128, 127]. However, Illium® corrodes about 150 times faster than unalloyed nickel at 527°C, which can be attributed to the high vapor pressures of the fluorides of chromium and molybdenum.

Hastelloy® C (Ni-17Cr-17Mo-6Fe-4W-1Mn-1Si) is attacked in fluorine between 297 and 397°C up to 5 times and between 397 and 497°C about 10 to 100 times faster than nickel [127].

The alloy INOR® 8 (Ni-17Cr-16Mo-5Fe-0.8Mn-0.35Cu-0.3Si-0.2Co) is not resistant at all in these temperature ranges [127, 129].

Attack by fluorine can also lead to inner fluorination corresponding to inner oxidation. This phenomenon was observed with a Ni-5.6Cr alloy between 800 and 1,000°C, which was embedded in a powder mixture of equal volume parts of Ni and

NiF_2 inside an evacuated quartz capsule [130]. In accordance with the diffusion of fluorine to the inside at a defined fluorine pressure, given by the equilibrium $Ni + F_2 = NiF_2$, CrF_3 is especially precipitated at the grain boundaries. The time dependence of the penetration of the inner reaction front from the surface into the inside of the alloy is shown in Figure 7. The reaction front advances by a parabolic rate law, since diffusion of the fluorine is rate-determining.

Incoloy® 825 (Ni-21.5Cr-30Fe-3Mo-2.25Cu) is recommended for hydrogen fluoride production plant [131].

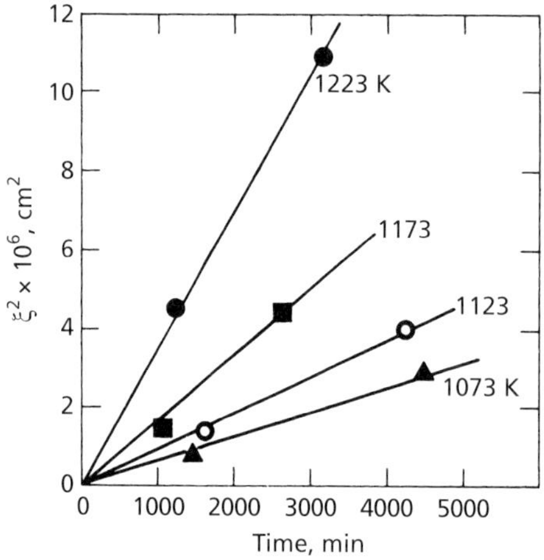

Figure 7: Time dependence of the square of the inner oxidation front ξ^2 in an Ni-5.6Cr alloy at different temperatures [130]

The corrosion behavior of Hastelloy® N in molten lithium tetrafluoroberyllate is dependent upon the chromium content of the alloy, since chromium has little resistance to fluoride melts. Its metal fluorides and oxides are readily dissolved by the melt so that no film formation and therefore no passivation can take place. To keep the corrosion of the alloy at a low level, the partial pressure of hydrogen fluoride and oxygen above the melt must be reduced to a minimum [132].

The behavior of Hastelloy® C (Ni-15.3Cr-16.5Mo-7.1Fe-3.8W-0.25Si-0.03C) in 50 to 100% hydrofluoric acid at 20 and 50°C, (impurities 0.08 to 0.13% of H_2SiF_6, 0.18% to 0.28% of H_2SO_4, 0.36 to 1.5% of SO_2) is shown in Figure 8 and Figure 9. As can be seen from Figure 8, despite considerable scattering of the test results, the weight change slightly decreases with increasing concentration of hydrofluoric acid from 50 to 90%, but decreases by about two orders of magnitude in anhydrous acid (see ordinate on the right of Figure 8).

In the defluorination of phosphoric acid and phosphates, when solutions containing 1 to 2% HF and 0.2 to 1% H_2SiF_6 at 80°C are present, the alloy KhN60-MB (Ni-30 to 35Cr-5Mo-1 to 2Nb) is more resistant than N54Ch22M7S4D4L (Ni-22Cr-7Mo-4Si-4Cu). The former alloy loses 0.02 mm/a at 2% HF at 20°C and 0.84 mm/a at 80°C [133]; however, it is recommended as a material for chemical pumps and injectors only at temperatures of up to 40°C.

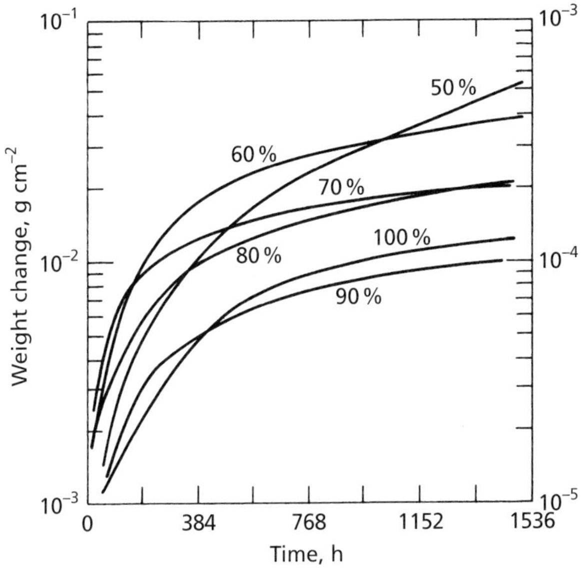

Figure 8: Time dependence of the weight change of Hastelloy® C in hydrofluoric acid (50 to 100%) at 20°C [134]. The ordinate on the right refers to anhydrous hydrofluoric acid (100 %)

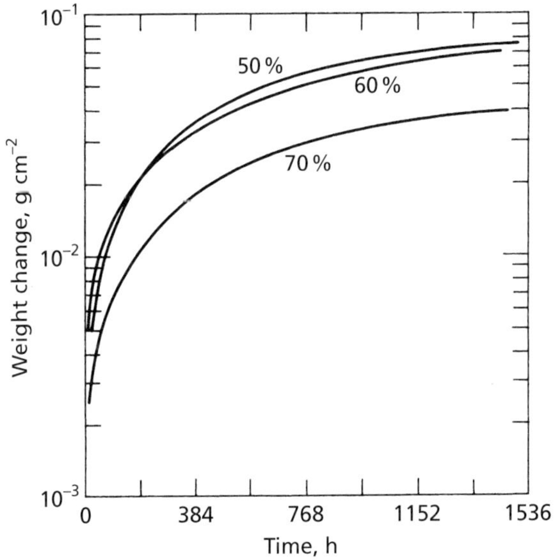

Figure 9: Time dependence of the corrosion of Hastelloy® C in hydrofluoric acid (50 to 70%) at 80°C [134]

A weldable alloy 03ChN60M5 (EP-758; 60% Ni, 5% Mo, 35% Cr), which is resistant to hydrofluoric acid up to 80°C and is not susceptible to intercrystalline corrosion, was developed as a material for the manufacture of phosphate fertilizers, during which HF-containing gases are also formed. An absorber for HF-containing waste gases employing 12 to 18% H_2SiF_6 as a falling-film liquid, showed no visible damage not even at the weld seams after 3 years under industrial conditions [135].

To recover uranium from spent fuel elements, the latter were treated with solutions containing hydrofluoric acid. Even high-alloy CrNiMo-steels are not resistant to these solutions. This is the reason why the following NiCr-alloys were tested as to their behavior under these conditions [136]:

Medium	*Hastelloy® C-276*	*Hastelloy® C-4*	*Hastelloy® S*	*Hastelloy® G*	*Inconel® 625*	*Incoloy® 825*
	Corrosion rates, mm/a					
8 mol/l HF	11.29	2.20	8.85	18	37.82	49.41
8 mol/l HF + 1.3 mol/l Zr	2.50	1.10	0.95	5.8	2.44	< 1.31
8 mol/l HF + 1.3 mol/l Zr + 0.5 mol/l HNO_3	18.61	14.03	14.95	31.11	3.05	10.37

Hastelloy® C-4 (Ni-10.4 to 18Cr-2Co-14 to 17Mo-3Fe-0.7Ti-0.08Si-0.02C)
Hastelloy® C-276 (Ni-14.5 to 16.5Cr-2.5Co-15 to 17Mo-4 to 7Fe-1Mn-3 to 4.5W-0.35V-0.08Si-0.02C)
Hastelloy® G (Ni-21 to 23.5Cr-2.5Co-5.5 to 7.5Mo-18 to 21Fe-1W-1.5 to 2.5Cu-1 to 2Mn-1Si-0.05C)
Hastelloy® S (Ni-15.5Cr-14.5Mo-1Fe-0.4Si-0.5Mn-0.2Al-0.02La-0.02C)
Incoloy® 825 (Ni-30Fe-19.5 to 23.5Cr-2.5 to 3.5Mo-1.5 to 3Cu-0.6 to 1.2Ti-1Mn-0.2Al-0.05C)
Inconel® 625 (Ni-20 to 23Cr-1Co-8 to 10Mo-5Fe-3.2 to 4.2Nb/Ta-0.5Si-0.5Mn-0.4Ti-0.1C)
Inconel® 671 (Ni-48Cr-0.35Ti-0.05C)
Inconel® 690 (Ni-30Cr-9.5Fe-0.03C)

Table 18: Corrosion rates for NiCr-alloys in solutions which are formed in the reprocessing of fuel elements, test duration 168 h [136]

In particular, the alloy Hastelloy® C-276 has proven to be suitable under these conditions. As shown in Table 18, admittedly considerable corrosion losses occur, but the following remarks have to be taken into account. These values have been extrapolated from short-term exposures; however, because of the short exposure time the effective stress remains low, therefore the total exposure times add up, for example, to only 8.5 weeks, from which a value of 3.0 mm/a is calculated. This appears to be acceptable under these conditions (see Table 19) [136].

For a distillation unit used for recovering nitric acid which is contaminated with hydrogen fluoride, a corrosion rate of 1.1 mm/a is given for Inconel® 690, which is too high compared to Hastelloy® C-276. Hastelloy® G-3 while having good weldability is also more resistant to hydrofluoric acid even in combination with nitric acid and is also said to be resistant to stress corrosion cracking [137].

*Solution**	*Exposure time weeks*	*Corrosion rate mm/a*
1	2.5	2.3
2	1	1.6
3	1	5.2
2	1	1.7
4	1	13.7
5	1	0.12
3	1	1.9
6	1	0.30

* 1) dissolution products of zirconium (8 mol/l HF, 1.36 mol/l Zr)
2) solution 1 + $Al(NO_3)_3$
3) 8 mol/l HF
4) solution 1 + 0.5 mol/l HNO_3
5) 2.2 mol/l $Al(NO_3)_3$
6) solution 1 + boron

Table 19: Corrosion of reaction vessels made of welded Hastelloy® C-276 in the media used in the reprocessing of fuel elements [136]

Two nickel alloys Ni-30 to 32Cr-8.5 to 9.5W-2 to 4Fe-2.2 to 2.5Si-0.7 to 0.9B-0.9 to 1C and Ni-7.5 to 10Cr-1.5 to 3.5Fe-2.3 to 4Si-1.5 to 2B-0 to 0.55C have good resistance to hydrofluoric acid at least at room temperature, see Table 20 [138].

The NiCrCoMo-alloy (Ni-22Cr-12.5Co-9Mo-1Al-0.07C) is also said to have good resistance to hydrogen fluoride (up to 20%) in the gas phase. However, quantitative data are not given [139].

In fluoride-containing phosphoric acid, pitting corrosion occurs on pipes made from NiCrFeMo-alloys, in the course of which iron is preferentially (13.6%) dissolved out of the alloy, while nickel and chromium at 3.3 and 2.4% are less severely attacked [140].

HF, %	*Material consumption rate, g/cm² h (Corrosion rate, mm/a)*		
	*Ni-alloy**	*Ni-alloy***	*Co-alloy****
6	1.46×10^{-5} (0.16)	1.46×10^{-5} (0.16)	8.5×10^{-4} (9.5)
40	3.96×10^{-5} (0.44)	4.13×10^{-5} (0.46)	2.0×10^{-3} (22.0)

* Ni-(30–32)Cr-(8.5–9.5)W-(2–4)Fe-(2.2–2.5)Si-(0.7–0.9)B-(0.9–1)C
** Ni-(7.5–10)Cr-(1.5–3.5)Fe-(2.3–4)Si-(1.5–2)B-(0–0.55)C
*** Co-(26–28)Cr-(4–5)W-(1.1–1.4)C

Table 20: Corrosion behavior of NiCr-alloys and a Co-hard alloy in hydrofluoric acid at 20°C; test duration 26 days [138]

The NiCrFe-alloys Ni-14 to 17Cr-6 to 10Fe-1Mn-0.5Cu-0.5Si, Inconel® 600, Nicrofer®, and Ferrochromin® are used as materials in the presence of hydrofluoric acid up to 40 and even up to 60% between room temperature and above 97°C. Thus, they can be used in areas 1 and 3 of Figure 5. NiCrMo-alloys, for example Ni-20Cr-15Mo, can also be used at even higher hydrofluoric acid concentrations up to 70% at 27°C [122, 141].

In particular the alloy 0ChN65MV (Ni-(14.5 to 16.5)Cr-(15 to 17)Mo-3W) is said to have proven its suitability in Russia as a material for mixers for the work-up of fluoride concentrates using sulfuric acid. The corrosion rate under production conditions was less than 1 mm/a and thus significantly less than the corrosion rate of 5 mm/a of the NiCrMo-alloy ChN40MDTYu (Ni-(14–17)Cr-(4.5–6)Mo-(2.7–3.3)Cu-(2.5–3.2)Ti-(0.7–1.2)Al); erosion corrosion was also more severe in the second alloy [142].

The corrosion behavior of alloy 47ChNM2 (Ni-47Cr-2Mo) in HNO_3-HF solutions of different compositions between 50 and 94°C can be seen in Table 21. As is evident, the corrosion rate at sufficiently low hydrogen fluoride contents of nitric acid is still acceptable even at 94°C [143]. The results summarized in Table 21 can be used to estimate by calculation the corrosion rates k to be expected at different temperatures and acid concentrations:

$$\ln k = -7.611 + 0.171[HNO_3] + 2.403[HF] + 0.046(T\text{-}273)$$

In order to maintain the corrosion rates of the alloys 46ChNM (Ni-46CrMo) and 03ChN60V (Ni-39CrV) below 0.5 mm/a, the HF concentration in 6 mol/l HNO_3 must not exceed 1 mol/l above 72°C as shown in Figure 10. The corrosion rate is, however, only slightly affected by the flow rate [144]. The corrosion rate k of the two NiCr-alloys 46ChNM and 03ChN60V can also be estimated by using the following formula:

for 46ChNM

$$\ln k = -7.155 + 0.194[HNO_3] = 1.629[HF] + 0.050(T\text{-}273)$$

for 03ChN60V

$$\ln k = -7.135 + 0.218[HNO_3] + 1.582[HF] + 0.047(T\text{-}273)$$

NiCr-alloys of the types EI-435 (Ni-20Cr-2Fe-0.5Mn-0.6Si-0.3Ti-0.1A1-0.09C) and EI-437 (Ni-20Cr-2.5Ti-0.8Fe-0.8A1-0.3Mn-0.55Si-0.05C) are susceptible to stress corrosion cracking and intercrystalline corrosion in solutions consisting of 20% nitric acid and 4% hydrofluoric acid even at room temperature, in which uniform corrosion can be ignored. The alloys which are arc-melted in a vacuum are said to be more susceptible than those prepared by induction-melting [145].

Inconel® 600 is used as a material for flexible metal tubing and compressors in steam turbines. In some cases, after fairly long operation of up to 12 years at 317°C and a pressure of about 1 MPa, defects due to stress corrosion cracking were observed. Heat-treated flexible tubes remained resistant at 327°C if previously pickled with solutions consisting of HNO_3-HF, since in that case pore-free films can be formed [146].

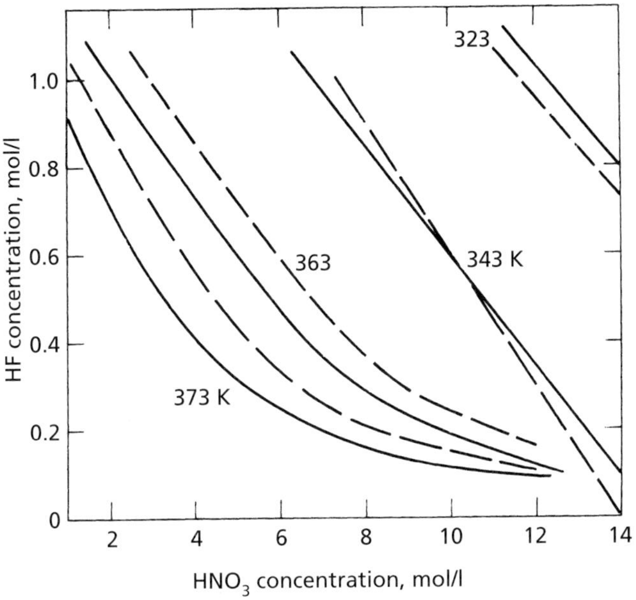

Figure 10: Isocorrosion lines (for 0.5 mm/a) of 46ChNM (– – –) and 03ChN60V (——) in the industrially interesting concentration range of HNO_3 and HF [144]

A NiCr-alloy (Ni-22Cr-1Al-12.5Co-9Mo-0.35Ti-0.03B-0.06 to 0.8C) having good high-temperature strength is said to have good resistance at room temperature to solutions consisting of H_3PO_4 and HF in the entire concentration range and good mechanical properties even at temperatures as high as 1,097°C. It is also very resistant to non-isothermal oxidation [147].

A NiCrW-alloy containing 39 to 41% chromium and 1% tungsten is said to have good mechanical and physical properties. It has remained for 4 years without damage as a container material for HNO_3-HF-containing pickling baths [148].

The effect of the method of melting and the heat treatment on the corrosion behavior of the alloy EI-437 B (Nimonic® 80 (Ni-20.5Cr-2.7Ti-0.48Fe-0.16Mn-0.70Al-0.11W-0.27Si-0.05C)) in a solution containing 15% HNO_3 and 4% HF at 60°C is shown in Figure 11, from which it can be seen that the type of melting (open crucible, then forged, melted in an arc under vacuum and forged or cast) has no noticeable effect on the extent of corrosion. On the other hand, the heat treatment between 950 and 1,100°C has a considerable effect on the time dependence of corrosion, which is less at lower temperatures. To avoid the increase of corrosion caused by the heat treatment at 1,100°C, which is characterized by structure-selective corrosion of the alloy, the concentration of nitric acid in the pickling solution can be reduced to 8 to 10%, which, however, requires longer pickling times, which in turn lead to increased dissolution of the metal. To prevent this, it is necessary to add inhibitors [149].

Concentration HNO_3	mol/l HF	Temperature °C	Corrosion rate mm/a
1.5	0.02	50	0.021 (0.008)
1.5	1.0	50	0.127 (0.046)
1.5	0.02	94	0.020 (0.056) (0.014)
1.5	1.0	94	0.358 (0.372)
4.0	0.2	50	0.011 (0.020)
4.0	1.0	50	0.083 (0.037)
4.0	0.2	94	0.221 (0.285)
4.0	1.0	94	0.314 (0.751)
6.0	0.2	70	0.103 (0.118)
6.0	1.0	70	0.286 (0.444)
6.0	0.2	94	0.374 (0.218)
6.0	1.0	94	1.30 (0.906)
14.0	1.0	50	0.457 (0.404) (0.704)
14.0	1.0	70	1.43
14.0	1.0	80	2.01
14.0	1.0	94	4.38 (4.58) (4.23)

Table 21: Corrosion rates of the alloy 47ChNM2 (Ni-47Cr-2Mo) in HNO_3-HF solutions at different temperatures, test duration 100 h. The values in parentheses show the range of variation of the results [143]

Corrosion-resistant fine-grain Hastelloy® S without a carbide phase is obtained by quenching in water at 1,220°C; this material is highly resistant to phosphoric acid which contains 3 to 4% sulfuric acid and 1% hydrogen fluoride. The resistance is said to increase further by a two-hour heat treatment between 497 and 547°C. However, if this material is slowly cooled starting at 1,217°C, carbide precipitates are formed and the resistance decreases [150].

The susceptibility of the NiCr-alloy ChN60V to intercrystalline corrosion in a solution consisting of 16 to 18% nitric acid and 4% hydrofluoric acid is not increased by the addition of 0.3 to 1.5% manganese but considerably intensified by the addition of 0.07 to 1% silicon, particularly because of carbide precipitates [151].

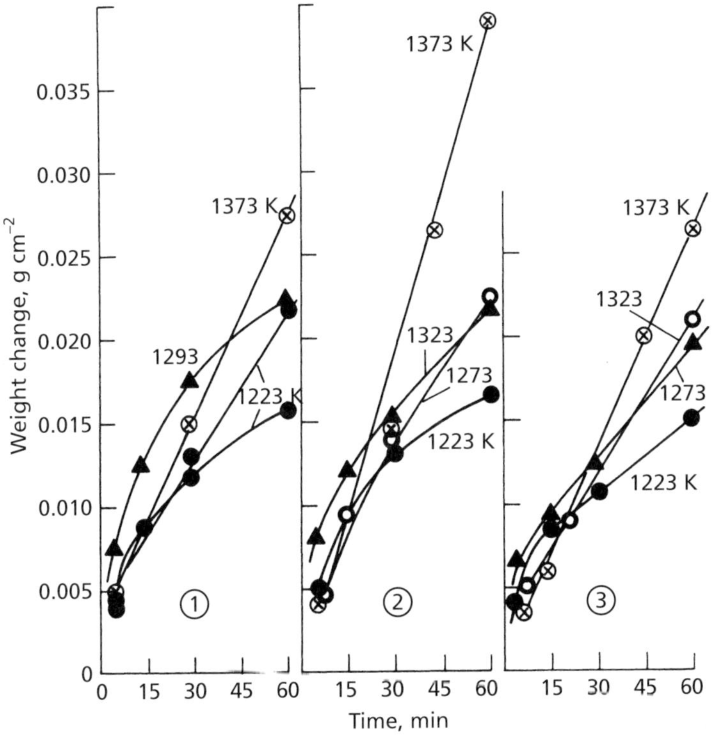

Figure 11: Effect of the production conditions on the time dependence of the corrosion of the Ni-alloy EI-437 B (Nimonic® 80) in 15% HNO3 + 4% HF at 60°C, following heat treatment at different temperatures [149]
① alloy melted in an open crucible and forged
② alloy melted in an arc under vacuum and forged
③ alloy melted in an arc under vacuum and cast

The alloy ChN58V (Ni-39 to 40Cr-0.5 to 1.5W) shows a remarkably low material consumption rate of about 0.00005 g/cm^2 h (0.11 mm/a) in a mixture consisting of 18% HNO_3 and 4% HF at 60°C. The corrosion rate of the steel 12Ch18N10T (c.f. 1.4541) is about 10 times greater than that of ChN58V [151].

A study of external corrosion rates of various nickel base alloys over wide ranges of concentrations and temperatures revealed a strong beneficial influence of molybdenum content [152]. However, tungsten, which is used as a partial replacement for molybdenum in some Ni-Cr-Mo alloys, appears to render the alloys more prone to internal attack. Copper appeared to be a beneficial addition to Ni-Cr-Mo alloys, as was found by weight loss measurement experiments in 5% and 10% hydrofluoric acid, as well as at temperatures of 66 and 79°C.

A series of immersion tests for eight high-alloy materials was performed in a variety of hydrofluoric acid (HF) environments [153]. Solution strengths of 0.05%, 1%, 10%, 24%, and 48% HF were examined at 24, 50, and 76°C. Coupons of each alloy were exposed fully immersed in the solution and partially exposed in the aerated vapor above each solution.

For the low acid concentration (0.05% HF), many alloys provided excellent corrosion resistance with only minor influence of temperature. Most prominent among these were alloys C-22, C-276, 825, and 20. As the acid concentration was increased (1%, 10% HF), corrosion resistance for fully immersed materials became more aggressively influenced by temperature. Under these conditions, alloys C-22, C-276, 200, and 400 exhibited the best overall performance. For fully immersed conditions, an increase in acid concentration above 10% (up to 48%) did not significantly affect the corrosion rate for the most resistant materials (alloys C-22 and 400).

Exposure to the vapor phase above the HF solutions confirmed the significance of relative aeration of the HF for corrosion resistance. Alloys 400 and 200 were found to be susceptible to extreme general corrosion rates (an order of magnitude higher than for fully immersed in the equivalent solution) and further susceptible to localized attack at the solution vapor interface. The other alloys examined revealed various degrees of increased general corrosion in the vapor space as acid concentration and temperature increased. For acid concentrations above 0.05% in solution, alloy C-22 exhibited the best vapor corrosion resistance, often by a substantial margin over other materials. Alloy C-276 also exhibited noteworthy vapor-phase resistance in many environments.

The corrosion behavior of eleven commercial nickel base alloys in wet hydrofluoric acid was quantified through laboratory testing [154]. Variables that were studied included testing time, acid concentration, temperature, vapor and liquid phases and the presence of residual stresses.

Results showed that for nickel alloys containing high chromium and which were exposed to the vapor phase, the corrosion rate decreased as the testing time increased. The corrosion rate was not highly dependent on the testing time when the coupons were exposed to the liquid phase. The corrosion rate of both alloy 400 and C-2000 alloy increased with the acid concentration and the temperature. However, both for increasing acid concentration and temperature, the corrosion rate of alloy 400 increased faster than the corrosion rate of C-2000 alloy, especially in the vapor phase. Considering factors such as corrosion rate by weight loss, internal penetration and susceptibility to stress cracking, C-2000 alloy had the best performance of all the tested alloys.

Nickel alloys containing niobium are prone to higher corrosion rates in HF. Table 22 includes corrosion data on wrought nickel alloys 625 (UNS N06625), C-276, C-22 (UNS N06022), and C-2000 in 5% HF [155]. Notably, from among these alloys, 625 had the highest corrosion rate at 79°C. This alloy has nominal 3.7% Nb.

Alloy	Corrosion rate at given temperature, mm/a	
	52°C	79°C
316L	15.4	98.5
904L	2.2	11.6
825	0.27	0.81
625	0.25	1.14
C-276	0.24	0.40
C-22	0.22	0.63
C-2000	0.06	0.27

Table 22: Wrought alloys exposed to 5% hydrofluoric acid [155]

Nickel-copper alloys

NiCu-alloys are distinguished by particularly high resistance in atmospheres and solutions containing fluorine and hydrogen fluoride. They are commercially available under the trade name Monel®, but do not have a uniform composition [156], for example

- Monel® 400 (Ni-31.5Cu-1.25Fe-1Mn-0.25Si-0.15C)
- Monel® 404 (Ni-44Cu-0.25Fe-0.05Mn-0.05Si-0.08C)
- Monel® K-500 (Ni-29.5Cu-2.7Al-1Fe-0.75Mn-0.6Ti-0.25Si-0.13C)

It is true that Monel® also has good resistance to fluorine, but between 348 and 437°C combustion-like reactions have been observed [157].

Monel® is used, amongst others, as a material for containers in which fluorine is prepared by means of graphite anodes [158]. The steel pipes which are concentrically fitted around the graphite anodes and serve as cathodes are separated from the anode by a diaphragm made of Monel®. Likewise, pumps containing Monel® casings and nickel pistons are used if fluorinated compounds and fluorine are present [159].

The following corrosion rates in Table 23 are given for Monel® (composition not disclosed) in fluorine:

Temperature, °C	Corrosion rate mm/a
400	0.15
500	0.60
600	18.00
700	45.00

Table 23: Corrosion behavior of Monel® in fluorine at various temperatures [157]

However, these values can serve only for orientation, since it is quite possible that the alloying additions present in the different Monel® types change the corrosion behavior. Thus, titanium drastically increases the corrosion rate of Monel® at higher temperatures [157]. Furthermore, corrosion is considerably intensified by moisture in the fluorine [110, 160].

In agreement with the above-mentioned values, the highest application temperature of Monel® alloys not only in fluorine but also in hydrogen fluoride is stated to be 537°C [161].

The fluorination of solid UF_4 to give gaseous UF_6 between 350 and 500°C is carried out in a reactor made of Monel®. Problems of severe corrosion did not arise [162].

By using Monel® as a material for valves which come into contact with fluorine or hydrogen fluoride, the service life could be prolonged by a factor of 3 to 4 and the costs reduced to about 1/10 [163].

In addition to nickel and platinum, Monel® is one of the metals having good resistance to hydrofluoric acid [103].

Monel® alloys have excellent resistance to hydrofluoric acid of all concentrations at room temperature, and fairly good resistance at temperatures up to 102°C [121, 122].

The positive effect of air removal through nitrogen purging on the corrosion rates of Monel® 400 for 24 hours exposure was investigated in [164]. Table 24 highlights the results. The results show that for a given set of conditions purging with nitrogen greatly reduces the corrosion attack.

Condition	Concentration, %	Temperature, °C	Corrosion rate, mm/a
saturated with air	25	30	1
	25	80	0.28
	50	30	0.2
	50	80	1
purged with nitrogen	25	30	0.005
	25	80	0.061
	50	30	< 0.003
	50	80	0.01

Table 24: Corrosion rates of Monel® 400 after 24 hours exposure to different concentrations of hydrofluoric acid with and without nitrogen purging [164]

In Figure 12 the isocorrosion curve for Monel® 400 is presented using data from immersion exposure to air-free and air-purged HF solutions [165]. A remarkable corrosion increase occurs in aerated HF.

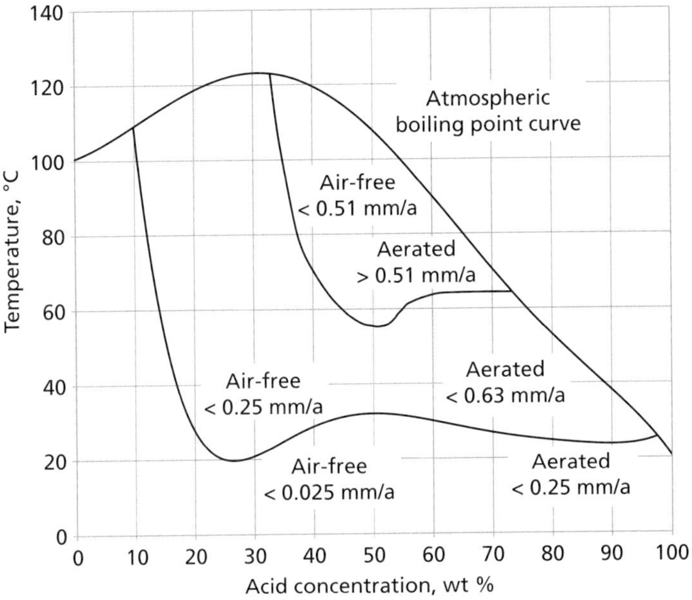

Figure 12: Isocorrosion curve for Monel® 400 in liquid HF with and without solution aeration [165]

In hydrogen fluoride which contains water vapor and air, Monel® is not as resistant as when air is excluded and is then also susceptible to stress corrosion cracking [166]. This is also true, for example, for 49% hydrofluoric acid at the boiling point [167].

Some values of the corrosion rates of a Russian Monel® alloy MHJMn28-2.5-1.5 (Ni-28Cu-2.5Fe-1.5Mn) in 6 to 60% hydrofluoric acid between 20 and 80°C are summarized in Table 25 [168].

As can be seen, the corrosion rate reaches a maximum at 40% HF and then decreases again at higher acid concentrations.

The fairly good corrosion behavior of the Monel® type NMJZ28-2.5-2.5 (Ni-28Cu-2.5Mo-1.5Fe) is also confirmed by results from the operation of gas purification units in magnesium and aluminum production (Table 26) [169].

Temperature, °C	*HF concentration, %*			
	6	*20*	*40*	*60*
	Corrosion rate, mm/a			
20	0.13	0.03	0.27	0.08
50	0.26	0.39	0.27	0.16
80	0.40	0.59	0.73	0.31

Table 25: Corrosion rates of Monel® NMJMn28-2.5-1.5 in hydrofluoric acid between 20 and 80°C [168]

Temperature, °C					
70	80	80	90	120	150
Concentration					
1% HF	1% HF	2% HF	2% HF	1 mg/m^3	HF
1)*	2)*	3)*	4)*	5)*	6)*
Corrosion rate, mm/a					
0.05	0.03	0.03	0.65	0.20	0.19

* 1) waste gas line, 2) discharge line from the scrubber, 3) and 4) line leading to the scrubber, 5) and 6) vessel in the foam zone

Table 26: Corrosion rates of Monel® NMJZ in waste gases containing hydrogen fluoride in various parts of the gas purification unit [169]

A special laboratory test system has been designed to develop stress corrosion cracking and general corrosion data for several nickel alloys in hydrogen fluoride and hydrofluoric acid environments at temperatures from 50 to 450°C [170]. Test results from this system and aqueous corrosion test results show Ni-Cu alloy 400 (UNS N04400) to be one of the best materials for high temperature applications and better than the ten other nickel base alloys examined in this study for resistance to general corrosion in aqueous hydrofluoric acid.

General corrosion data has been collected for several nickel containing alloys in dilute to concentrated aqueous HF solutions at temperatures up to the boiling point. Aqueous HF resistance for the alloys is summarized in Figure 13. The 0.51 mm/a isocorrosion curves are given for the various alloys.

The relative corrosion resistance of the materials examined in all concentrations and temperatures up to boiling in aqueous hydrofluoric acid liquid phase exposure was as follows:

Ni+Cu (alloy 400) > Ni+Mo (alloy B-2) > Ni+Cr+Mo+W (alloys C-276, 686, C-22, 59) > Ni+Cr+Fe (alloys 600, 690) = Ni+Cr+Fe+Mo (alloys 825, G-30) >> Fe+Cr+-Ni+Mo (316L).

UNS N04400 and UNS N24135 have been used at temperatures up to at least 149°C in many HF alkylation unit applications for distillation columns, heaters, recovery towers, pumps, and valves where process conditions include small amounts of water under air-free immersion conditions [171].

The two most useful alloys for hydrofluoric acid applications seem to be C-2000 and alloy 400. The corrosion behavior of these two nickel alloys in hydrofluoric acid has been characterized regarding the effect of temperature, acid concentration, and testing time [171]. Table 27 shows the uniform corrosion rate calculated by weight loss and the depth of attack by internal penetration, as a function of the temperature, and vapor versus liquid phase of exposure, for a testing time of 240 h. As the testing time increased from 24 to 240 h, the corrosion rate of C-2000 alloy in the vapor phase for 20% HF at both 79 and 93°C markedly decreased, while in the liquid

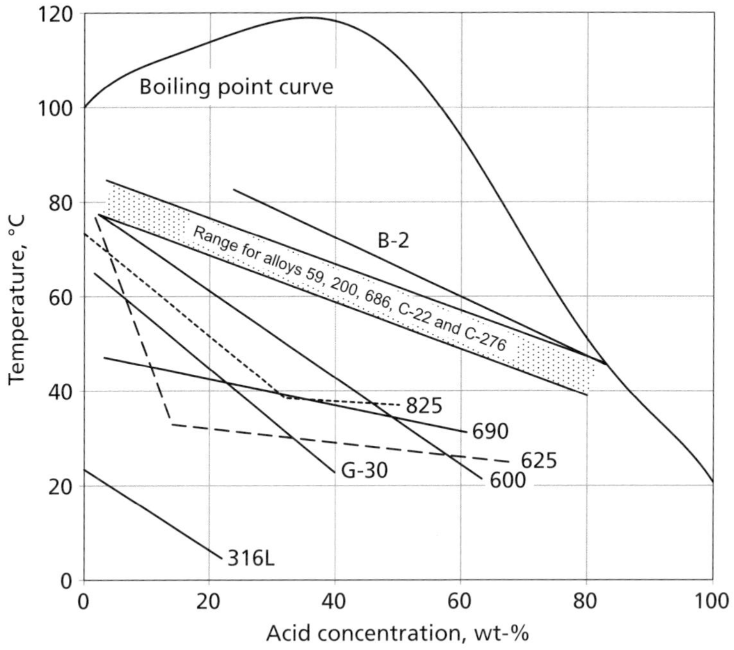

Figure 13: Summary of hydrofluoric acid isocorrosion curves showing 0.51 mm/a curves for various alloys [170, 172]

phase, the corrosion rate was almost independent of the testing time. On the other hand, under the same test conditions, the corrosion rate of alloy 400 in the liquid phase slowly increased as the test time increased. In the vapor phase, the corrosion rate of alloy 400 was high (2.5 to 10 mm/a) and independent of the testing time. As would be expected, as the acid concentration increased from 1 to 20%, the corrosion rates of both alloys increased; however, the effect was more pronounced for alloy 400, especially in the vapor phase. Both materials had similar corrosion rates at the lower temperatures of 38 and 52°C; however, as the temperature was raised to higher values, the corrosion rate of alloy 400 increased faster than the corrosion rate of C-2000 alloy, both in the vapor and liquid phases. Table 27 also indicates that the internal penetration in alloy 400 is high at all temperatures in the vapor phase. On the other hand, in the liquid phase, the internal penetration was higher for C-2000 alloy, mainly at 79 and 93°C. From these results, it is evident that chromium is a beneficial element for vapor phase applications. It is also apparent that copper is the most beneficial alloying element for nickel alloys, while iron seems to be detrimental.

Common name	UNS No.	Temperature, °C	Phase (a)	Average corrosion rate, mm/a	Internal penetration, μm	Observations
400	N04400	38	L	0.27	13	Uniform corrosion, (b)
		38	V	1.15	241	IGA. Cracks at the perforated hole and edges
		52	L	0.46	3	Uniform corrosion, (b), GB etching
		52	V	3.31	622	Severe IGA. Some cracks from edges
		66	L	0.22	13	Uniform corrosion, (b)
		66	V	1.76	191	IGA. Cracks along drilled hole
		79	L	1.1	25	Uneven corrosion. GB etching
		79	V	7.59	356	Deep IGA
		93	L	2.06	25	Uneven corrosion. GB etching
		93	V	12.3	368	Deep IGA and uneven corrosion
C-2000	N06200	38	L	0.21	13	Uniform corrosion, (b)
		38	V	0.19	241	Uniform corrosion, (b), trace of copper on surface
		52	L	0.48	13	Uniform corrosion, (b), copper on surface
		52	V	0.36	13	Uniform corrosion, (b)
		66	L	0.85	38	Uneven general corrosion. Small de-alloyed layer. Sample plated with copper and probably nickel
		66	V	0.85	13	Minimal corrosion. Sample plated with nickel and traces of copper

(a) L, liquid; V, vapor. (b) Roughness reported as internal penetration. IGA, intergranular attack; GB, grain boundary

Table 27: Effect of temperature on the corrosion behavior of unstressed UNS N04400 and UNS N06200 coupons exposed in 20% hydrofluoric acid for 240 h [172]

Table 27: Continued

Common name	*UNS No.*	*Temperature, °C*	*Phase (a)*	*Average corrosion rate, mm/a*	*Internal penetration, μm*	*Observations*
C-2000	N06200	79	L	0.67	102	Thin, lacelike penetration. Light copper color
		79	V	0.92	38	Shallow, isolated, lacelike penetration
		93	L	0.74	216	Thin, lacelike penetration
		93	V	0.69	64	Isolated, lacelike penetration. Green corrosion products

(a) L, liquid; V, vapor. (b) Roughness reported as internal penetration. IGA, intergranular attack; GB, grain boundary

Table 27: Effect of temperature on the corrosion behavior of unstressed UNS N04400 and UNS N06200 coupons exposed in 20% hydrofluoric acid for 240 h [172]

Nickel-molybdenum alloys

NiMo-alloys are resistant to hydrofluoric acid and anhydrous hydrogen fluoride in the temperature range from 17 to 57°C.

Even in the particularly aggressive solution of 10 mol/l HCl and 0.2 to 1.0 mol/l HF, the NiMo-alloy EP-496 (Ni-25 to 29Mo- < 4Fe-1.4 to 1.7V- < 0.5Mn- < 0.5Si- < 0.05C) is highly resistant at a maximum corrosion rate of 0.52 mm/a. However, oxidizing compounds, for example Fe^{3+}, accelerate the corrosion, since they shift the potential to more positive values. One possible protection method consists of the addition of a strong and rapidly acting reducing agent which, for example, reduces the Fe^{3+}-ions, thus making them ineffective [173]. As shown in Figure 14, the potential decreases steeply at the beginning and reaches a stationary value at 6 g/l Fe^{3+}, which characterizes the transpassive range in which the corrosion is, however, still considerable. The corrosion rate in a solution containing 10 mol/l HCl + 0.002 mol/l F^- + 5 g/l Fe(III) salt can be reduced from 24 to 0.54 mm/a, which corresponds to the values of the acid free oxidizing agent, by means, for example, of 10 g/l $SnCl_2$. Aluminum chloride is also effective. The best reagent is titanium chloride, which reduces the corrosion rate by about a factor of 10 to 0.04 mm/a.

In the presence of Fe^{2+}-ions, the titanium ions, due to the high overvoltage, prevent the formation of the corrosion-promoting Fe^{3+}-ions. The redox system Ti^{2+}/Ti^{3+} is probably responsible for this effect. However, in the presence of 1 to 5 g/l Fe^{3+}-ions 10 to 20 g/l titanium ions are necessary to inhibit corrosion effectively [173].

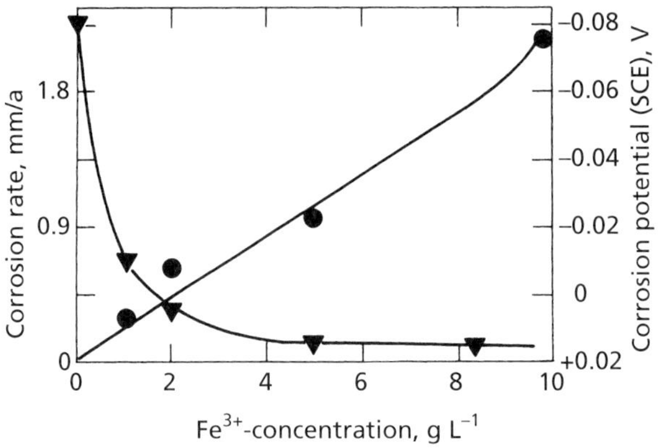

Figure 14: Corrosion rates (left axis) and corrosion potentials (right) of the alloy EP-496 in a solution consisting of 10 mol/l HCl + 0.2 mol/l F^- as a function of the Fe^{3+}-concentration [173]

A NiMoFe-alloy (Ni-27.9Mo-6Fe-0.99Cr-0.02Si-0.03C) has good resistance in 50 to 100% hydrofluoric acid at 20°C (Figure 15).

In anhydrous hydrofluoric acid, the weight change decreases by about two orders of magnitude (see the ordinate on the right of Figure 15). This demonstrates the corrosion-promoting effect of water quite clearly [134]. At 50°C, the attack on this alloy corresponds to that on the NiCrMoFe-alloy in Figure 9.

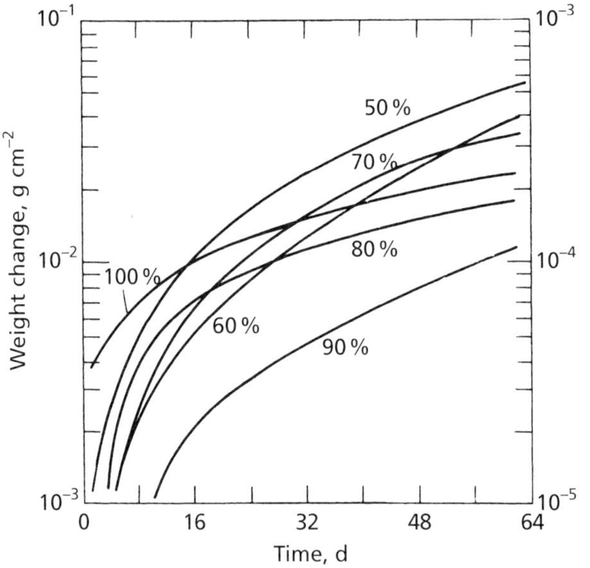

Figure 15: Time dependence of the corrosion of an NiMoFe-alloy at 20°C in hydrofluoric acid of different concentrations. The ordinate on the right refers to anhydrous 100% hydrofluoric acid [134]

Hastelloy® B (Ni-26 to 30Mo-4 to 7Fe-2.5Nb) and Hastelloy® B-2 (Ni-26 to 30Mo-2Fe-1Co) are stated to be resistant to hydrofluoric acid, although only in a non-oxidizing environment [174]. Hastelloy® B-2 alloys are not preferentially attacked due to their low carbon and silicon contents. Finally an alloy (Hy-Mu80 (Ni-16Mo-7Cr-5Fe-0.8Mn-0.35Cu-0.3Si-0.2Co)) which is resistant to gaseous fluorine should be mentioned [114].

The good resistance of NiMo-alloys to acids containing hydrogen fluoride is also confirmed by Figure 16.

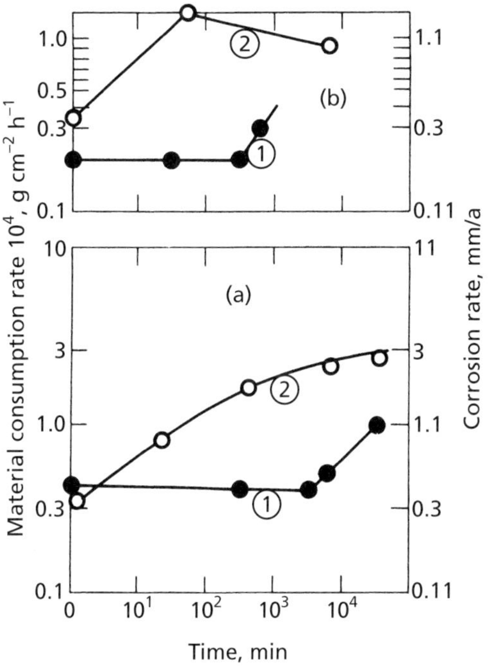

Figure 16: Dependence of the corrosion behavior of a monocrystalline (1) and polycrystalline (2) Ni28Mo alloy in 18% HNO_3 + 4% HF at 60°C on the duration of the heat treatment at 700°C (a) and 800°C (b) [175]

Single crystals of these alloys are clearly more resistant than polycrystalline materials. Furthermore, heat treatment at 800°C is more favorable than treatment at 700°C [175].

Other nickel alloys

Of the other nickel alloys, the NiFe- and NiAl-alloys are fairly resistant materials to fluorine and hydrogen fluoride. Figure 17 shows the time dependence of the corrosion of a Ni15Fe alloy (79 NM) at different temperatures in a gas mixture consisting of 93% fluorine, 4 to 5% hydrogen fluoride and 3 to 4% inert gas (containing F_2O)

at a flow rate of 20 cm/s. However, above 497°C the corrosion rate of the alloy is very high at 2.5 mm/a [176].

Comparison of the corrosion behavior of a NiAl-alloy (NA-1) containing 1 to 1.5% aluminum with that of unalloyed nickel NP-2 (Figure 18) shows that nickel is attacked in a mixture consisting of 94% fluorine + 5% HF + 0.3% O_2 + 0.2% CO_2 + 0.04% F_2O at test durations of 10 to 100 h more severely than the NiAl-alloy [177], while the difference in the corrosion rates below 597°C can be disregarded. A fine-crystalline fluoride film which has protective properties is formed on the alloy, while the film which grows on nickel due to corrosion and has a clearly coarser structure does not act as protection. Alloys which have higher aluminum contents and contain titanium and carbon, for example Duranickel® (Ni-4.5Al-0.5Ti-0.15C), are attacked at 500°C about 3 times faster than unalloyed nickel [114].

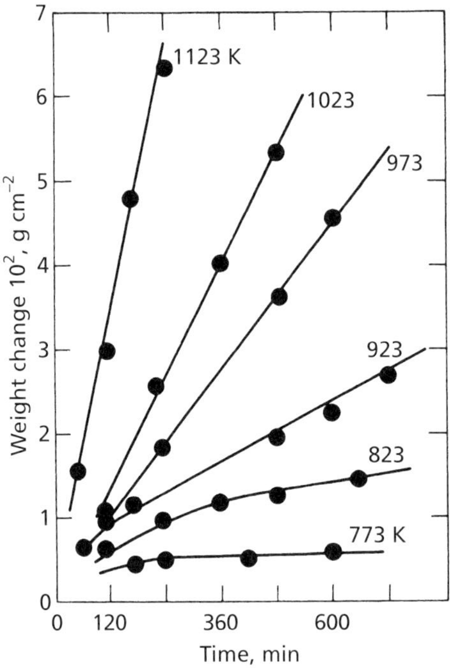

Figure 17: Time dependence of the corrosion of 79NM (Ni15Fe) at different temperatures in a gas consisting of 93% fluorine, 4 to 5% HF and 3 to 4% inert gas at a gas velocity of 20 cm/s [176]

NiMn-alloys (D nickel and nickel 211) containing 4.5% manganese and 0.1% carbon corrode in fluorine at 600°C about 6 times faster than pure nickel [127].

For nickel 61 (Ni-3Ti-1.5Al-1Mn-1Fe-0.75Si-0.25Cu-0.15C), which is used for the welding of nickel, different corrosion values are given. In one case, the corrosion rate is said to be 800 to 1,000 times greater at 477°C than that of pure nickel [178], while in another case the corrosion rate of the two materials at 500°C is said to be approximately the same [129]. As a result of the relatively high titanium content in nickel 61, the higher value is most likely to represent the true situation.

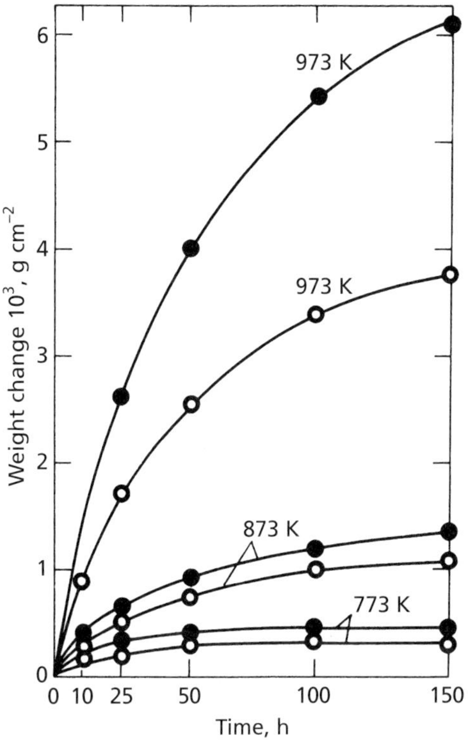

Figure 18: Progress in time of the corrosion of nickel NP-2 (●) and nickel-aluminum NA-1 (○) in fluorine containing (percent by volume) 5 HF, 0.3 O_2, 0.2 CO_2 and 0.04 F_2O of 0.12 to 0.13 MPa between 500 and 700°C [177]

Alloys of the type Ni-17Fe-2Mo (Permalloy®), which are used in induction coils and for integrated circuits, have to be coated with protective films of high electrical resistance. This is done by exposing the metal powder to fluorine between 197 and 447°C, preferably 347°C, for 15 minutes to 8 h; the fluoride films produced are 0.6 μm thick and have a specific resistance of 2×10^{11} Ω cm [179]. This avoids the complicated process of applying insulating films consisting of $Mg(OH)_2$-Na_2SiO_3. Apart from fluorine, NH_4F is also said to be suitable for producing this type of insulating fluoride film.

In a new process, Permalloy® films (Ni-16Fe-3Mo) are sprayed onto partially oxidized hot silicon (3 μm thick) at 240°C coated with titanium about 0.05 to 0.10 μm thick and then etched in H_2SO_4-HF solutions, in the course of which HF concentrations of 0.2 mol/l and more corrode the duplex film Permalloy®/titanium in an oriented fashion at an angle between 6 and 40°, as is necessary for integrated circuits [180].

In a solution consisting of 3.9 mol/l H_2SO_4, 1.12 mol/l H_2O_2, 0.2 to 3.1 mol/l HF, the etching rate – independent of the HF concentration – is 24 μm/h at room temperature.

Formic Acid

Author: *H. Leyerzapf* / Editor: *R. Bender, R. Durham*

Nickel

Results from laboratory and operational tests on the corrosion behavior of nickel differ very widely [181–184]. According to the results of a 48 hour laboratory test, on exposure to boiling, deaerated 10% formic acid there is no difference between the corrosion rates of nickel and copper, both being 0.15 mm/a [181]. Deaerated 90% boiling formic acid has a more marked effect on nickel than on copper, whereas in the presence of air, nickel is attacked substantially less than copper. At the same time, however, the corrosion rates of 0.38 to 0.99 mm/a for nickel in the liquid phase and 0.30 to 0.53 mm/a in the vapor space are such as to exclude their use [181]. According to the results of an operational test, in this case the material loss of nickel on exposure for several days is about 47% of that of copper, although the corrosion rate of 0.46 mm/a means that nickel cannot be used [184].

Nickel-chromium alloys

Nickel-chromium-iron alloys

Nickel-chromium-molybdenum alloys

Nickel-copper alloys

Nickel-molybdenum alloys

Other nickel alloys

Laboratory tests were carried out on the corrosion behavior of various nickel alloys in boiling formic acid. The corrosion rates determined are listed in Table 28 [181].

The nickel alloys are equivalent to the DIN materials listed in Table 29 below or contain the most important alloying elements listed there.

The behavior of the nickel-based alloys Hastelloy® C-276 and Hastelloy® B-2 on exposure to boiling 40% and 88% formic acid in comparison with the behavior of the Ferralium® alloy and the austenitic CrNiMo-steel SAE 316L has been examined elsewhere [185].

The results of laboratory tests with the nickel-based alloys Hastelloy® C-4 and Hastelloy® B after exposure to 50% boiling formic acid with the exclusion of air are given in Table 30. Air was excluded by gassing with "special" high-purity nitrogen (99.996% N_2) [186].

HCOOH-Concentration %	Material	Corrosion rates, mm/a			
		Non-aerated		Aerated	
		Liquid	Vapor phase	Liquid	Vapor phase
10	Monel® 400	0.025	0.08	–	–
50	Hastelloy® C	–	–	0.13	0.05
50	Monel® 400	0.025	0.10	–	–
50	Hastelloy® B	–	–	0.56	0.13
90	Inconel® 600	0.53–0.99	1.30	1.47	0.23
90	Hastelloy® C	0.05	–	0.18	0.025
90	Monel® 400	0.025–0.08	0.025	7.62	0.23
90	Hastelloy® B	0.05	–	0.15	0.05

Table 28: Corrosion rates of nickel alloys in boiling formic acid of various concentrations (test duration 48 hours) [181]

Nickel alloy	DIN-Material No.	Alloy composition, %
Inconel® 600	2.4816	Ni-15.5Cr; 8Fe; 0.05C
Hastelloy® C	2.4602	Ni-16-18Mo; 13-17.5Cr; 3.7-5.3W; 4.5-7.0Fe
Hastelloy® C-276	2.4819	Ni-16Cr; 16Mo; 5Fe; 4W; 2.5Co; 0.35V
Hastelloy® C-4	2.4610	Ni-14-17Mo; 14-18Cr; ≤ 0.015C; ≤ 2.0Co; ≤ 0.7Ti; ≤ 3.7Fe
Monel® 400	2.4360	Ni-0.12C; 0.9Mn; 1.35Fe; 0.005S; 0.15Si; 31.5Cu
Hastelloy® B	2.4800	Ni-0.6Cr; 28Mo; 5.5Fe; 0.3V; 1.25Co; 0.1 C
Hastelloy® B-2	2.4615	Ni-26-30Mo; ≤ 0.02C; ≤ 1.0Co; ≤ 1.0Cr; ≤ 2.0Fe

Table 29: DIN-Material No. and alloying elements of the nickel alloys investigated in Table 28

Alloy	DIN-Material No.	Corrosion rate
		mm/a
NiMo 30 (Hastelloy® B)	2.4810	0.02
NiCr 21 Mo (Incoloy® 825)	2.4858	0.07
NiCr 16 Mo 15 (Hastelloy® C-4)	2.4610	0.11

Table 30: Corrosion rates of nickel-based alloys in 50% boiling formic acid (test duration 14 days) [186]

The relatively moderate corrosion of the Hastelloy® C-4 alloy in 50% boiling formic acid is due to the complete exclusion of air [186]. According to information provided by a manufacturer of Hastelloy® C-276, the maximum corrosion rate for this material in formic acid is < 0.05 mm/a [187]. The use of Hastelloy® B under severe conditions highly-stressed in the manufacture of formic acid has already been mentioned [188].

Hydrochloric Acid

Author: *P. Drodten* / Editor: *R. Bender, R. Durham*

Nickel

The corrosion resistance of nickel and nickel-based alloys covers a wide range of very aggressive media. While pure nickel is predominantly used in situations where there is exposure to very alkaline media, the NiCrMoFe and NiMo alloys are particularly suitable for acidic media.

Pure nickel (nickel 200, UNS N02200, Ni99.6 DIN-Mat. No. 2.4060, Ni99.2 DIN-Mat. No. 2.4066) exhibits some resistance in dilute hydrochloric acid at room temperature. The attack is enhanced by aeration and the presence of oxidizing agents in the acid.

Nickel-chromium alloys

Nickel-chromium-iron alloys (without Mo)

The molybdenum-free NiCr or NiCrFe alloys, e.g. NiCr16Fe (UNS N06600, Inconel® 600, Alloy 600, DIN-Mat. No. 2.4816) or NiCr21Fe (Inconel® 800), are not used in hydrochloric acid because in dilute solutions they are susceptible to pitting corrosion and in concentrated solutions at elevated temperatures they are also susceptible to general surface corrosion. The corrosion resistance of NiCr and NiCrFe is lower than that of pure nickel or alloy 400. More information on the resistance of NiCr or NiCrFe alloys in hydrochloric acid is given in reference [189].

Nickel-chromium-molybdenum alloys

The nickel-chromium alloys used in chemical process engineering generally contain molybdenum as well to increase the pitting and crevice corrosion resistance in acidic media. The most commonly used materials are given in Table 31 with the frequently used and familiar manufacturer's designations.

As the molybdenum content increases, the resistance of the NiCrMo alloys increases in hydrochloric acid so that alloy 625 has better resistance than alloy G-3, which in turn is more resistant than alloy 825.

As the isocorrosion diagram in Figure 19 shows for alloy 825, it can be expected that this material has insufficient resistance at low acid concentrations and temperatures [190].

Manufacturer's designation	UNS	DIN-Mat. No.	Abbreviation
Incoloy® 825 Nicrofer® 4221 Alloy 825 Uranus® 825	N08825	2.4858	NiCr21Mo
Hastelloy® G-3 Alloy G-3 Nicrofer® 4823 hMo	N06985	2.4619	NiCr22Mo7Cu
Inconel® 625 Alloy 625 Nicrofer® 6020 hMo Uranus® 625	N06625	2.4856	NiCr22Mo9Nb
Hastelloy® C-276 Alloy C 276 Nicrofer® 5716 hMoW	N10276	2.4819	NiMo16Cr15W
Hastelloy® C-4 Alloy C-4 Nicrofer® 6616 hMo	N06455	2.4610	NiMo16Cr16Ti
Nicrofer® 5923 hMo Alloy 59	N06059	2.4605	NiCr23Mo16Al
Hastelloy® C-22 Alloy C-22 Nicrofer® 5621	N06022	2.4602	NiCr21Mo14W
Inconel® 686 Alloy 686 Nicrofer® 6032	N06686	2.4606	NiCr21Mo16W

Table 31: Nickel-chromium-molybdenum alloys frequently used in chemical process engineering

The isocorrosion diagrams for alloy G-30 (UNS N06030, a modified G-3 variant) in Figure 20, alloy C-276 in Figure 21, alloy C-4 in Figure 22 and alloy 59 in Figure 23 demonstrate the considerably improved behavior of these alloys [190].

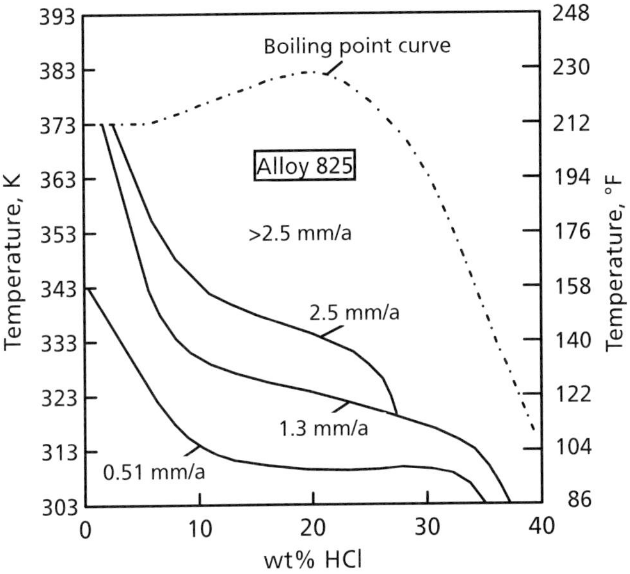

Figure 19: Isocorrosion diagram for alloy 825 in hydrochloric acid [190]
> 2.5 mm/a (> 100 mpy)
2.5 mm/a (100 mpy)
1.3 mm/a (50 mpy)
0.51 mm/a (20 mpy)

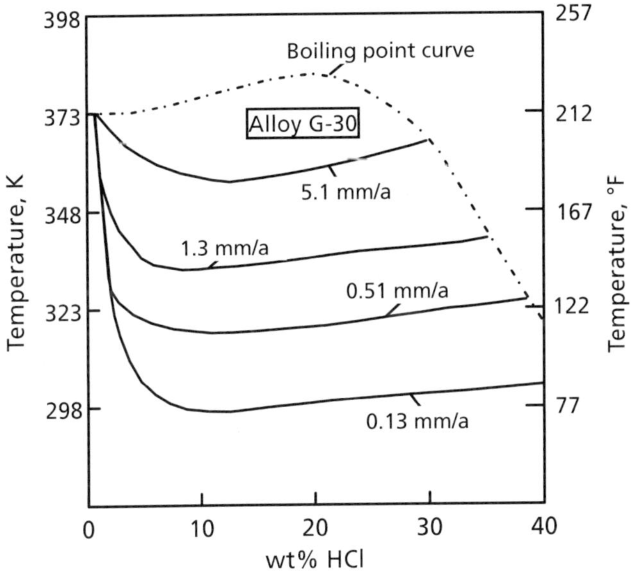

Figure 20: Isocorrosion diagram for alloy G-30 in hydrochloric acid [190]
5.1 mm/a (200 mpy)
1.3 mm/a (50 mpy)
0.51 mm/a (20 mpy)
0.13 mm/a (5 mpy)

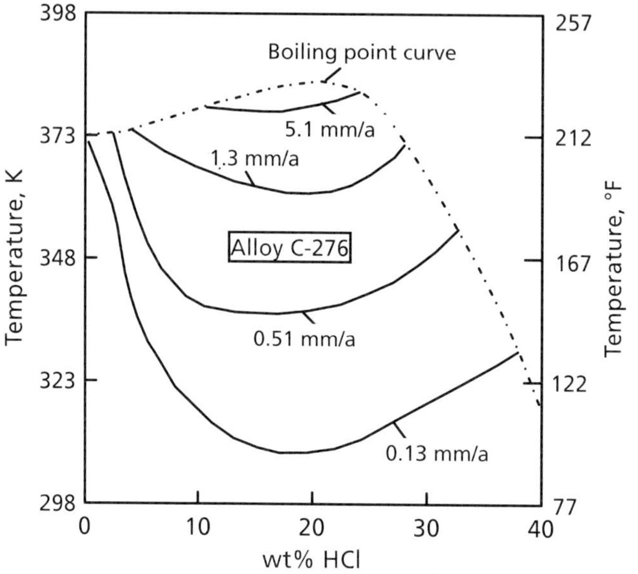

Figure 21: Isocorrosion diagram for alloy C-276 in hydrochloric acid [190]
5.1 mm/a (200 mpy)
1.3 mm/a (50 mpy)
0.51 mm/a (20 mpy)
0.13 mm/a (5 mpy)

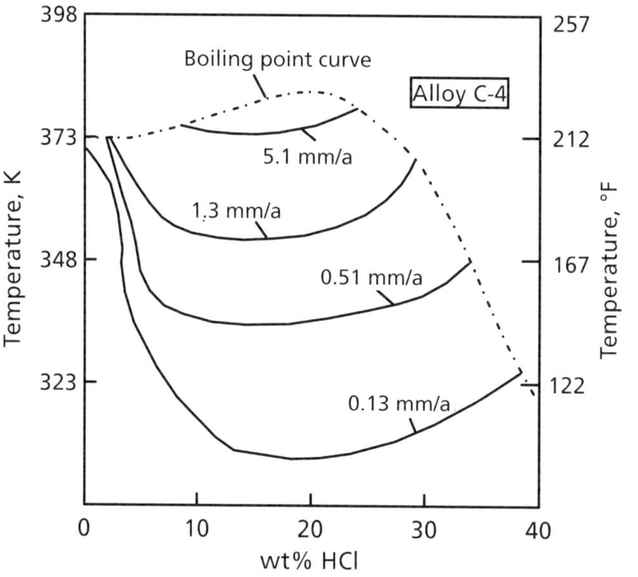

Figure 22: Isocorrosion diagram for alloy C-4 in hydrochloric acid [190]
5.1 mm/a (200 mpy)
1.3 mm/a (50 mpy)
0.51 mm/a (20 mpy)
0.13 mm/a (5 mpy)

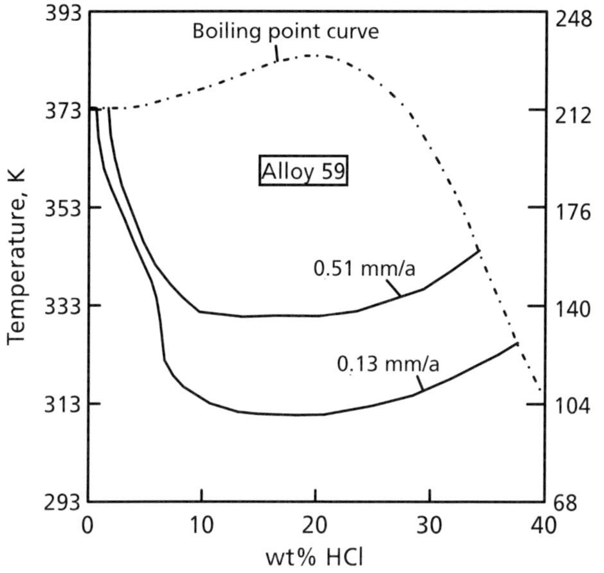

Figure 23: Isocorrosion diagram for alloy 59 in hydrochloric acid [190, 191]
0.51 mm/a (20 mpy)
0.13 mm/a (5 mpy)

Table 32 is taken from [192] and summarizes data for various Ni-Cr-Mo alloys. At higher acid concentrations alloy 625 showed satisfactory corrosion resistance. At low concentrations of boiling acid C-276 showed better corrosion resistance than C-22.

Alloy	*HCl Concentration, %*	*Temperature, °C*	*Tested Time*	*Corrosion rate, mm/a*	*Reference*
G-30	5	50		0.49	[193]
	10	20		0.5	[193]
	20	20		0.5	[193]
	30	20		0.5	[193]
C-276	1	boiling		0.25	[194]
	1	boiling		0.34	[194]
	1.5	boiling		0.74	[194]
	2	90		0.025	[194]
	2	boiling		1.55	[194]
	3	boiling		1.78	[194]

* – welded with 625

Table 32: Corrosion rates of selected nickel alloys in hydrochloric acid [192]

Table 32: Continued

Alloy	HCl Concentration, %	Temperature, °C	Tested Time	Corrosion rate, mm/a	Reference
C-22	1	boiling		0.076	[194]
	1.5	boiling		0.28	[194]
	1	boiling	5 x 48 h	0.36	[195]
	2	90		nil	[194]
	2	boiling		1.55	[194]
	3	boiling		< 1	[194]
	3	boiling		2.13	[194]
625	1.3	40	28 days	< 0.01	[196]
	1.3	40	28 days*	0.09	[196]
	5	66		1.8	[197]
	10	66		2.1	[197]
	15	66		1.7	[197]
	20	66		1.3	[197]
	25	66		1.0	[197]
	30	66		0.9	[197]
	concentrated	66		0.4	[197]

* – welded with 625

Table 32: Corrosion rates of selected nickel alloys in hydrochloric acid [192]

Reference [198] discusses the factors that led to the development of the Ni-Cr-Mo-W alloy Hastelloy® C-22 (UNS N06022). Compared to other nickel-based alloys, it exhibits very high corrosion resistance both in very aggressive reducing and oxidizing acids.

Table 33 gives the composition of alloys in the development sequence of alloys developed prior to alloy C-22. During welding of alloy C with max. 0.08% C and max. 1.0% Si, chromium carbides were precipitated on the grain boundaries in the substrate material next to the weld seam, making this zone susceptible to intergranular corrosion. Therefore, the utilization of this material was greatly limited because a subsequent solution annealing and quenching to eliminate the susceptible zones was only possible to a limited extent.

Alloy	*Cr*	*Mo*	*W*	*Fe*	*Nb*	*Ti*	*V*	*Mn*	*C*	*Si*	*Ni*
Alloy C	15	16	4	5	–	–	0.35(1)	1.0(1)	0.08(1)	1.0(1)	balance
Alloy C-276	16	16	4	5	–	–	0.35(1)	1.0(1)	0.01(1)	0.08(1)	balance
Alloy C-4	16	16	–	3(1)	–	0.70(1)	–	1.0(1)	0.01(1)	0.08(1)	balance
Alloy C-22	22	13	3	3	–	–	0.35(1)	0.50(1)	0.015(1)	0.08(1)	balance
Alloy 625	21.5	9	–	5.0(1)	3.5	0.40(1)	–	0.5(1)	0.10(1)	0.50(1)	balance
Alloy HC new	20	15	–	2	–	–	–	–	0.01	0.03	balance

(1) Maximum

Table 33: Compositions of the Ni-Cr-Mo/W alloys, mass-% [198]

By reducing the carbon content to max. 0.01% and the silicon content to 0.08%, intergranular susceptibility was eliminated in alloy C-276 [199]. In spite of its excellent general corrosion resistance, this alloy is still somewhat susceptible to preferred attack in the welding seam and in the heat-affected zone of the welded material because in the temperature range from 700 to 1,100°C a chromium and molybdenum-rich intermetallic phase (μ-phase) can be precipitated that is also susceptible to intergranular corrosion in aggressive acids.

Alloy C-4 was developed at a later date. This tungsten-free alloy with a reduced iron content is particularly thermally stable. An intermetallic phase is precipitated only after 100 h in the temperature range from 650 to 1,100°C. The intergranular susceptibility due to precipitation of carbides is avoided by the low carbon content and by stabilization with titanium. However, due to the lack of tungsten, the resistance to local corrosion (pitting and crevice corrosion) is reduced. Additionally, the resistance of alloy C-4 to oxidizing acids is limited.

Increasing understanding of the effect of the individual alloying elements on the corrosion behavior allows the development of alloys with optimum corrosion properties. In nickel-based alloys, chromium enhances the resistance in oxidizing acids, whereas molybdenum and tungsten enhances that in reducing acids. Based on this knowledge, the atomic percent factor (APF) was developed in order to compare the corrosion resistance of alloys with differing compositions. It expresses the opposing effect of chromium compared to that of molybdenum and tungsten. The ratios of the atomic weights of the relevant elements are

$$\text{Cr (52) : Mo (96) : W (184)} = 1 : 2 : 4$$

The APF is defined as the ratio of four times the weight-% of chromium to the sum of twice the weight-% of molybdenum + one times the weight-% of tungsten, namely

$$\text{APF} = 4\ \text{Cr}/(2\ \text{Mo} + 1\ \text{W}).$$

A high APF indicates high resistance to oxidizing acids, whereas a low APF indicates a high resistance to reducing acids. For an intermediate value of 2.5 to 3.3, the resistance to oxidizing as well as reducing acids is high. The ratios are represented graphically in Figure 24. Alloy C-22 has an APF ≈ 3, which is the most favorable composition with respect to the corrosion behavior in oxidizing and reducing acids.

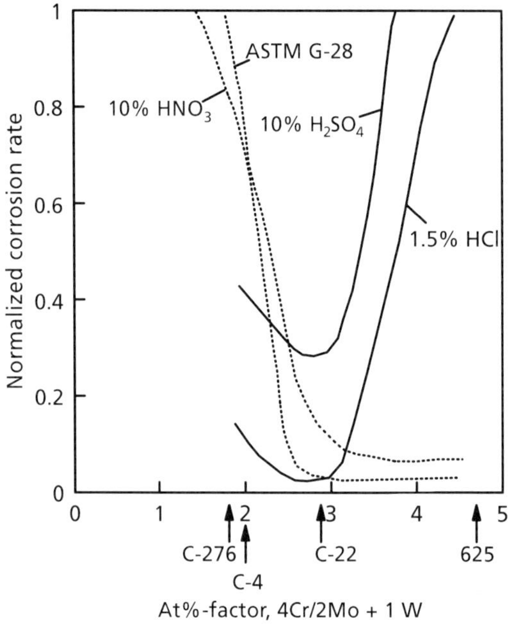

Figure 24: Range of optimum resistance in reducing and oxidizing environments [198]

C-22 has a higher thermal stability than C-276. On tempering at 927°C, the temperature at which the fastest precipitation of the intermetallic phase occurs, intergranular susceptibility starts in less than 2 min for C-276, and in less than 15 min for C-22.

Acid mixture	*Temperature °C*	*C-22*	*C-276*	*C-4*	*625*
5% HNO_3 + 6% HF	60	1.68	5.18	5.1	1.825
5% HNO_3 + 1% HCl	boiling	0.0125	0.2	0.275	0.025
5% HNO_3 + 25% H_2SO_4 + 4% NaCl	boiling	0.3	1.6	2.425	17.83
23% H_2SO_4 + 1.2% HCl + 1% $FeCl_3$ + 1% $CuCl_2$	boiling	0.175	1.375	57.35	96.18

Table 34: Corrosion rates under oxidizing conditions, mm/a [198]

Table 34 gives comparative corrosion rates of the alloys listed in Table 33 in oxidizing and reducing acidic aggressive mixed media. Alloy C-22 exhibited the highest corrosion resistance in all media [198].

Alloy C-22 also exhibited the highest resistance in pitting and crevice corrosion tests in the FeCl3 test with increasing temperature [201]. The lower molybdenum content compared to alloy C-276 is more than over compensated for the higher chromium content.

Even in the stress corrosion cracking tests, C-22 was generally better than alloys C-276 and 625 in aggressive media that trigger stress corrosion cracking.

However, as the values in Table 35 show for pure hydrochloric acid, the corrosion rates of alloy 59 and alloy C-276 were lower than those of alloy C-22 [190].

Acid concentration %	***Temperature °C***	***Corrosion rate mm/a***		
		C-22	**C-276**	**59**
1.5	Boiling	0.87	0.69	0.38
2	90	0.051	0.53	0.00
2	Boiling	2.18	1.14	0.99
3	90	1.41	0.96	0.053
3	boiling	4.55	2.06	2.05
5	93	5.07	1.99	2.81
10	55	0.77	0.44	0.55
10	82	3.03	1.56	2.47
10	Boiling	9.26	6.07	7.50

Table 35: Corrosion rates of the NiCrMo alloys C-22, C-276 and 59 in hydrochloric acid of differing concentrations and temperatures [190]

Acid mixtures

Acid mixtures are frequently used in chemical and metallurgical processes. For example, mixtures of nitric acid and hydrofluoric acid are used to pickle stainless steels. Although there have been many publications on the corrosion behavior of many alloys in pure acids, not enough is known about the corrosion behavior in mixed acids. Reference [202] reports on the corrosion behavior of nickel-based alloys in mixed acids. The results are based on the determination of the mass loss rates in static solutions that have not been deaerated.

Alloy	UNS No.	Fe	Cr	Ni	Mo	C	others
SAE 316L (c.f. DIN-Mat. No. 1.4404)	S31603	balance	17	12	2.5	0.93	–
Ferallium® Alloy 255 (c.f. DIN-Mat. No. 1.4492)	S32550	balance	26	5.5	3.0	0.04[(1)]	Cu = 1.7 N = 0.2
Alloy C-276	N10276	5.0[(1)]	16	balance	16	0.01[(1)]	W = 4.0
Alloy C-22 (c.f. DIN-Mat. No. 2.4602)	N06022	3.0	22	balance	13	0.015[(1)]	W = 3
Alloy G-3 (c.f. DIN-Mat. No. 2.4619)	N06985	20	22	balance	7	0.015[(1)]	Cu = 2.0 Nb = 0.5[(1)]
Hastelloy® Alloy G-30 (c.f. DIN-Mat. No. 2.4603)	N06030	15	29.5	balance	5	0.3[(1)]	Nb = 0.7 W = 2.5 Cu = 1.7
Alloy B-2 (c.f. DIN-Mat. No. 2.4617)	N10665	2.0[(1)]	1.0[(1)]	balance	28	0.02[(1)]	–

[(1)] Minimum

Table 36: Compositions of the investigated alloys, mass-% [202]

Table 36 gives the maximum chemical composition of the investigated alloys in mass-%. For comparison purposes, the stainless steel SAE 316L (1.4404) and the duplex steel Ferallium® Alloy 255 were also included.

In general, the corrosion rate in reducing acids may increase if oxidizing media are also present in the mixture. Therefore, it is particularly important to determine the corrosion behavior of engineering alloys in acid mixtures.

Figure 25 gives the isocorrosion curves for alloy Hastelloy® C-276 at 80°C in the three-component representation H_2O – HCl – HNO_3. The data are limited to one corner of the triangular representation because the maximum concentrations of the analytical grade acids used was 71% for HNO_3 and 38% for HCl. This form of representation shows the corrosion behavior of an alloy in a wide range of acid combinations; however, it is not very suitable for a comparison of the corrosion behavior of different alloys. Further results are thus given as a function of the proportion of an acid in another acid of constant concentration.

Figure 26 gives the corrosion rates of alloys C-22, C-276 and B-2 in 15% H_2SO_4 at 80°C with increasing HCl concentration. Due to the reducing effect of hydrochloric acid, the corrosion rate increases by differing amounts, and increases more strongly with decreasing Mo content in the alloy. The attack is in the form of pitting corrosion.

On the other hand, with respect to the favorable behavior of B-2 in this mixture of two reducing acids (Figure 26), it must be pointed out that the corrosion resistance of B-2 alloy is reduced in the presence of oxidizing media, e.g. Fe^{3+} ions which are frequently present in industrial acids.

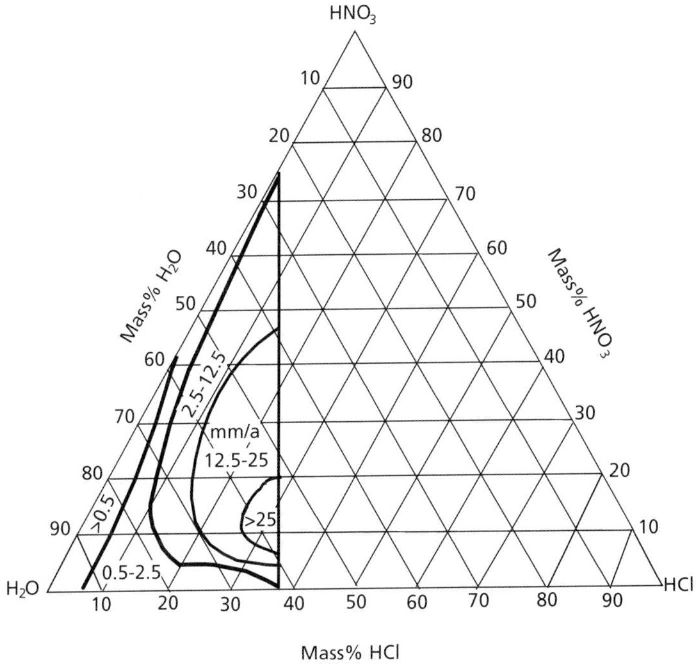

Figure 25: Isocorrosion curves for the alloy Hastelloy® C-276 at 80°C in the three-component representation H_2O – HCl – HNO_3 [202]
< 0.5 mm/a (< 20 mpy)
0.5–2.5 mm/a (20–100 mpy)
2.5–12.5 mm/a (100–500 mpy)
12.5–25 mm/a (500–1000 mpy)
> 25 mm/a (1000 mpy)

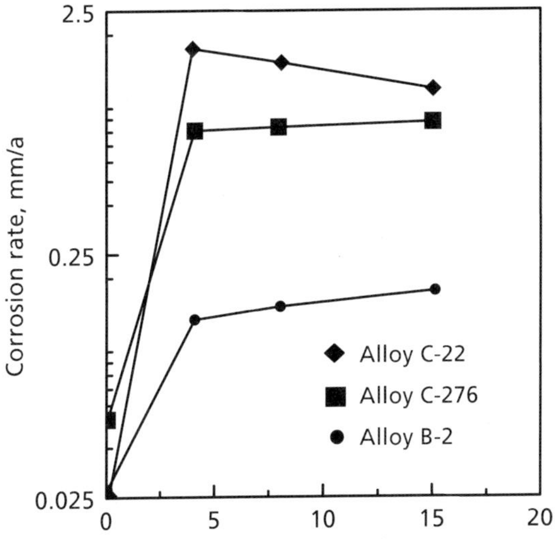

Figure 26: Corrosion rates of the alloys in 15% H_2SO_4 at 80°C [202]
(1 mm/a ≈ 39.37 mpy)

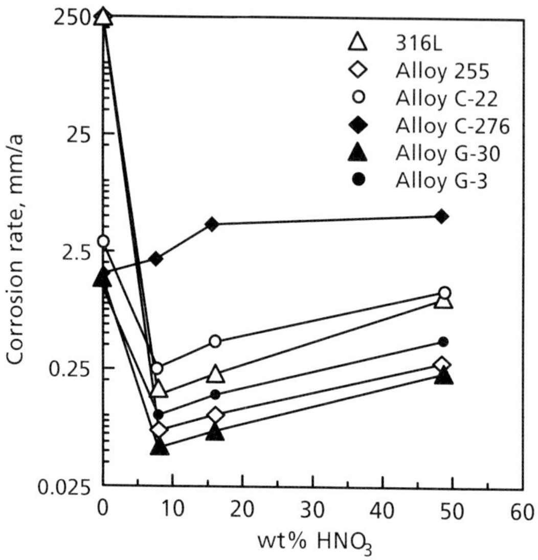

Figure 27: Corrosion rates in boiling 30% H_2SO_4 with increasing concentration of HNO_3 [202] (1 mm/a ≈ 39.37 mpy)

With the exception of C-276, the addition of approx. 10% oxidizing nitric acid to boiling 30% H_2SO_4, a reducing acid, led to a large reduction in the corrosion rate, as shown in Figure 27. The effect is particularly large for the stainless steels SAE 316L (1.4404) and Ferralium® Alloy 255, which are strongly attacked by boiling 30% H_2SO_4. The corrosion rate is decreased by almost four orders of magnitude. If the HNO_3 concentration is increased further, then the corrosion rate increases again. In Figure 27, the value of the corrosion rate is mainly determined by the amount of chromium and molybdenum contained in the alloy. The corrosion rate decreases with increasing chromium content and decreasing molybdenum content. Alloy C-276, with the lowest chromium (16%) and highest molybdenum contents (16%) has the highest corrosion rate and shows abnormal behavior. Alloy B-2 with 28% Mo and no Cr was not included in the investigation.

The addition of HNO_3 to 4% HCl at 80°C leads to results that are comparable to those in systems with boiling 30% H_2SO_4 + HNO_3 (Figure 27 and Figure 28). There is a corrosion minimum for 10% HNO_3. For further additions there is a stronger increase, whereby the level of the corrosion rate corresponds to the sequence in Figure 27, namely, the same alloying relationships are decisive for the corrosion rate. The attack is of the pitting type.

The corrosion rates for the nitric acid – hydrofluoric acid system are also given (Figure 29). This system is important for the pickling of stainless steels. The addition of HF to 20% HNO_3 at 80°C gives rise to a large continual increase in the corrosion rate already commencing at 1% HF. Again, high Cr and low Mo alloys offered the best resistance.

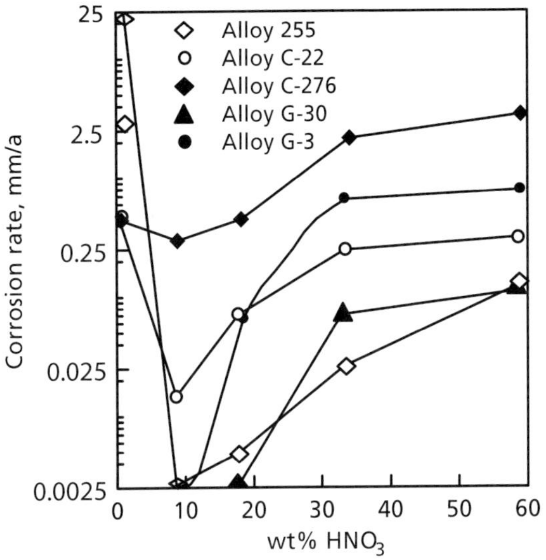

Figure 28: Corrosion rate in a 4% HCl solution with increasing HNO_3 concentration at 80°C [202] (1 mm/a ≈ 39.37 mpy)

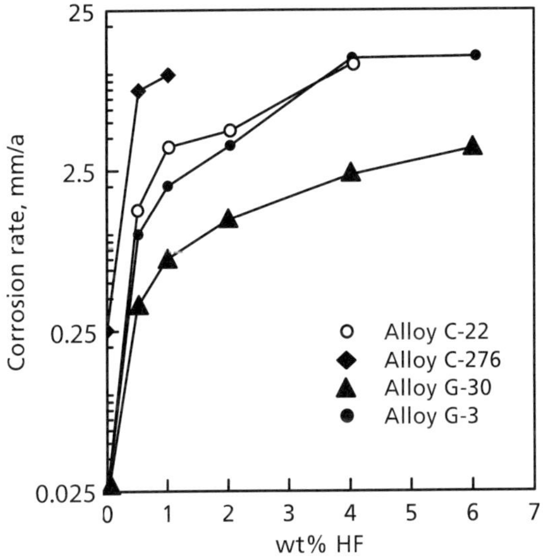

Figure 29: Corrosion rate in a 20% HNO_3 solution with increasing HF concentration at 80°C [202] (1 mm/a ≈ 39.37 mpy)

High-performance alloys

Although extensive corrosion data is available for long-used stainless steels, there is still a lack of corresponding data for the "high-performance alloys". These include corrosion-resistant NiCr alloys with relatively high molybdenum contents up to approx. 30% for the chemical process industry and superalloys for the gas turbine industry. The work in reference [204] reports on the nickel-based alloys given in Table 37, the corrosion resistant alloys Ferralium® Alloy 255, Haynes® No. 20-Mod. and, for purposes of comparison, the stainless steels SAE 304 (X5CrNi18-10, DIN-Mat. No. 1.4301) and SAE 316L (X2CrNiMo17-12-2, DIN-Mat. No. 1.4404).

	Ni	Fe	Cr	Mo	W	Co	others
Hastelloy® Alloy G	balance	19.5	22	6.5	< 1	< 2.5	Cu-2, Nb + Ta-2
Haynes® Alloy No. 625	balance	5	21.5	9	–	< 1	Nb + Ta-3.5
Hastelloy® Alloy C-276	balance	5	16	16	4	< 2.5	V-0.35
Hastelloy® Alloy B	balance	5	< 1	28	–	< 2.5	–
Hastelloy® Alloy B-2	balance	< 2	< 1	28	–	< 1	low carbon content
Ferralium® Alloy 255	5	balance	25.5	3.5	–	–	Cu-1.7, N-0.17
Haynes® Alloy No. 20-Mod.	26	balance	22	9	–	–	Ti-4×C Min.
SAE 304	9	balance	19	–	–	–	–
SAE 316L	12	balance	17	2.5	–	–	low carbon content

Table 37: Composition of the alloys, mass-% [204]

The corrosion rates in various acids obtained from measurements of the mass losses are summarized in Table 38. As expected, the highest corrosion rates were found in boiling 10% HCl, the most strongly reducing acid. By far the highest corrosion resistance was exhibited by B-2 in hydrochloric acid. In contrast, it had the lowest resistance by far in boiling 10% HNO_3. As already found in other work, the level of molybdenum is decisive for the corrosion behavior in hydrochloric acid, whereas the level of the Cr content is decisive in nitric acid.

The degree of corrosion resistance of Hastelloy® Alloy B-2 is not negatively influenced by cold deformation and welding. Table 39 shows that for cold forming of up to 50%, cold forming with subsequent welding as well as cold forming after welding, the corrosion rate remained < 0.25 mm/a.

Alloys	10% HCl	10% H_2SO_4	10% HNO_3	99% CH_3COOH	40% HCOOH	88% HCOOH	55% H_3PO_4	85% H_3PO_4
Hastelloy® Alloy G	28.85	0.635	0.02	0.041	0.13	0.109	0.099	0.686
Haynes® Alloy No. 625	12.9	0.168	0.025	0.01	0.185	0.236	0.114	1.803
Hastelloy® Alloy C-276	5.7	0.584	0.432	0.01	0.074	0.046	0.071	0.508
Hastelloy® Alloy B-2	0.175	0.046	490.5	0.03	0.011	0.015	0.091	0.102
Ferralium® Alloy	478.2 (1)	7.04	0.048	0.03	0.196	0.965	0.130	44.35
Haynes® Alloy No. 20-Mod.	281.2	4.45	0.015	0.018	0.356	0.711	0.279	10.62
SAE 316L	449.2 (1)	60.88	0.053	0.097	1.067	0.483	0.201	16.38

(1) Sample dissolved

Table 38: Corrosion rates (in boiling media), mm/a [204]

State	mm/a	State	mm/a	State	mm/a
0% cold-formed	0.165	0% cold-formed + welded	0.165	welded + 0% cold-formed	0.155
10% cold-formed	0.133	10% cold-formed + welded	0.188	welded + 10% cold-formed	0.138
30% cold-formed	0.145	30% cold-formed + welded	0.235	welded + 30% cold-formed	0.14
50% cold-formed	0.108	50% cold-formed + welded	0.155	welded + 50% cold-formed	0.15

Table 39: Corrosion rates for Hastelloy® alloy B-2 in 10% HCl [204]

Furthermore, this work also investigated the behavior of the alloys with regard to types of local corrosion, namely pitting, crevice and stress corrosion cracking [204]. For passive materials, the levels of the chromium and molybdenum contents are decisive for the resistance to pitting and crevice corrosion. These types of corrosion are caused by chloride ions, and the presence of any oxidizing agents enhances the effect. A 10% $FeCl_3$ solution is generally used to test for pitting and crevice corrosion of stainless and chemically resistant steels. However, this test solution is not aggressive enough for nickel-based alloys that often contain large amounts of molybdenum. To test the pitting corrosion of nickel-based alloys, a solution consisting of 7 vol% H_2SO_4 + 3 vol% HCl + 1% $FeCl_3$ + 1% $CuCl_2$ was used to simulate the conditions in scrubbers. Table 40 gives the results of samples with artificial crevices and of samples that are not in contact with another body. In both test solutions, and also at the higher temperatures of 50 and 102°C, C-276 did not exhibit either pitting or crevice corrosion because of its high pitting index of almost 70 (calculated according to W = %Cr + 3.3 x %Mo). It must be mentioned at this point that in the presence of crevices, crevice corrosion always occurs before pitting corrosion. The samples become active in the crevice, and pitting corrosion on the surface is prevented because of cathodic protection by the anodic crevice region. The results in Table 40 confirm this fundamental corrosion mechanism for passive materials.

	10% $FeCl_3$; Sample with artificial crevice		***Smooth samples 24 h 7 vol% H_2SO_4 + 3 vol% HCl + 1% $FeCl_3$ + 1% $CuCl_2$***	
Alloy	***25 °C 240 h***	***50 °C 100 h***	***25 °C***	***102 °C***
Hastelloy® Alloy G	0.4	0.575 (crevice corrosion)	0.01	43.95 (pitting corrosion)
Haynes® Alloy No. 625	0.038	3.1 (crevice corrosion)	0.008	47.3 (pitting corrosion)
Hastelloy® Alloy C-276	0.005	0.005 (crevice corrosion)	0.008	0.6
Ferralium® Alloy	0.015	19.95 (crevice corrosion)	0.003	75.4 (pitting corrosion)
Haynes® Alloy No. 20-Mod.	0.105	2.35 (crevice corrosion)	0.058	57.95 (pitting corrosion)
SAE 316L	7.8 (crevice corrosion)	11.5 (crevice corrosion)	2.53 (pitting corrosion)	97.73 (pitting corrosion)

Table 40: Corrosion rates, mm/a [204]

The author of [204] gives a very detailed report on the testing of local corrosion of nickel-based alloys in reference [205]. In addition to the materials given in Table 37 [204], the alloy Incoloy® 825 with the nominal percentage composition 42% Ni, 39% Fe, 21% Cr, 3% Mo, max. 1% Mn, max. 0.5% Si, max. 0.05% C, 0.8% Ti, 2% Cu was included in the investigation. The main aim was to correctly determine the "critical pitting corrosion potential" using electrochemical methods because this potential (if correctly measured) would allow a reliable comparison of the pitting corrosion behavior of various materials in one medium. The difficulty concerned with the premature appearance of crevice corrosion must be pointed out as it prevents the correct determination of the critical pitting corrosion potential. In addition, it is shown that the potentiodynamic quick test with a ramping rate of 1000 mV/min, which is often used, does not allow differentiation of the pitting corrosion behavior, particularly of the nickel-based alloy.

The work focused on potentiostatic holding tests with holding times of 24 h for each potential step of 100 mV, from −200 mV to + 800 mV, measured against a saturated calomel electrode. The sample was checked for local corrosion (pitting and/or crevice corrosion) after every holding test. For practical applications, this is a damage limit, regardless of whether one or the other type of corrosion occurs. Table 41 summarizes the results of the 24 h holding tests in five different solutions at 70°C for Hastelloy® G. The shaded area represents the region of the free corrosion potential. The acidic and at the same time oxidizing solutions of columns 1 and 3 are suitable as test solutions for nickel-based alloys because their corrosion potential, due to a high redox potential of the solution, is polarized to the damage potential, whereas the corrosion potentials of the three other solutions without oxidizing agents are more or less negative and lie below the corresponding danger potential for localized corrosion.

U_{SCE} mV	7 vol% H_2SO_4 + 3 vol% HCl + 1% $CuCl_2$ + 1% $FeCl_3$ (pH ≈ 0.2)	1 vol% HCl + 3.8% NaCl (pH ≈ 1.2)	3.8% $FeCl_3$ (pH ≈ 1.3)	4% NaCl (pH ≈ 5.5)	3% Na_2SO_3 + 3% $Na_2S_2O_5$ + 4% NaCl (pH ≈ 12)
+ 800	*	*	*	*	o
+ 700	*	*	*	*	o
+ 600	*	*	*	*	o
+ 500	*	*	*	*	o
+ 400	*	*	*	*	o
+ 300	o	*	*	*	o
+ 200	o	*	o	*	o
+ 100	o	o	o	*	o
0	o	o	o	o	o
– 100	o	o	o	o	o
– 200	o	o	o	o	o

* = local corrosion; o = no local corrosion

Table 41: Hastelloy® alloy G; Results of the 24 h holding tests in five different solutions at 70°C [205]

Table 42 gives the potential ranges for damage or resistance of all investigated materials in a 3.8% $FeCl_3$ solution at 70°C, determined with long-term potentiostatic tests. The resistance ranges are extended to more noble potentials as the molybdenum content in the alloys increases. This result agrees with theoretical and practical findings and confirms the serviceability of this test method.

In reference [206], the comparative corrosion resistance of alloys 625, C-276, Hastelloy® alloy C-22 and ALLCORR® was determined in HCl, H_2SO_4 and formic acid, the last two partly with small additions of HCl, HF, $FeCl_3$ and $CuCl_2$, at differing concentrations and temperatures over a test period of 48 h. The tolerance range for the composition of the alloys and the actual contents of the elements are summarized in Table 43. Welded samples were used that were prepared using the filling wires and welding conditions recommended by the manufacturers. The reason for the investigation was to select suitable materials for handling highly radioactive waste solutions, some of which are exceptionally corrosive.

U_{SCE} mV	2 to 3% Mo SAE 316	3% Mo Incoloy® Alloy 825	6.5% Mo Hastelloy® Alloy G	9% Mo Haynes® Alloy 625	16% Mo Hastelloy® Alloy C-276
+ 800	*	*	*	*	o
+ 700	*	*	*	*	o
+ 600	*	*	*	*	o
+ 500	*	*	*	o	o
+ 400	*	*	*	o	o
+ 300	*	*	*	o	o
+ 200	*	*	o	o	o
+ 100	*	*	o	o	o
0	*	*	o	o	o
– 100	*	o	o	o	o
– 200	o	o	o	o	o

* = localized corrosion; o = no localized corrosion

Table 42: 24 h exposure at constant voltage in 3.8% $FeCl_3$ solution at 70°C [205]

	Alloy 625 (UNS N06625)	Alloy C-276 (UNS N10276)	Hastelloy® Alloy C-22 (UNS N06022)	ALLCORR® (UNS N06110)
Ni	60.20 (balance)	56.42 (balance)	57.52 (balance)	56.03 (balance)
Cr	22.34 (20.0–23.0)	16.11 (14.5–16.5)	21.31 (20.0–22.5)	30.93 (27.0–33.0)
Mo	8.48 (8.0–10.0)	15.99 (15.0–17.0)	13.55 (12.5–14.5)	9.97 (8.0–12.0)
W	– (–)	3.85 (3.0–4.5)	3.17 (2.5–3.5)	1.97 (4.0 max.)
Fe	4.53 (5.0 max.)	5.76 (4.0–7.0)	3.52 (2.0–6.0)	0.11 (–)
C	0.04 (0.10 max.)	0.002 (0.02 max.)	0.003 (0.015 max.)	0.030 (0.15 max.)

Numbers in brackets give the maximum content or the possible fluctuation

Table 43: Compositions, mass-% [206]

Table 43: Continued

	Alloy 625 (UNS N06625)	*Alloy C-276 (UNS N10276)*	*Hastelloy® Alloy C-22 (UNS N06022)*	*ALLCORR® (UNS N06110)*
To	0.25 (0.40 max.)	– (–)	– (–)	0.24 (1.5 max.)
Nb (+ Te)	3.51 (3.15–4.15)	– (–)	– (–)	0.42 (2.0 max.)
P	0.010 (0.015 max.)	0.004 (0.030 max.)	0.007 (0.02 max.)	0.005 (–)
Al	0.14 (0.40 max.)	– (–)	– (–)	0.23 (1.50 max.)
Si	0.34 (0.50 max.)	0.03 (0.08 max.)	0.021 (0.08 max.)	0.02 (–)
V	0.001 (0.015 max.)	0.002 (0.030 max.)	0.002 (0.02 max.)	0.001 (–)
Mn	0.16 (0.50 max.)	0.47 (1.0 max.)	0.21 (0.50 max.)	0.01 (–)
Co	– (–)	1.22 (2.5 max.)	0.52 (2.5 max.)	0.01 (12.0 max.)
Cu	– (–)	– (–)	0.07 (–)	0.01 (–)
V	– (–)	0.14 (0.35 max.)	0.01 (0.35 max.)	0.01 (–)

Numbers in brackets give the maximum content or the possible fluctuation

Table 43: Compositions, mass-% [206]

Table 44 gives the results in HCl. The corrosion resistance is essentially determined by the amount of molybdenum contained in the alloy. The addition of 17% HNO_3 considerably lowers the corrosion rate, particularly that of alloy ALLCORR®, which has the highest Cr content of 31%. For the formic acid-sulfuric acid mixture, the small additions of HCl, HF, $FeCl_3$ and $CuCl_2$ (not shown) increase the corrosion rate to different extents. It must be pointed out that, after the tests, many cases of intergranular corrosion and strong attack of the weld seam were found on the samples.

Solution [(1)]	*%*	*Temperature °C*	*Alloys*	*Corrosion rate (µm/a)*
HCl	35	84	625 C-276 C-22 ALLCORR®	28,000 4,100 (2) 6,800 (2) 7,900 (2), (3)
HCl	15	108	625 C-276 C-22 ALLCORR®	25,000 5,600 (2) 7,100 (2), (3) 15,000 (2), (3)
HCl	15	55	625 C-276 C-22 ALLCORR®	840 280 360 (3) 710 (3)
HCl	10	104	625 C-276 C-22 ALLCORR®	36,000 6,100 (3) 11,000 (3) 22,000 (2), (3)
HCl + HNO_3	28 17	109	625 C-276 C-22 ALLCORR®	> 99,000 17,000 (3) 2,000 (3) 1,400 (2)

(1) Balance water (2) Intergranular corrosion (3) Welding seam corrosion

Table 44: Corrosion rates in HCl solutions [206]

Investigations of various NiCrMo alloys for their possible use in supercritical water oxidation (SCWO) of organic waste is reported in [207]. This process is typically operated above the critical point of water (374°C and 221 bar for pure water). In the presence of compounds containing halogens and sulfur, the oxidation products are the corresponding acids, e.g. HCl, HF and H_2SO_4, in addition to CO_2 and H_2O. The materials listed in Table 45 were included in the investigation.

Material	*UNS code*	*Cr*	*Ni %*	*Fe %*	*Al %*	*Mo %*	*W %*	*others*
G-30	N06030	29.4	40.1	14.5	–	5.1	2.9	Co, Cu
214	N07214	16.0	75.4	3.7	4.4	0.1	–	Zr, Cs*
602 CA	N06025	25.3	62.4	9.5	2.1	–	–	Ti
625	N06625	21.6	62.7	2.3	0.1	9.0	–	Ti
686	N06686	20.4	58.1	1.0	–	16.2	3.9	Ti

* in the commercially available variant, alloy 214 contains yttrium instead of caesium

Table 45: Composition of the investigated materials [207]

The tests were carried out in autoclaves in a mixture of water with 10 g/l dichloromethane (CH_2Cl_2) and 16 g/l H_2O_2 as the oxidizing agent at a temperature of 420°C and a pressure of 400 bar. The test duration was 24 h. The mass losses of the samples for the 2 test runs are given in Table 46.

Material	*Mass loss mg/cm²*	*Mass loss mg/cm²*
G-30	2.8	2.9
214	26.6	–
214 preoxidized	50.8	39.9
602 CA	46.0	–
602 CA preoxidized	15.5	–
625	–	33.5
686	73.3	23.8

Table 46: Mass losses of the samples for two test runs after a 24 h test period [207]

According to this, only alloy G-30 can be expected to exhibit satisfactory corrosion behavior under SCWO conditions. The other alloys show unacceptably high corrosion rates after 24 h.

Nickel-copper alloys

Nickel-copper alloys, e.g. Monel® alloy 400 (UNS N04400, c.f. DIN-Mat. No. 2.4360, 2.4366) exhibit lower resistance in hydrochloric acid than nickel itself. Some older data on the behavior of Monel® in different hydrochloric acid solutions at different temperatures is given in [189].

Table 47 is taken from [192] and shows data for alloy 400 in boiling hydrochloric acid at various concentrations.

Alloy	*Notes*	*Concentration, %*	*Temperature, °C*	*Time, d*	*Corrosion rate, mm/a*	*References*
400	no aeration	0.5	boiling	10	0.74	[208]
	no aeration	1	boiling	10	1.07	[208]
	no aeration	5	boiling	10	1.12	[208]

Table 47: Corrosion rates of alloy 400 in boiling hydrochloric acid at various concentrations [192]

Nickel-molybdenum alloys

Molybdenum has very good corrosion resistance against non-oxidizing solutions of hydrochloric acid at most temperatures and concentrations. NiMo alloys which generally contain 26 to 30 wt-% molybdenum belong to the few metallic materials that are resistant against non-aerated hydrochloric acid at all concentrations and temperatures [191]. As shown in Figure 30 the corrosion rates of such high molybdenum containing alloys in 10% boiling hydrochloric acid solutions are extremely low.

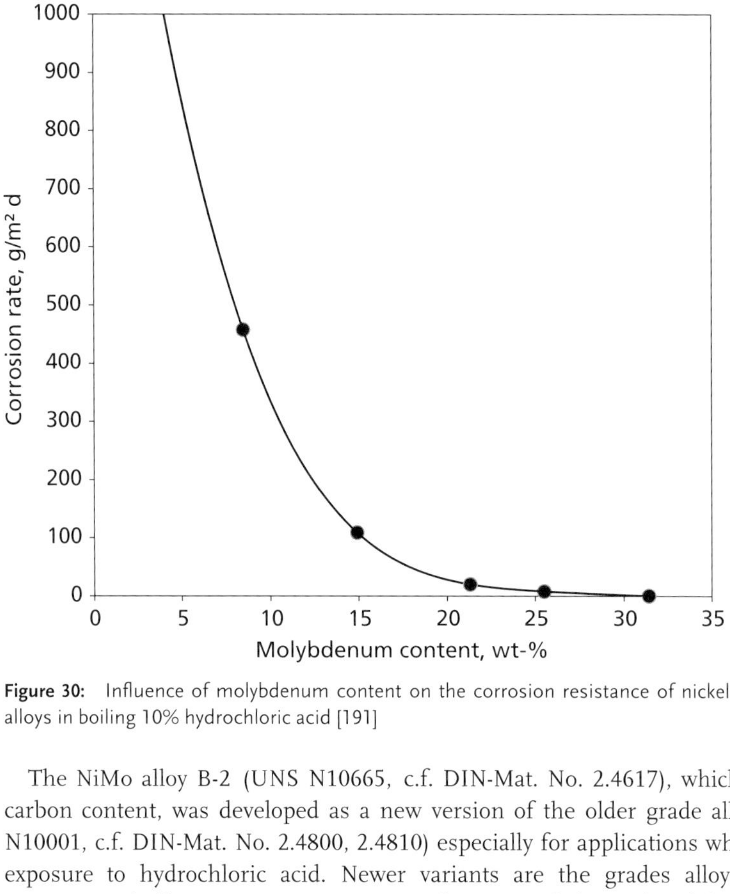

Figure 30: Influence of molybdenum content on the corrosion resistance of nickel-molybdenum alloys in boiling 10% hydrochloric acid [191]

The NiMo alloy B-2 (UNS N10665, c.f. DIN-Mat. No. 2.4617), which has a low carbon content, was developed as a new version of the older grade alloy B (UNS N10001, c.f. DIN-Mat. No. 2.4800, 2.4810) especially for applications where there is exposure to hydrochloric acid. Newer variants are the grades alloy B-3 (UNS N10675) and alloy B-4 (UNS N10629) with improved thermal stability. Figure 31 shows an isocorrosion diagram for the grade B-2 [190] As the solubility of oxygen decreases as temperature increases and thus the corrosion rate decreases at higher temperatures, there is a second line for the corrosion rate of 0.13 mm/a at higher temperatures close to the boiling point.

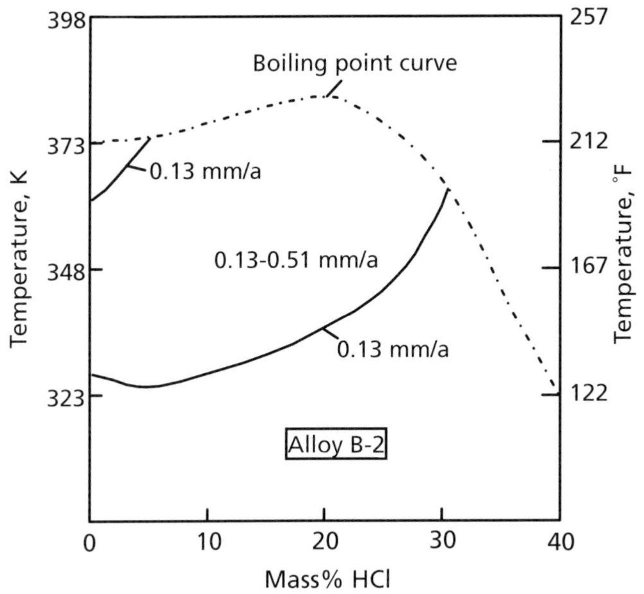

Figure 31: Isocorrosion diagram for alloy B-2 in hydrochloric acid [190]
0.13 mm/a (5 mpy)
0.13–0.51 mm/a (5–20 mpy)

Because of its high molybdenum content of 28%, alloy B-2 and its variants exhibit the best resistance of all nickel alloys in hydrochloric acid. In non-aerated hydrochloric acid and in the absence of oxidizing agents, alloy B-2 is one of the few metallic materials with a corrosion rate of < 0.5 mm/a for all acid concentrations and temperatures. Data from [209] shown in Table 48 also confirms this.

Alloy	*Concentration, %*	*Temperature, °C*	*Time, h*	*Corrosion rate, mm/a*	*Reference*
B-2	1	boiling	120	0.02	[209]
	2	boiling	120	0.08	[209]
	5	boiling	120	0.13	[209]
	10	boiling	120	0.18	[209]
	15	boiling	120	0.28	[209]
	20	boiling	120	0.38	[209]

Table 48: Corrosion rates of alloy B-2 in boiling hydrochloric acid at various concentrations [209]

Oxidizing agents in the acid can have a noticeable detrimental effect on the resistance of the alloy, as shown in Figure 32 by the isocorrosion diagram for alloy B-2 in hydrochloric acid purged with oxygen [190].

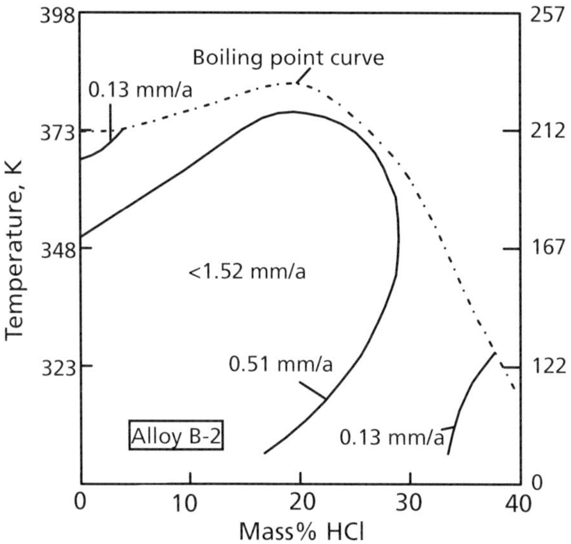

Figure 32: Isocorrosion diagram for alloy B-2 in hydrochloric acid purged with oxygen [190]
0.13 mm/a (5 mpy)
0.51 mm/a (20 mpy)
< 1.52 mm/a (60 mpy)

The presence of a strong oxidizing agent, such as Fe^{3+} or Cu^{2+} ions can lead to even higher corrosion rates in the acid, as shown in Figure 33 [190].

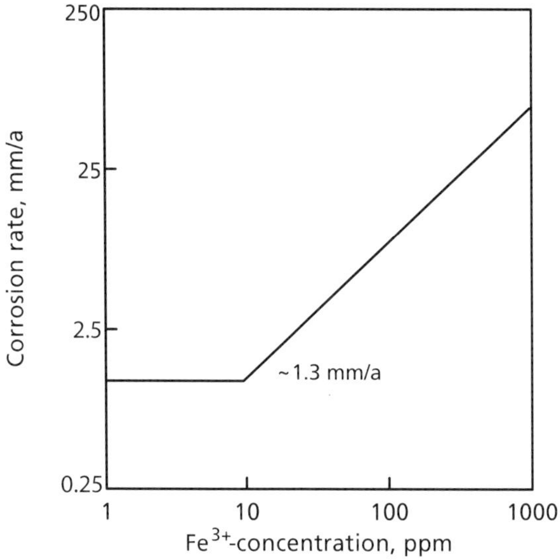

Figure 33: Influence of Fe^{3+} ions in boiling 20% hydrochloric acid on the corrosion of alloy B-2 [190]
1.3 mm/a (~50 mpy)

Therefore, before NiMo alloys are used in hydrochloric acid, it is strongly recommended that welded samples are tested under near-service conditions.

Reference [210] mentions that melts of NiMo alloy B-2 (UNS N10665) that are within the permissible analysis limits can greatly differ in their mechanical properties and that some melts cause problems during production whereas others can be processed without problems. Table 49 gives the chemical composition of the NiMo alloys. The decisive difference between Hastelloy® B-2 and alloy B (UNS N10001) is the reduction of the carbon and silicon contents as well as the reduction of the iron content to max. 2%. Due to the low carbon content of max. 0.01%, it is not necessary to stabilize alloy B-2 with vanadium as it is for alloy B. Since the knife-line-corrosion and the attack of the heat-affected zone is prevented in B-2, in contrast to alloy B, alloy B-2 can be used in the welded state. Therefore, B-2 can be used for a much wider range of applications in modern process engineering. However, occasional problems during processing (cracking) with individual melts necessitated a detailed investigation to elucidate this phenomenon.

Alloys (UNS No.)	*Ni*	*Mo*	*Fe*	*Cr*	*Mn*	*Si*	*C*
Alloy B (N10001)	balance	26–30	4–6	1 (A)	1 (A)	1 (A)	0.05 (A)
Alloy B-2 (N10665)	balance	26–30	2 (A)	1 (A)	1 (A)	0.08 (B)	0.01 (A)
Alloy B-3 (N10675)	65 (B)	27–32	1–3	1–3	3 (A)	0.10 (A)	0.01 (A)
Alloy B-4 (N10629)	balance	27–30	2–5	0.5–1.5	1.5	0.05 (A)	0.01 (A)
Alloy B-4 (C)	balance	28	3	1.3	0.6	0.03	0.006

(A) Maximum
(B) Minimum
(C) Typical composition

Table 49: Composition of the NiMo alloys, mass-% [210]

As can be seen in Figure 34, the two brittle intermetallic phases Ni_4Mo and Ni_3Mo exist in the two-phase Ni-Mo system in addition to the face-centred cubic Ni-Mo matrix. Embrittlement and thus processing problems occur if intermetallic phases are formed, particularly Ni_4Mo. Earlier publications report that the formation of Ni_4Mo in Hastelloy® B depends on the iron content [211, 212]. The formation of this phase is suppressed at 4% Fe. There is a significant reduction in the amount of Ni_4Mo for 2% Fe in Hastelloy® B-2. Chromium has a similar effect on the formation of Ni_4Mo.

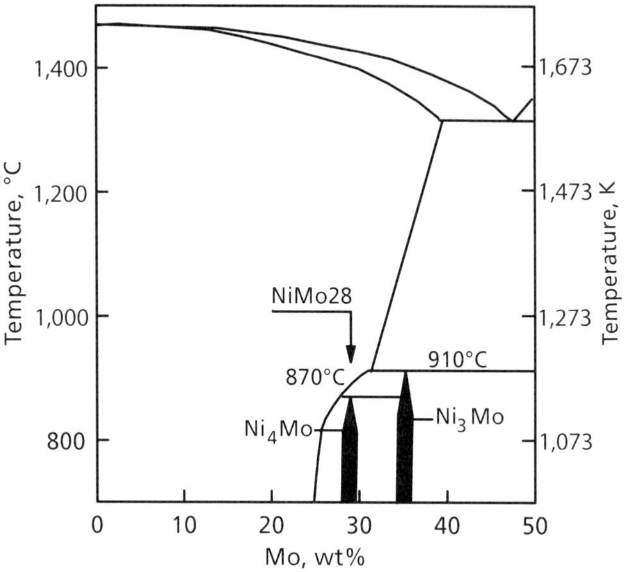

Figure 34: Phase diagram of the Ni-Mo system [210]

The influence of iron and chromium on the precipitation kinetics of Ni_4Mo was investigated with the melts summarized in Table 50. Melts 1–3 lie within the composition range of Hastelloy® B-2, whereas melts 4–6 lie outside it. The results show that embrittlement due to Ni_4Mo is prevented for a controlled content of 2–5% Fe and 0.5–1.5% Cr (see Table 49) and thus cracking during manufacture is avoided.

Melts	*Mo*	*Fe*	*Cr*	*Mn*	*Si*	*Al*	*C*
1	28.15	0.11	0.02	0.12	0.02	0.40	0.002
2	27.6	1.13	0.47	0.42	0.01	0.30	0.003
3	26.9	1.75	0.68	0.68	0.01	0.24	0.003
4	26.9	3.17	1.42	0.61	0.03	0.26	0.006
5	26.9	3.23	0.72	0.61	0.03	0.27	0.006
6	26.9	5.68	0.78	0.60	0.03	0.28	0.006

Table 50: Compositions of the investigated melts of NiMo alloys, mass-% [210]

Within these limits, the corrosion resistance is not negatively influenced as compared to that of Hastelloy® B-2 (Table 51). The resistance to stress corrosion cracking, which is influenced in NiMo alloys by the precipitation of intermetallic phases, was investigated by means of the ASTM G 30 test (boiling 10% H_2SO_4) in the tempered state (1 h at 700°C and 6 h at 700°C). There was an improvement of a few hours endurance for melts 1–3, which corresponds to B-2, whereas the endurance for melts 4–6, which correspond to B-4, increased to > 100 h (Table 52).

Melts	Corrosion rate mm/a	
	boiling 10% HCl 24 h (+)	boiling 20% HCl 24 h (+)
1	0.29	0.58
2	0.31	not detectable
3	0.37	0.60
4	0.25	0.56
5	0.24	0.46
6	0.38	0.58

(+) Intergranular attack was less than 50 μm for all melts

Table 51: Corrosion rates in boiling HCl solutions [210]

Melts	Content mass%		Time to fracture h		
	Fe	Cr	solution-annealed	tempered 1 h, 700°C	tempered 6 h, 700°C
1	0.08	≤ 0.01	> 100	3	2
2	1.13	0.47	> 100	3	3
3	1.75	0.68	> 100	45	3
4	3.17	1.42	> 100	> 100	> 100
5	3.23	0.72	> 100	> 100	> 100
6	5.86	0.78	> 100	> 100	> 100

(A) Investigations according to ASTM G 30 in 10% boiling H_2SO_4

Table 52: Results of the stress corrosion cracking tests on the NiMo alloys(A) [210]

The corrosion rates of melt 3 (corresponds to B-2) and melt 4 (corresponds to B-4) in boiling 10% HCl are given in Table 53 dependent upon the tempered state. The results show that the higher iron and chromium contents in B-4 have a favorable effect on the resistance after tempering.

Tempering temperature °C	*Corrosion rate, mm/a*					
	Melt 3 (1.75% Fe, 0.68% Cr)			*Melt 4 (3.17% Fe, 1.42% Cr)*		
	0.5 h	*1 h*	*8 h*	*0.5 h*	*1 h*	*8 h*
600	0.32	0.29	0.39	0.18	0.33	0.31
650	0.33	0.38	0.33	0.27	0.37	0.31
700	0.35	0.36	0.87[A]	0.27	0.32	0.26
750	0.35	0.35	0.39	0.24	0.35	0.29
800	0.33	0.31	0.33	0.25	0.23	0.25
850	0.32	0.32	0.31	0.34	0.20	0.25
900	0.30	0.27	0.34	0.21	0.24	0.18

[A] Intergranular attack was less than 50 μm; and less than 25 μm for all other samples

Table 53: Corrosion rates of melt 3 and melt 4 in boiling 10% HCl with respect to the tempered state [210]

The alloy Hastelloy® B-4 (UNS N10629) represents a further improvement in the series of Ni-28 Mo alloys. It has advantages with regard to production and practical utilization compared to Hastelloy® B-2.

In reference [214], comparative tests with non-heat-treated welded samples of Hastelloy® B and Hastelloy® B-2 proved the superiority of B-2 in boiling 20% HCl for a test duration of 16 days. Mass loss measurements gave a corrosion rate of 1.5 mm/a for Hastelloy® B and 0.6 mm/a for Hastelloy® B-2. The corrosion of alloy B was mostly in the form of deep penetrating knife-line corrosion at the transition between the welding seam and the parent material. Mass was lost uniformly from the surface of B-2 as it does not have any knife lines. Figure 35 gives the isocorrosion curve of B-2 in hydrochloric acid. From this it follows that the mass loss rate is < 0.5 mm/a in the entire concentration and temperature range. It is pointed out that Hastelloy® B-2 also reacts sensitively to oxidizing auxiliary substances in HCl. It is not recommended for use in direct contact with iron and copper in chloride-containing environments. The presence of these metals can lead to the formation of iron(III) and copper(II) chlorides which increase the corrosion rate.

The susceptibility of NiMo alloys to intergranular corrosion in the strongly reducing acids HCl and H_2SO_4 is essentially due to the precipitation of molybdenum carbides on the grain boundaries and the associated molybdenum depletion in the region of the grain boundary. Thus the behavior is similar to that of CrNi steels that are also susceptible to intergranular corrosion due to chromium carbide precipitation and the associated chromium depletion.

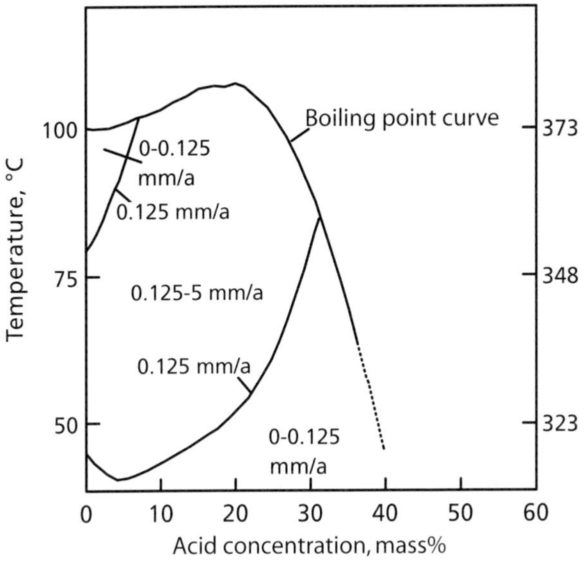

Figure 35: Isocorrosion curve of alloy B-2 in HCl [214]
0–0.125 mm/a (0–5 mpy)
0.125 mm/a (5 mpy)
0.125–5 mm/a (5–200 mpy)

In reference [216], the carbide precipitation and the associated intergranular susceptibility of NiMo alloys is discussed on a thermodynamic basis. Here, the solubility product of the carbide and the carbon activity are the decisive factors. The latter does not only depend on the main components, but also on other elements that are present in small amounts and which can be regarded as impurities. The tendency towards carbide precipitation increases as the solubility product of the carbide decreases and the carbon activity increases. Therefore, the carbon content must be kept as low as possible to counteract the tendency to precipitate carbides. Because impurities, particularly silicon, increase the carbon activity, they must also be reduced to a minimum. The theoretical results of this work are in agreement with the practical measures taken during the development of Hastelloy® B-2.

In [216], the corrosion behavior of NiMo alloys was investigated with respect to the molybdenum content of up to 40% in boiling 21% HCl (Figure 36). At 40% Mo, the corrosion rate was approx. 20 times lower than that of the industrially used alloy with 28% Mo. The alloy with 40% Mo had a corrosion rate of 0.01 g/m^2 h, which is the same resistance as that of pure molybdenum. In the technical alloy, the level of molybdenum was limited to 28% Mo because the intermetallic phases that occur above this value lead to precipitates, particularly during welding, that impair the corrosion resistance.

A Ni alloy with 25% Mo and 8% Cr (UNS N10242) was developed for use at higher temperatures. It has a low thermal expansion and high strength over a wide temperature range. The behavior of this material in various acids of differing con-

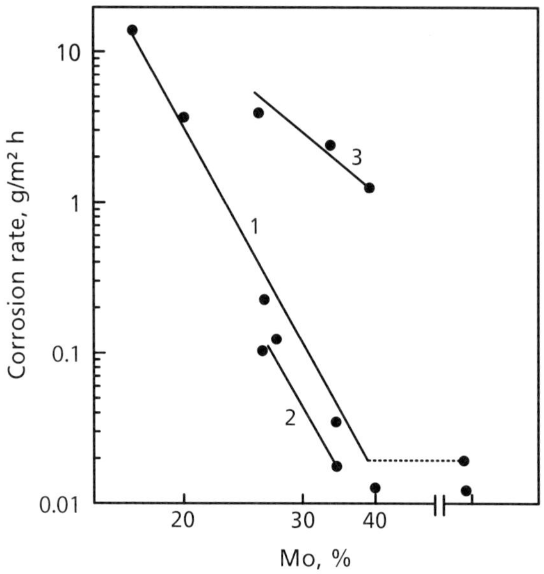

Figure 36: Corrosion behavior of NiMo alloys dependent upon the molybdenum content in the following aqueous acids:
1) 21% boiling HCl;
2) 40% boiling H_2SO_4;
3) 70% boiling H_2SO_4;
Exposure time 200 h [216]

centrations was investigated with respect to the heat-treatment and the results were compared to those from the NiMoCr alloy B-2 and the NiCrMo alloy C-2000 [218]. The compositions of the three materials are given in Table 54.

Alloy	*UNS No.*	*Ni %*	*Cr %*	*Mo %*	*C max. %*	*Others max. %*
alloy 242	N10242	65	8	25	0.03	2.5 Co; 2 Fe; 0.8 Mn; 0.8 Si; 0.5 Al; 0.5 Cu
alloy B-3	N10675	65	1.5	28.5	0.01	3 Co; 1.5 Fe; 3 W; 3 Mn; 0.1 Si; 0.2 Cu
alloy C-2000	N06200	59	23	16		1.6 Cu

Table 54: Nominal compositions of the investigated alloys [218]

The strength values of alloy 242 can be noticeably increased by a corresponding heat treatment. They are then two to three times higher than those for the other two alloys, as shown in Figure 37.

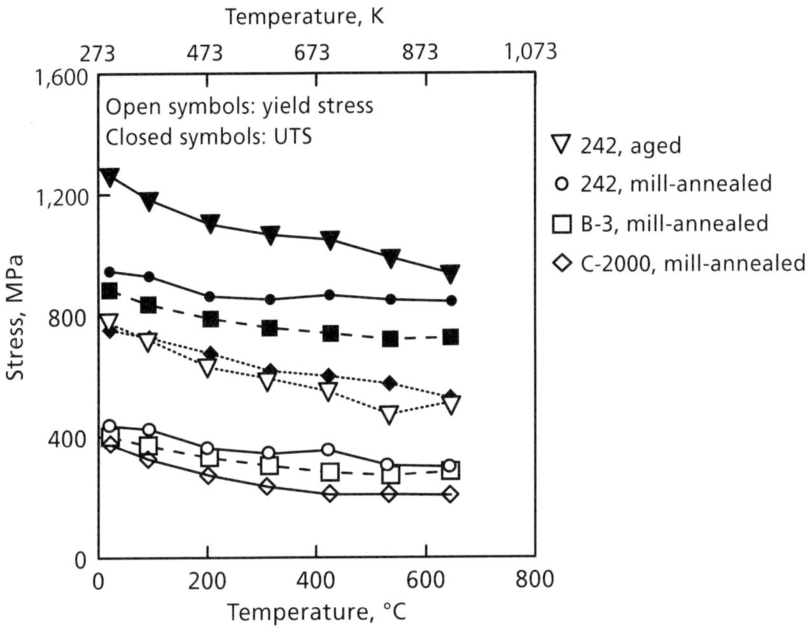

Figure 37: Values of the tensile strengths and yield strengths of the investigated alloys dependent upon the temperature [218]

All materials were tested in the as-delivered state. The strength of alloy 242 was further increased by tempering at 650°C for 24 h with subsequent cooling in air. Electrochemical investigations were carried out as well as immersion tests according to ASTM G 3 at 66°C, 79°C and at the respective boiling point of the test solutions.

The following conclusions can be drawn from the results:

- Alloy 242 exhibits good corrosion resistance in reducing hydrochloric and sulfuric acids
- The corrosion resistance of alloy 242 in hydrofluoric acid is similar to that of alloy B-3
- The corrosion rates of alloy 242 in phosphoric acid, formic acid and acetic acid are very low and comparable to those of alloy B-3 under the same conditions
- Under moderately oxidizing conditions, e.g. in aerated strong acids, alloy 242 exhibits better corrosion resistance than alloy B-3
- The use of alloy 242 is not recommended in strongly oxidizing acids (e.g. HNO_3)
- The heat-treated variant of alloy 242 exhibited higher corrosion rates than the alloy in the as-delivered state. This is particularly noticeable under strongly reducing conditions
- In both heat-treated states, alloy 242 is not susceptible to stress corrosion cracking in boiling acids and in boiling 45% $MgCl_2$ solutions

- Alloy 242 appears to be susceptible to hydrogen embrittlement, particularly in the heat-treated state

Information on the concentration- and temperature-dependent corrosion behavior is given in Figures 38 to 41.

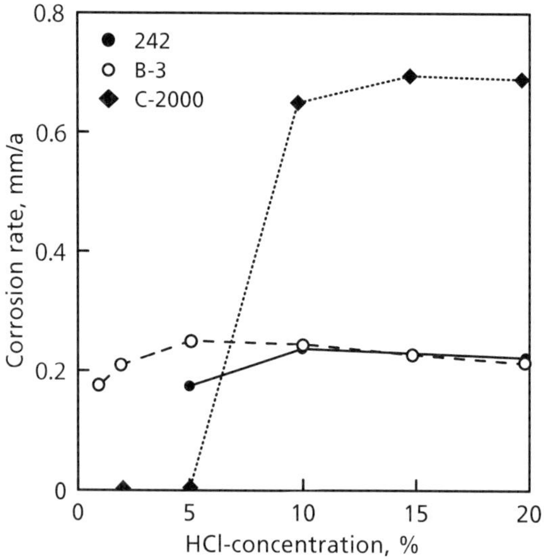

Figure 38: Corrosion rates dependent upon the HCl concentration at 66°C [218]
(1 mm/a ≈ 39.37 mpy)

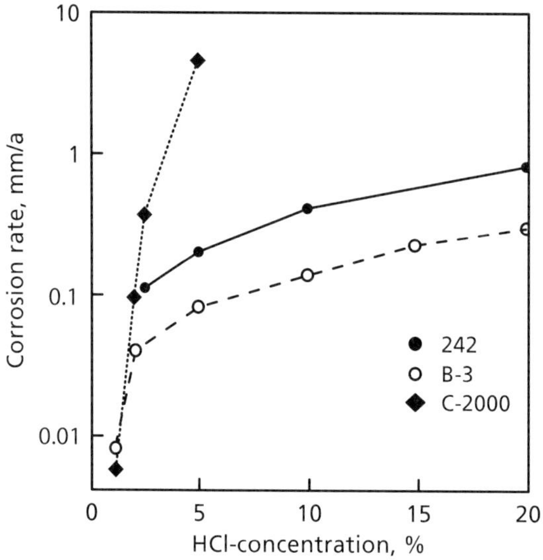

Figure 39: Corrosion rates of the alloys dependent upon the HCl concentration at the boiling point of the individual solution [218]
(1 mm/a ≈ 39.37 mpy)

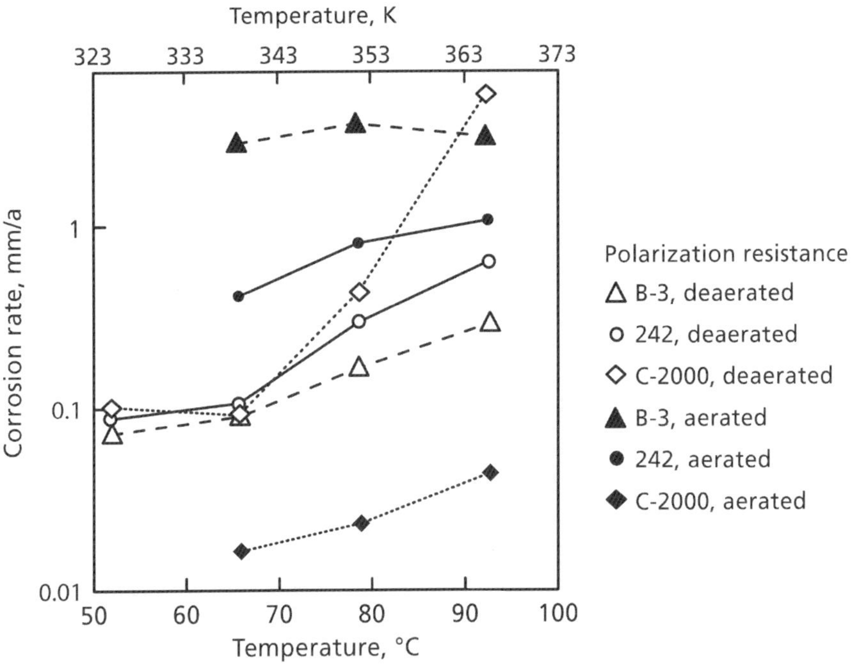

Figure 40: Corrosion rate of the alloys dependent upon the temperature in aerated and deaerated 1 M HCl [218]
(1 mm/a ≈ 39.37 mpy)

Reference [219] reports on a case of damage to alloy B-2 in a Venturi scrubber of a catalytic reformer unit. The scrubber is used to neutralize the acidic process gases (CO_2, HCl, Cl_2) with an alkaline solution (pH 8 to 9). Leakages appeared in the upper region of the scrubber after an operating time of 9 months.

The damage was caused by the combined effect of corrosion and cavitation. A large temperature drop from 230 to 70°C in the affected area led to condensation of the very acidic gas components with simultaneous collapse of the gas bubbles. The cavitation effect of the bursting gas bubbles destroyed the protective surface coatings, enabling corrosive attack by the acidic condensates.

In addition to hydrochloric acid, to which alloy B-2 has high resistance, high proportions of strong oxidizing agents, such as oxygen, chlorine and Fe^{3+} ions were also present in the condensates, so that the corrosion resistance of the material would not have been sufficient under these conditions even without the cavitation effect. Subsequent laboratory investigations confirmed that a corrosion rate of 0.9 mm/a must be expected for the material under these conditions. Investigations with the NiCr alloys C-276 (N10276; c.f.: DIN-Mat. No. 2.4819), C-22 (N06022; c.f.: DIN-Mat. No. 2.4602), C-2000 (N06200) Alloy 59 (N06059; c.f.: DIN-Mat. No. 2.4605) and Alloy 686 (N06686; c.f.: DIN-Mat. No. 2.4606); which usually exhibit better resistance in oxidizing acids, also exhibited unsatisfactory behavior under these critical conditions. Thus, the same damage occurred in a replacement scrubber made of C-276 after an operating time of one year. Recommended countermeasures are either a strict control of temperature to prevent the temperature dropping below the dew point thus avoiding the formation of the aggressive condensates, or to test the use of acid-resistant ceramic coatings.

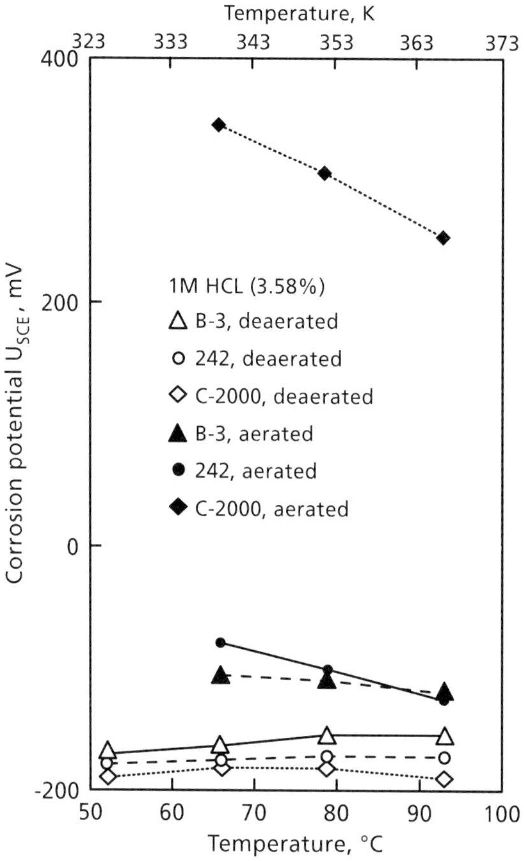

Figure 41: Corrosion potential of the alloys dependent upon the temperature in aerated and deaerated 1 M HCl [218]

Mixed Acids

Author: *M. B. Rockel* / Editor: *R. Bender, R. Durham*

Nickel

Pure nickel is not resistant in inorganic acids and their mixtures.

The use of pure nickel in apparatus and plant engineering is based on its excellent resistance in alkaline solutions, even at the highest concentrations and temperatures.

Nickel-chromium alloys

NiCr alloys have only little resistance in inorganic acids and mixed acids. They usually do not have any advantages over austenitic steels, which are generally more resistant and less expensive.

Nickel-chromium-iron alloys (without Mo)

Molybdenum-free NiCrFe alloys have only little resistance in inorganic acids and mixed acids. They usually do not have any advantages over austenitic steels.

Nickel-chromium-molybdenum alloys

The most resistant nickel materials include those with high-Mo and high-Cr, i.e. those with 16–23% Cr and 10–16% Mo (so-called nickel alloys of the C series). They are resistant if the concentration is not too high, but even then at higher temperatures (however, below the boiling point). They are strongly attacked by HNO_3 – HF mixtures, as well as by mixed acids with higher amounts of hydrochloric acid. However, they exhibit excellent resistance in many technical process media and are always used in those cases in which the austenitic steels – even the so-called superaustenites – fail due to general corrosion and in particular, by pitting corrosion, crevice corrosion and stress corrosion cracking.

The following NiCrMo alloys, also commonly designated as just "alloy", are listed with their international trade names. For purposes of simplification and to explain their corrosion behavior with respect to the alloying elements, their approximate compositions as well as the DIN-Mat. No. are summarized in Table 55 below.

Name	DIN-Mat. No.	Abbreviation	Ni	Cr	Mo	Cu	Other
alloy 20	2.4660	NiCr20CuMo	38	20	2.4	3.4	0.2 Nb
alloy 825	2.4858	NiCr21Mo	42	22	3.2	1.5	0.8 Ti
alloy 625	2.4856	NiCr22Mo9Nb	62	22	9		3.4 Nb
alloy G	2.4618	NiCr22Mo6Cu	47	22	6	1.8	20 Fe
alloy G-3	2.4619	NiCr22Mo7Cu	44	22	7	1.8	20 Fe
alloy G-30	2.4603	NiCr30FeMo	46	30	5	1.5	15 Fe; 2.5 W
alloy 690	2.4642	NiCr29Fe	61	29			9 Fe; 0.25Ti
alloy C-276	2.4819	NiMo16Cr15W	57	16	16		3.5 W
alloy C-22	2.4602	NiCr21Mo14W	60	20	14		3 W; 3 Fe
alloy C-4	2.4610	NiMo16Cr16Ti	66	16	16		1 Fe
alloy 59	2.4605	NiCr23MoTi	59	23	16		<1 Fe
alloy 686	2.4606	NiCr21Mo16W	57	21	16		3.5 W

Table 55: Approximate compositions of NiCrMo alloys

The Metals Handbook reports the results of investigations in some mixed acids [220, 221]. It is pointed out that, in general, the corrosion behavior in the individual acids only provides qualitative indications of the behavior in mixed acids. However, there may also be abnormal, i.e. strongly accelerating or inhibiting effects, in the case of acid mixtures.

Alloys of type Hastelloy® C have good resistance towards hot mixed acids [222]. Chlorimet® 3 is suitable for tower scrubbers operating with mixtures of sulfuric, nitric and hydrochloric acids as well as selenic acid and sodium chloride at 60–65°C [223]. Ni-O-Nel® (alloy 825, DIN-Mat. No. 2.4858) also has good resistance in mixed acids of various compositions and temperatures [223].

At this point, new results in the most important pure inorganic acids should be briefly mentioned because they provide a general indication of the possible applications with regard to concentration and temperature. New test results are available for alloy 59 (DIN-Mat. No. 2.4605) in sulfuric and hydrochloric acids. This nickel-based alloy is essentially resistant in phosphoric acid; the resistance is moderate in nitric acid of the highest concentration. A corrosion rate of 0.05 mm/a was found in boiling 10% HNO_3, while it was 1.02 mm/a in boiling 65% HNO_3. Under these conditions, only the much more resistant and less expensive, fully austenitic high-chromium steels can be used. Therefore, only data on their behavior in sulfuric and hydrochloric acids is given below [224].

The material is resistant up to 100°C in dilute H_2SO_4 up to 10%. For concentrations of 10–80%, the material is resistant up to 80°C; however, the corrosion rate increases steeply at 80%. At a temperature of 100°C, the material has no resistance at a concentration of 20%. As the isocorrosion diagram given in Figure 43 shows, alloy 59 is resistant up to 80°C for all sulfuric acid concentrations between 90 and 100%. Above 98% the range of resistance increases up to 200°C [224].

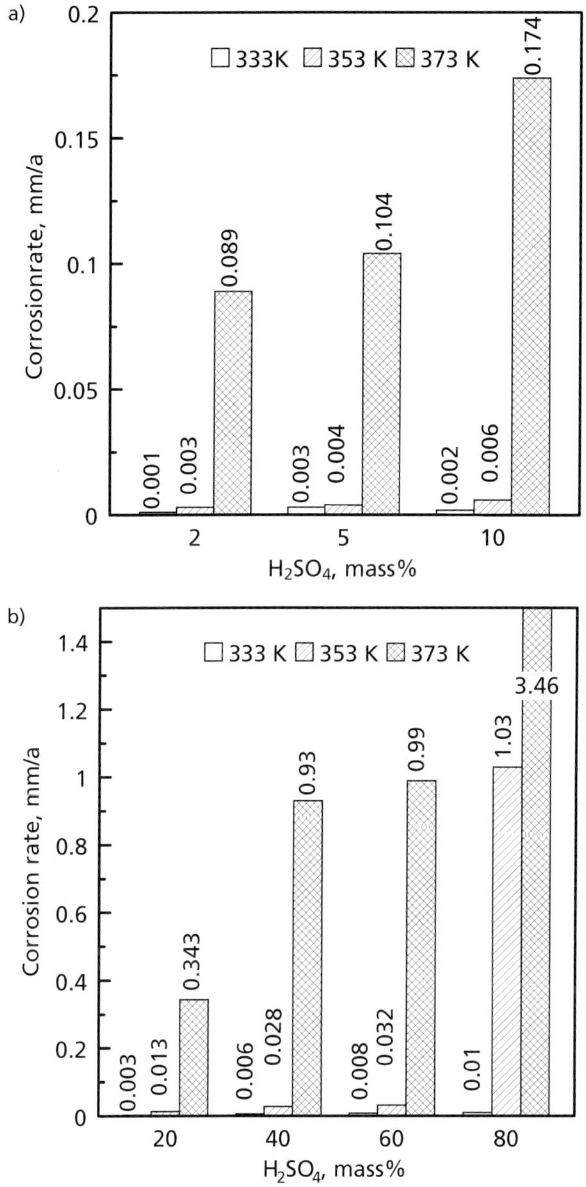

Figure 42: Behavior of Nicrofer® 5923 hMo – alloy 59 (DIN-Mat. No. 2.4605) in sulfuric acid a) dilute b) medium concentration; corrosion rate in mm/a [224]

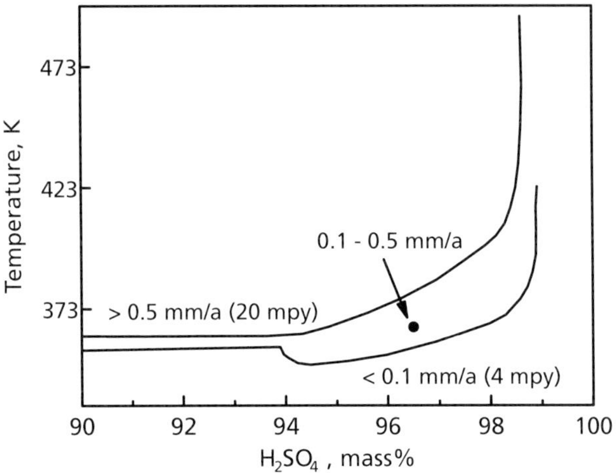

Figure 43: Isocorrosion diagram of alloy 59 (DIN-Mat. No. 2.4605) in highly concentrated sulfuric acid above 90% [224]

The corrosion behavior of alloy 59 (DIN-Mat. No. 2.4605) in pure hydrochloric acid is noticeably less favorable. Figure 44 shows that the material is resistant in all concentrations of acid only up to 40°C. If corrosion rates up to 0.5 mm/a are tolerated, then the limiting temperature lies at 60°C. In very dilute acids up to 10%, none of the nickel-based alloys listed in Table 56 are resistant in boiling acid.

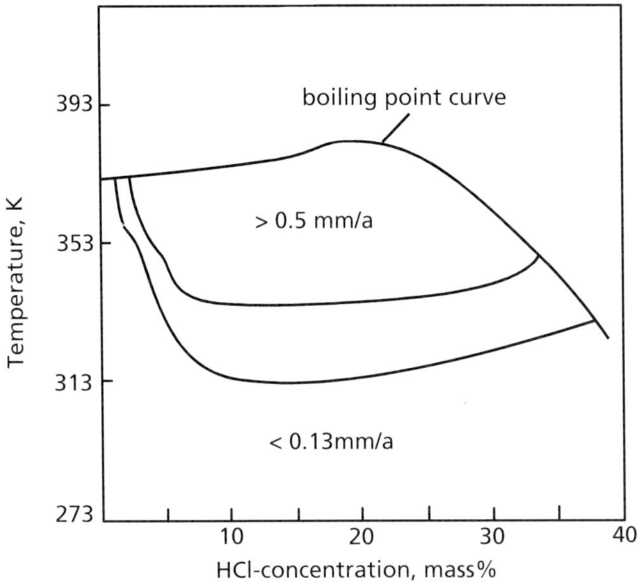

Figure 44: Isocorrosion diagram of alloy 59 (DIN-Mat. No. 2.4605) in hydrochloric acid [224, 225]

Test solution	Temperature, °C	Corrosion rate, mm/a		
HCl		alloy C-22	alloy C-276	alloy 59
1.5%	boiling	0.87	0.69	0.38
2%	90	0.05	0.53	0
	boiling	2.20	1.14	0.99
3%	90	1.41	0.96	0.05
	boiling	4.55	2.06	2.05
5%	93	5.07	1.99	2.81
10%	66	0.77	0.44	0.81
	82	3.04	1.56	2.47
	boiling	9.26	6.07	7.50

Table 56: The corrosion behavior of nickel alloys of the C series in dilute hydrochloric acid [224]

Only alloy 59 is resistant (0.05 mm/a) in 3% solution up to 90°C. In spite of this rather unfavorable behavior of the high-alloy nickel materials in hydrochloric acid, it must be pointed out that in the process media used in industrial practice, hydrochloric acid is very often present only in trace amounts and that high-alloy steels and most nickel alloys rapidly fail due to pitting, crevice and strong uniform corrosion. In this case, alloys such as alloy 59, exhibit the best behavior and generally very good resistance.

Sulfuric and nitric acids

NiCrMo alloys containing sufficient chromium and which also exhibit active-passive behavior are more resistant on the addition of nitric acid to sulfuric acid (Figure 45).

It can be seen that all listed materials are not resistant in sulfuric acid. Additions of approx. 10% nitric acid obviously have a passivating effect since the corrosion rates decrease to 0.25 mm/a and below, even for 18Cr10Ni steel, and then increase again at higher nitric acid concentrations. Only the very high-chromium alloys, such as alloy G-30, Ferralium® 255 as well as alloy G-3, are still resistant at 50% nitric acid addition. In contrast, the well known nickel-based alloy C-276 (NiMo16Cr15W, DIN-Mat. No. 2.4819) generally becomes less resistant with increasing nitric acid concentration. This can be explained due to its relatively low chromium content (16% Cr) [221].

In 25% H_2SO_4 + 5% HNO_3 + 4% NaCl (boiling), alloy C-22 exhibits the best behavior, with a corrosion rate of 0.3 mm/a, compared to the other nickel alloys C-276, C-4 and 625.

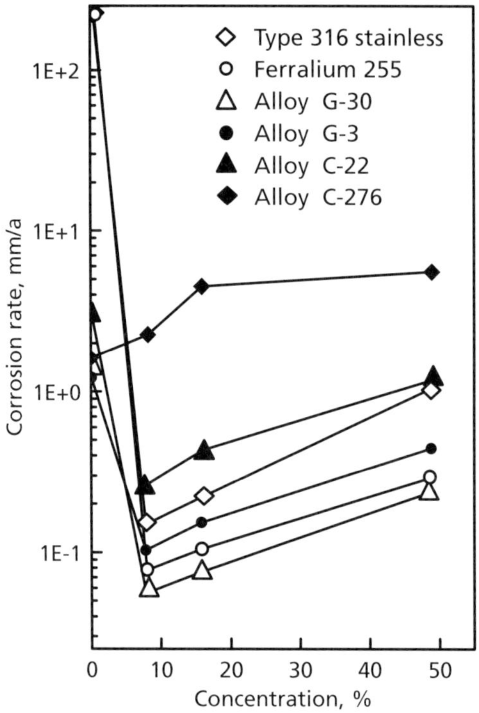

Figure 45: Influence of nitric acid on the corrosion of different alloys in boiling 30% sulfuric acid [220]

In a mixed acid of 50% H_2SO_4 + 50% HNO_3, Langley® alloy 20V (29Ni, 20Cr, 3Cu, 3Mo) and alloy 7R (54Ni, 23Cr, 6Cu, 6Mo, 2W, 3Si) as well as alloy C-276 (DIN-Mat. No. 2.4819) exhibit good resistance up to the boiling point with < 0.15 mm/a; however, in a mixture of 25% HNO_3 + 75% H_2SO_4 (boiling) only alloy 20V is resistant [226].

In a ternary mixture of 72% H_2SO_4 + 8% HNO_3 + 4% HF, the nickel alloy G-30 is very resistant with a corrosion rate of only 0.02 mm/a at 54°C.

In a boiling solution of 25% H_2SO_4 + 5% HNO_3 + 4% NaCl, alloy C-276 (1.5 mm/a), alloy C-4 (2.5 mm/a) and particularly alloy 625 (18 mm/a) are not resistant, while alloy C-22 is the most resistant with a corrosion rate of 0.3 mm/a [227].

Alloy C-276 is absolutely not resistant (> 12 mm/a) in 12.5-47.1% H_2SO_4 with 21.5-55.9% HNO_3 (boiling); alloy C-22 also cannot be used (approx. 2 mm/a). Only alloy G-30 has limited suitability with corrosion rates of 0.24 to 0.36 mm/a (Table 58).

In a mixture of 50% H_2SO_4 + 22% HNO_3 + 19% H_2O, alloy 825 (NiCr21Mo, DIN-Mat. No. 2.4858) shows good resistance with 0.01 mm/a at 66°C and 0.11 mm/a at 83°C [228]. Likewise in a mixture of 36% H_2SO_4 + 28% oxalic acid + 32% water + 4% ash at 60°C (0.06 mm/a).

Medium	*Testing temperature, °C*	*Alloy C-22*	*Alloy C-276*	*Alloy C-4*	*Alloy 625*
5% HNO_3 + 6% HF	60	1.70	5.26	5.18	1.85
5% HNO_3 + 1% HCl	boiling	0.01	0.20	0.28	0.03
5% HNO_3 + 25% H_2SO_4 + 4% NaCl	boiling	0.30	1.63	2.46	18.1
23% H_2SO_4 + 1.2% HCl + 1% $FeCl_3$ + 1% $CuCl_2$	boiling	0.18	1.40	58.3	97.7

Table 57: Uniform corrosion rates in mm/a; oxidizing media [227]

H_2SO_4 / HNO_3, vol%	*Corrosion rate (4 to 24 h periods), mm/a*					
	SAE 316	*alloy 625*	*alloy G-3*	*alloy G-30*	*alloy C-276*	*alloy C-22*
12.5 / 55.9	1.27	0.51	0.41	0.24	12.65	1.57
24.9 / 41.9	1.42	0.48	0.48	0.28	16.00	2.03
38.5 / 21.5	1.45	0.61	0.48	0.25	17.45	2.39
47.1 / 26.3	1.96	0.91	0.69	0.36	12.78	2.31

Table 58: Corrosion rates in boiling H_2SO_4 / HNO_3 mixtures [227]

Sulfuric and hydrochloric acids

Additions of chlorides and hydrochloric acid generally increase the corrosion rates in sulfuric acid [220] (see also Figure 46). Alloy C-22 (DIN-Mat. No. 2.4602) and alloy C-276 (DIN-Mat. No. 2.4819) exhibit corrosion rates of approximately 1 mm/a above a HCl addition of 4%. The positive influence of molybdenum (in alloy C-276 approx. 3% higher than in alloy C-22), is recognizable, particularly in the behavior of the chromium-free NiMo alloy B-2 (2.4617) that contains a very high amount of molybdenum [221].

In dilute mixtures of sulfuric and hydrochloric acids, the nickel alloy G-30 (< 0.01 mm/a) exhibits excellent behavior, followed by alloy C-22 (< 0.03 mm/a) (Table 59). Alloy C-276 and, particularly alloy G-3, exhibit higher corrosion rates if the hydrochloric acid content exceeds 1% [227].

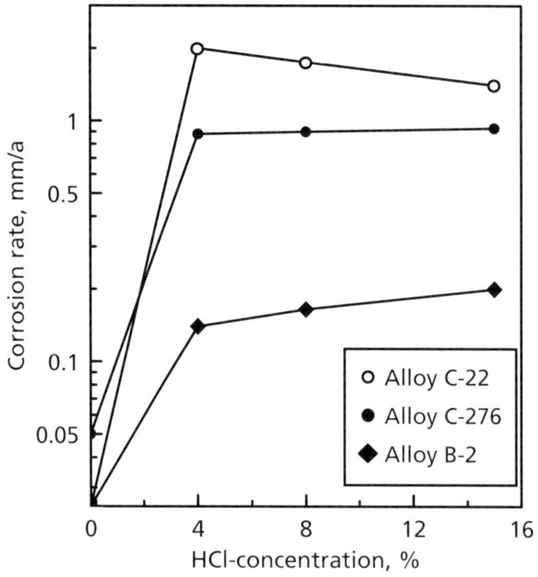

Figure 46: Influence of hydrochloric acid on the corrosion of some nickel-based alloys in 15% sulfuric acid 80°C [220]

	Corrosion rate (4 to 24 h periods), mm/a					
H_2SO_4 / HCl, vol%	***SAE 316***	***alloy B-2***	***alloy G-3***	***alloy G-30***	***alloy C-276***	***alloy C-22***
1.8 / 0.45	2.06	0.17	0.008	0.003	0.03	0.003
1.8 / 1.7	14.68	0.20	1.80	0.003	0.33	0.018
3.5 / 0.45	3.12	0.16	0.48	0.003	0.05	0.008
5.0 / 1.3	12.17	0.18	0.91	0.005	0.64	0.030
6.8 / 0.45	3.94	0.15	0.23	0.008	0.36	0.023

Table 59: Corrosion data in dilute H_2SO_4/HCl solutions at 80°C [227]

Alloy 625, C-276 and C-22 are not resistant in 10% H_2SO_4 + 1% HCl at 70°C. Under these conditions, only ALLCORR® (UNS N06110, see also below under HNO_3 + HCl) has the most favorable resistance with a corrosion rate of 0.4 mm/a [229].

Alloy C-22 (DIN-Mat. No. 2.4602) exhibited the best behavior in mixed acids based on H_2SO_4 + HCl or with additional chlorides as compared to the other C-type nickel alloys (Table 60) [230].

In 10% H_2SO_4 + 1% HCl, the material has good resistance with a corrosion rate of 0.02 mm/a, while alloy C-276 is more strongly attacked (0.3 mm/a) and alloy 625 shows extreme attack (3 mm/a). In a boiling mixed acid with additional oxidizing iron(III) and copper(II) ions, this material is the most resistant compared to the three other nickel materials (Table 60).

Medium	Test temperature, °C	Corrosion rate, mm/a			
		alloy C-22	alloy C-276	alloy C-4	alloy 625
Non-oxidizing acids					
1.5% HCl	boiling	0.279	0.737	1.626	8.966
10% HCl	boiling	10.16	7.32	5.79	16.31
10% H_2SO_4	boiling	0.279	0.584	0.787	1.168
10% H_2SO_4 + 1% HCl	70	0.013	0.279	0.610	3.073
Oxidizing acids					
5% HNO_3 + 6% HF	60	1.70	5.26	5.18	1.85
5% HNO_3 + 1% HCl	boiling	0.01	0.20	0.28	0.03
5% HNO_3 + 25% H_2SO_4 + 4% NaCl	boiling	0.30	1.62	2.46	18.1
23% H_2SO_4 + 1.2% HCl + 1% $FeCl_3$ + 1% $CuCl_2$	boiling	0.17	1.40	58.3	97.7
1% HCl + 42 g/l $Fe_2(SO_4)_3$	93	0.05	1.04	–	6.05

Table 60: A comparison of the corrosion behavior of NiCrMo alloys [230]

The corresponding corrosion rates for the three nickel alloys C-22, C-276 and alloy 59 in mixtures of 10–85% H_2SO_4 + 1–7% HCl and at higher test temperatures, are given in Table 61 [231]. Up to approx. 50°C, the alloys exhibit good resistance; however, alloy C-276 shows the best behavior because the acid mixtures are strongly reducing. Under these conditions, the markedly higher Cr content (approx. 23% Cr) of the other two alloys is rather disadvantageous.

The behavior of the three high-alloy nickel alloys C-276, C-22 and 59 in standard tests in dilute sulfuric acid with additional HCl, in mixtures of 10% H_2SO_4 + 1% HCl as well as in some other pure acids is shown in Table 62. Of the five very highly alloyed NiCrMo materials, alloy 59 generally exhibited the best behavior. However, the good behavior in the tests ASTM-G-28B and green death is due to the addition of heavy metal ions (as they actually occur in practice, see below under FGD plants) which have an oxidizing effect. In their absence, the corrosion rates are much higher, as shown by the values for 10% H_2SO_4 + 1% HCl. The corrosion rate of alloy 59 only decreases to below 0.1 mm/a above 90°C [232].

Test solution	Temperature, °C	Corrosion rate, mm/a		
		alloy C-22	alloy C-276	alloy 59
10% H_2SO_4 + 1% HCl	70	0.025	0.020	0.025
	90	2.34	1.05	0.07
	boiling	8.98	2.20	1.77
20% H_2SO_4 + 4% HCl	52	0.21	0.15	0.008
	79	1.37	0.84	1.02
30% H_2SO_4 + 4% HCl	52	0.19	0.11	0.23
	79	1.49	0.77	1.10
80% H_2SO_4 + 7% HCl	52	0.07	0.04	0.08
85% H_2SO_4 + 3% HCl	52	0.71	0.05	1.32

Table 61: Comparison of the corrosion rates of the nickel alloys C-22, C-276 and 59 in mixed acids [231]

Corrosion rate, mm/a					
Medium	alloy C-276	alloy C-22	alloy 686	alloy C-2000	alloy 59
ASTM 28A	4.27	0.91	2.62	0.69	0.61
ASTM 28B	1.40	0.18	0.97	0.10	0.10
Green death	0.66	0.10	0.20	–	0.13
10% HNO_3	0.48	0.05	–	–	0.05
65% HNO_3	19.05	1.32	5.87	–	1.02
10% H_2SO_4	0.58	0.46	–	–	0.20
50% H_2SO_4	6.10	7.82	–	–	4.47
1.5% HCl	0.28	0.36	0.13	0.04	0.08
10% HCl	6.07	9.96	–	–	4.55
10% H_2SO_4 + 1% HCl	2.21	8.99	–	–	1.78
10% H_2SO_4 + 1% HCl *	1.04	2.34	–	–	0.08

*) 90°C

Table 62: Comparison of the corrosion of some NiCrMo alloys in various boiling acids in mm/a [232]

Comparative corrosion data for several nickel alloys of the C series have been published by a manufacturer of semi-finished products, and are given in Table 63 [233].

Medium	Testing temperature, °C	Average corrosion rate, mm/a			
		alloy C-22	alloy C-276	alloy C-4	alloy 625
5% H_2SO_4 + 0.1% HCl	boiling	0.66	0.84	1.24	3.84
5% H_2SO_4 + 0.5% HCl	boiling	1.55	1.24	2.31	11.02
10% H_2SO_4 + 1% HCl	70	< 0.03	0.28	0.61	3.07
10% H_2SO_4 + 1% HCl	90	2.39	1.14	1.68	8.28
10% H_2SO_4 + 1% HCl	boiling	5.72	2.95	4.88	22.1
10% H_2SO_4 + 2% HF	boiling	0.74	0.56	0.66	1.40
25% H_2SO_4 + 200 ppm Cl^-	70	0.28	0.30	0.94	2.79
25% H_2SO_4 + 200 ppm Cl^-	boiling	5.46	4.72	4.62	8.26
11.5% H_2SO_4 + 1.2% HCl + 1% $FeCl_3$ + 1% $CuCl_2$	boiling	0.08	1.07	21.3	46.1
23% H_2SO_4 + 1.2% HCl + 1% $FeCl_3$ + 1% $CuCl_2$ (ASTM G28B)	boiling	0.20	1.40	54.7	69.1
50% H_2SO_4 + 42 g/l $Fe_2(SO_4)_3$ (ASTM-G-28B)	boiling	1.02	6.35	3.63	0.58

Table 63: Comparative corrosion data in acid mixtures [233]

The individual data shows that alloy C-22, which contains a higher amount of chromium, exhibits better resistance compared to alloys C-276 and C-4 only under oxidizing conditions, i.e. with addition of heavy metal ions, such as Fe^{3+} and Cu^{2+}, which have an oxidizing effect, and not in primarily reducing media.

Nickel alloys have been tested for their suitability in mixed acids based on sulfuric acid and traces of hydrochloric acid (due to the addition of chlorine gas in the process, etc.), as they occur in the paper and pulp industry. While alloy C-276 (DIN-Mat. No. 2.4819) and alloy C-22 (DIN-Mat. No. 2.4602) do not show local attack, their corrosion rates of 0.5 and 0.3 mm/a are very unfavorable and noticeably higher than those of the superaustenites. The only nickel alloy which is not only free of local corrosion but also has an extremely low corrosion rate (< 0.01 mm/a), is alloy G-30. With respect to the unfavorable behavior of this alloy in hydrochloric acid – the corrosion rate is 0.5 mm/a in 10% HCl at room temperature. In contrast, according to the information given there, its behavior in phosphoric and sulfuric acids is more favorable [234].

Very dilute and high-chloride mixtures based on sulfuric and hydrochloric acids are produced during industrial flue gas cleaning in conventional power station engineering (flue gas desulfurization plant, FGDP). In hard coal-fired and brown coal-fired power stations, the sulfur dioxide formed during the combustion process is absorbed by a limestone suspension, and a sulfuric acid solution is formed [224, 235]. Because of the chlorides brought in by the coal or recirculated and highly enriched via the scrubbing process, this acidic solution is corrosive, especially at elevated temperatures (range 60–120°C). In waste incineration plants (WIP), there are additional traces of hydrochloric acid from the combustion of plastic waste (PVC etc.) Heavy metal ions, which have an oxidizing effect, as well as oxygen from the air increase the risk of crevice corrosion. Particularly in steels which are insufficiently alloyed, local corrosion occurs in the form of pitting corrosion or crevice corrosion not only in the metal-metal crevices caused by construction, but also underneath the large amounts of gypsum deposits. Above 50°C, for components which are under mechanical stress and which are not sufficiently highly alloyed with nickel (at least 25% Ni), susceptibility towards stress corrosion cracking (SCC) must be expected. In order to qualify stainless steels and nickel alloys for this, extensive laboratory tests have been carried out – partly in simulated FGD media and partly in enhanced laboratory media which provide a relative and differentiating estimation of the materials. Since the uniform corrosion under FGD conditions is small and can be neglected, the aim of the investigations was the estimation of the resistance of the nickel materials to pitting and crevice corrosion. For this the pitting corrosion or breakthrough potentials were determined electrochemically. For practicians, the so-called critical pitting corrosion temperatures and critical crevice corrosion temperatures are very meaningful – local corrosion can be definitely expected above these temperatures.

Material designation	*Testing temperature °C*	*Breakthrough potential, mV*	*Repassivation potential, mV*	*Corrosion rate, mm/a*	*Surface evaluation*
Nicrofer® 5716 hMoW (DIN-Mat. No. 2.4819)	60	1,125	935	–	–
	80	745	800	< 0.001	no corrosive attack
Nicrofer® 5923 hMoW (DIN-Mat. No. 2.4605)	60	1,155	1,015	< 0.001	no corrosive attack
	80	925	850	< 0.001	no corrosive attack

Table 64: Local corrosion behavior of Nicrofer® 5716 hMoW (alloy C-276, DIN-Mat. No. 2.4819) and Nicrofer® 5923 hMo (alloy 59, DIN-Mat. No. 2.4605) under electrochemically controlled conditions in a simulated FGD solution: H_2SO_4 (pH 1) + 15% chlorides + 0.01% fluorides + 15% FGD gypsum; aerated, stirred, at 80°C [235]

Table 64 gives the electrochemical behavior in a typical simulated FGD solution [235]. The values of the potential indicate very good resistance of the nickel materials. The breakthrough and repassivation potentials lie relatively close together, both at 60°C as well as at 80°C. The corrosion rates are very low, and no corrosive attack could be observed by microscopy/metallography. When studies were carried out in the same FGD solution on samples of the base material, on welded samples as well as those equipped with slotted teflon blocks (test duration 30 days), then even these highly alloyed materials exhibited crevice corrosion under the artificially enhanced conditions (see Table 65).

Testing medium	***Material*** ***1) base material*** ***2) weldment, manual TIG with filler*** ***3) plasma weldment***	***Corrosion rate, mm/a***	***Depth of crevice corrosion, mm***
Medium B: 7% Cl^-, 0.01% F^-, pH 1 (H_2SO_4), 15% FGD gypsum aerated, stirred	1) Nicrofer® 5923 hMo (DIN-Mat. No. 2.4605) 2) Nicrofer® S 5923 (DIN-Mat. No. 2.4607) 3) no filler	not measurable not measurable not measurable	< 0.1 – > 0.3
Medium C: 15% Cl^-, 0.01% F^-, pH 1 (H_2SO_4), 15% FGD gypsum aerated, stirred	1) Nicrofer® 5923 hMo (DIN-Mat. No. 2.4605) 2) Nicrofer® S 5923 (DIN-Mat. No. 2.4607) 3) no filler	not measurable not measurable not measurable	0.1–0.5 0.1–0.5 0.1–0.5
	1) Nicrofer® 5716 hMoW (DIN-Mat. No. 2.4819) 2) Nicrofer® S 5716 (DIN-Mat. No. 2.4886) 3) no filler	not measurable not measurable not measurable	0.1–0.5 0.1–0.5 0.1–0.5

Table 65: Local corrosion behavior of Nicrofer® 5716 hMoW (alloy C-276, DIN-Mat. No. 2.4819) and Nicrofer® 5923 hMo (alloy 59, DIN-Mat. No. 2.4605) in the unwelded and welded state in the simulated FGD solutions B and C after 30 days at 80°C [235]

In medium B, alloy 59 is obviously not susceptible to local corrosion since only isolated crevice corrosion depths of < 0.1 mm were observed, although a depth of > 0.3 mm was observed at one location in the welded material. In contrast, a measurable corrosion rate was not determined for any of the samples. The limiting range for the suitability of both nickel materials is obviously reached in the simulated FGD solution C. For continuous exposure in this enhanced medium and in the presence of extreme crevice conditions that were generated artificially in the laboratory, crevice corrosion depths of up to 0.5 mm were determined. This is only valid for the assumption of continuous loading, which is, however, not the case in practice since the operating conditions in continuous operation are generally less corrosive; this also applies to the process temperature. The simulated extreme case only occurs very rarely during the service life of a power station. On account of the pronounced repassivation ability of the high-alloy nickel materials (c.f. Table 64), it can be assumed that the corresponding extra wall thickness ensures reliability against corrosion during practical use. While the sheet thicknesses are 2–4 mm for the other plant regions of a FGD scrubber, thicknesses of 6 mm are chosen for particularly endangered regions, such as the raw gas inlet ducts for reasons of safety.

With regard to the behavior of high-alloy nickel materials used as metallic linings (so-called "metal sheeting" concept in which large sheets are tack-welded to carrier materials of carbon steel by hole spot welding) in scrubbers, hole spot-welds were also tested. This was carried out in order to test that their corrosion behavior was the same as the unwelded sheet material as well as the influence of iron migration from the carbon steel [235]. The test was performed on two-layer inert-gas weldments in pre-punched holes for a period of 30 days at 80°C.

Welding process: MIG	*Pulsed arc welding*	*Short-spray and pulsed arc welding*	*Pulsed arc welding*	*Short-spray and pulsed arc welding*
Base material	*Nicrofer® 5716 hMoW (DIN-Mat. No. 2.4819)*		*Nicrofer® 5923 hMo (DIN-Mat. No. 2.4605)*	
Filler	*Nicrofer® S 5923 (DIN-Mat. No. 2.4607)*			
Weld orientation:	*V*	*H*	*V*	*H*
Simulated FGD solution A	0.0007	0.0007	0.0008	0.001
	0.0008	0.0006	0.0008	0.001
Simulated FGD solution B	0.002	0.002	0.001	0.001
	0.002	0.002	0.001	0.01
Simulated FGD solution C	0.05	0.05	0.02	0.03
	0.07	0.06	0.02	0.04

FGD medium A: 3% Cl^- + 0.05% F^- + 15% FGD gypsum, pH 1 with H_2SO_4
FGD medium B: 7% Cl^- + 0.01% F^- + 15% FGD gypsum, pH 1 with H_2SO_4
FGD medium C: 15% Cl^- + 0.01% F^- + 15% FGD gypsum, pH 1 with H_2SO_4

Table 66: Investigation of the corrosion behavior of nickel materials after welding with the wall-papering process using hole spot welding. Welding method: metal inert gas (MIG) weld cladding; two-layer, in pre-punched holes, test duration 30 days at 80°C; weld orientation: vertical (V) and horizontal (H); corrosion rates in mm/a [235].

As can be seen from Table 66, independent of the weld orientation (vertical and horizontal), the corrosion rates lay on the boundary of experimental error. Local corrosion was not observed. Even the much more aggressive FGD solution C, containing a higher chloride content of 15%, only led to slightly increased corrosion rates, which, however, still lay clearly below 0.1 mm/a.

In flue gas desulfurization plants and waste incineration plants, high-alloy nickel materials are also used in combination with the carrier material carbon steel. While explosive cladding is used for small pieces of apparatus, hot-roll cladding is preferred for reasons of cost for large-scale sheets used in large-volume plants, such as scrubbers.

Equivalent corrosion resistance of the material must also be ensured in the cladded state (influence of heat during production can change the microstructure). Table 67 gives examples of corrosion results of a cladded layer of Nicrofer® 5716 hMoW (alloy C-276, DIN-Mat. No. 2.4819) [235].

Medium	Specimen	Corrosion rate, mm/a
A	cladding material	0.016
A	sample with weld seam	0.040
C	cladding material	0.042
C	sample with weld seam	< 0.1*

*) Increase in the mass loss rates due to local corrosion of an incorrectly executed sealing seam without a weld filler
FGD medium A: 3% Cl^- + 0.05% F^- + 15% FGD gypsum, pH 1 with H_2SO_4
FGD medium C: 15% Cl^- + 0.01% F^- + 15% FGD gypsum, pH 1 with H_2SO_4

Table 67: Corrosion behavior of a cladded layer of Nicrofer® 5716 hMoW (alloy C-276, DIN-Mat. No. 2.4819) in a simulated FGD solution, test conditions: 30 days at 80°C, aerated and stirred [235]

In both media, the unwelded material of the cladding exhibited corrosion rates of < 0.1 mm/a. The samples with welded joints behaved accordingly, with one exception: an incorrectly welded sealing seam (without weld filler) exhibited a higher corrosion rate as a result of local corrosion. Exposure tests in FGD scrubbers for 300 days did not show any measurable losses and no local corrosive attack of these nickel alloys in practice.

The behavior of some nickel alloys in sulfuric acid solutions with very high chloride contents at the boiling temperature and after 21 days exposure are given in Table 68 [224]. While the known nickel alloys alloy C-4 and alloy 625 exhibited clear susceptibility with relatively high corrosion rates, the C variants containing higher amounts of molybdenum were markedly more resistant; in particular Nicrofer® 5923 hMo (alloy 59, DIN-Mat. No. 2.4605), which exhibited an extremely low corrosion rate of 0.003 mm/a.

In order to obtain an even greater differentiation between the high-alloy materials with regard to their microstructural states after heat treatment and welding, even more enhanced test solutions have been introduced to simulate the most extreme corrosion conditions in FGD plants and WIPs. These solutions have proven useful in the selection of materials [224, 235], particularly the very aggressive test developed in the USA with the composition: 7% H_2SO_4 + 3% HCl + 1% $CuCl_2$ + 1% $FeCl_3$, which is called "green death" on account of its green color. Stainless steels and also superaustenites do not pass the test. To test the resistance towards pitting and crevice corrosion which is the most important factor in practice, the test does not determine uniform corrosion but rather the critical temperature of local corrosion [224], which is the temperature above which pitting or crevice corrosion is observed. Thus, the use of the material is not recommended above this temperature.

Alloy	Pitting corrosion/ crevice corrosion	Corrosion rate, mm/a
Nicrofer® 5716 hMoW (NiMo16Cr15W, DIN-Mat. No. 2.4819)	no	0.12
Nicrofer® 6616 hMo (NiMo16Cr16Ti, DIN-Mat. No. 2.4610)	no	0.28
Nicrofer® 5621 hMoW (NiCr21Mo14W, DIN-Mat. No. 2.4602)	no	0.08
Nicrofer® 5923 hMo (NiCr23Mo16Al, DIN-Mat. No. 2.4605)	no	0.003
Nicrofer® 6020 hMo (NiCr22Mo9Nb, DIN-Mat. No. 2.4856)	no	1.33

Table 68: The corrosion behavior of nickel alloys of the C series in sulfuric acid solution at pH 1 and 1.7% chloride addition at 105°C (boiling), after 21 days exposure [224]

Alloy	CPT, °C	CCT, °C
Nicrofer® 5716 hMoW (DIN-Mat. No. 2.4819)	115–120	105
Nicrofer® 6616 hMo (DIN-Mat. No. 2.4610)	100	85–95
Nicrofer® 5621 hMoW (DIN-Mat. No. 2.4602)	120	110
Nicrofer® 5923 hMo (DIN-Mat. No. 2.4605)	> 120	> 110
Nicrofer® 6020 hMo (DIN-Mat. No. 2.4856)	100	85–96

Table 69: Critical pitting (CPT) and critical crevice corrosion temperature (CCT) of nickel alloys of the C series in the solution "green death": 7% H_2SO_4 + 3% HCl + 1% $CuCl_2$ + 1% $FeCl_3$. After each period of 24 h exposure, the temperature was increased by a further 5°C [224]

It can be seen from Table 69 that the nickel alloys containing less chromium, such as alloy C-4 (NiMo16Cr16Ti, DIN-Mat. No. 2.4610) and alloy 625 (NiCr22-Mo9Nb, DIN-Mat. No. 2.4856), have markedly lower critical temperatures of local corrosion. In contrast, the best behavior was shown by the high-Mo alloys with chromium contents of 22–23%. Alloy 59 (NiCr23Mo16Al, DIN-Mat. No. 2.4605) gave the most favorable results in this very enhanced test.

A further enhanced method is the Method B test in the standard ASTM G-28 test. While Test Method A (50% H_2SO_4 + 42 g/l $Fe_2(SO_4)_3$, boiling) is a purely qualitative test of the resistance to intergranular corrosion, which is used to validate the perfect microstructural state of manufactured semi-finished products, Test Method B also provides a statement on the general corrosion resistance in strong acids and media containing high amounts of chloride. The boiling test solution is similar to the "green death", and is a mixed acid, but with a three-fold sulfuric acid content: 23% H_2SO_4 + 1.2% HCl + 1% $FeCl_3$ + 1% $CuCl_2$. It can be seen from Figure 47 that C variants containing relatively low amounts of chromium, approx. 16%, namely alloy

C-276 (DIN-Mat. No. 2.4819) and alloy C-4 (DIN-Mat. No. 2.4610), exhibited unfavorable corrosion behavior in both test solutions. Very good behavior is shown by alloy 59 (DIN-Mat. No. 2.4605), which can be explained by the high chromium content of 23% as well as the very stable, essentially non-segregated microstructure.

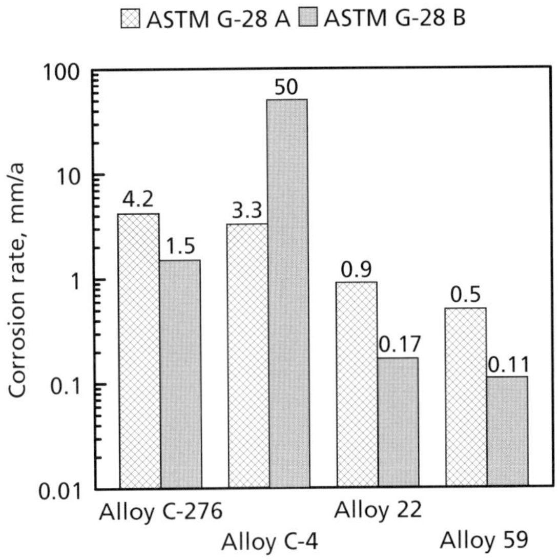

Figure 47: Corrosion rate of nickel alloys of the C series in the sulfuric acid standard test ASTM G-28 Methods A and B. Test duration 120 (A) and 24 hours (B); boiling solutions [224]

State	*Unwelded*	*Welded*		
		TIG	*plasma plug welding*	*MAW*
CPT (°C)	b) > 120	a) 115 HAZ + welded material b) > 120	a) > 120	b) > 120
CCT (°C)	a) 110 b) > 120	a) 110 b) > 120	a) 105 root of welded seam	b) > 120 root of welded seam

Table 70: Influence of welding on the critical pitting and crevice corrosion temperatures (CPT and CCT) of alloy 59 (DIN-Mat. No. 2.4605) in the solution "green death" (see also Table 69)
a) Test series 1 b) Test series 2 [235]
TIG: tungsten inert gas welding; MAW: manual arc welding

The behavior of practice-oriented welded joints was also investigated in short-term tests [235]. As the results in Table 70 show, the scattering range of the critical local corrosion temperatures of approx. 5°C is relatively low and indicates that the welding processes used in practice, namely tungsten inert gas welding (TIG), plasma-plug welding and manual arc welding (MAW) do not influence the generally good resistance to local corrosion of this material in the welded state. In the case of plasma plug welding, a slight decrease of the crevice corrosion temperature was found on the weld root. This indicates inclusion of oxides and thus the gas protection during welding was not 100%. The second test series also gave better values because the welding parameters had been carefully optimized. This shows once again how decisively dependent the corrosion resistance of weld seams is on the welding parameters, weld filler, welding process, type of inert gas and control of inert gas and, in particular, on the welding speed and the heat input (drawing energy). These parameters must be optimized in every case by preliminary trials. This is the only way to minimize the risk that the welding seam of a component becomes the weak point of the plant. If possible, a slightly overalloyed welding filler should always be selected. It should be overalloyed in those elements which decisively influence the corrosion resistance in the specific medium in use. In the case discussed here, these are chromium and molybdenum.

Samples subjected to TIG welding, plasma plug welding and MAW were also tested in a simulated FGD solution of the composition sulfuric acid pH 1 + 1.7% Cl^- at the boiling point 105°C) for 21 days. They gave corrosion rates of less than 0.01 mm/a. Local corrosion was not observed [235].

Alloy	**Corrosion rate, mm/a**	
	20% H_2SO_4 + 1.5% Cl^- 80°C	**50% H_2SO_4 + 1.5% Cl^- 50°C**
alloy C-276	0.007	0.42
alloy 22	0.004	0.45
alloy 59	0.003	0.38
alloy 625	0.003	0.75

Table 71: Corrosion behavior of alloy 59 (DIN-Mat. No. 2.4605) compared to other nickel materials in sulfuric acid of medium concentration with additions of chloride [224]

It is clear from Table 71, that nickel alloys have limited suitability in mixtures of sulfuric acid and traces of hydrochloric acid or chlorides [224]. The corrosion rates at 80°C are still less than 0.01 mm/a in up to 20% H_2SO_4 and 15,000 mg/l Cl^-. They already increase to considerably higher values at 50°C in 50% H_2SO_4 and the same chloride concentration. According to experience gained in FGD plants, such conditions only occur in a few regions, for example in the wet-dry zones of the raw gas inlet zone or in the heat transfer system. If necessary, greater wall thicknesses must be used.

During chlorine-alkaline electrolysis, a mixture of sulfuric acid and hydrochloric acid (formed from damp chlorine gas) is produced, which is a very corrosive medium, even for high-alloy nickel materials. Considerable corrosion problems occurred in the drying towers of such a plant, which operated with 65–95% H_2SO_4 to dry damp chlorine gas at 40°C. Particularly the cooling spirals in the lower region of the drying tower are subjected to extremely corrosive conditions. Crevice corrosion or uniform corrosion rates of up to 1 mm/a were observed on the nickel alloys of the C series used here. For this reason, the operator of a European chlorine-alkali electrolysis plant carried out laboratory studies of nickel alloys of the C series at 40°C in 75% H_2SO_4 saturated with chlorine gas [224]. After 32 days, the weight loss and behavior with regard to crevice corrosion of sheet samples of alloy C-276 (DIN-Mat. No. 2.4819), alloy C-4 (DIN-Mat. No. 2.4610), alloy 22 (DIN-Mat. No. 2.4602) and alloy 59 (DIN-Mat. No. 2.4605) were determined by the ASTM-Teflon thread test. Intermediate weighings were carried out to provide information on any changes in the corrosion behavior over the total test period. The corrosion rates were plotted against time. The initial corrosion rates of alloy C-4 and alloy 22 were very low; however, after 8 and 16 days, they increased to 0.04 and 0.13 mm/a, respectively. This indicates a continuous activation. Alloy C-276 exhibited high corrosion rates of 0.62 to 0.96 mm/a right from the start along with pronounced surface roughness. Nicrofer® 5923 hMo (DIN-Mat. No. 2.4605) exhibited an extremely low corrosion rate of 0.007 mm/a from the start, which remained almost constant during the entire test period and levelled out at a value of 0.003 mm/a after 32 days. The results of the test for crevice corrosion confirmed the sequence found for the resistance of the nickel alloys. Marked crevice corrosion was observed on alloy C-276, and slight crevice corrosion on alloy C-4 and alloy 22. Only the sample of alloy 59 retained its shiny surface and was free of local attack. On the basis of these results, the plant operator was recommended to use only the high-alloy nickel material Nicrofer® 5923 hMo (alloy 59, DIN-Mat. No. 2.4605).

Tests were carried out under laboratory conditions on three nickel alloys of the C series with regard to weight loss by means of immersion tests in a mixture of sulfuric and hydrochloric acids of various mixing concentrations and temperatures [236]. As a first result, the materials showed good resistance in all mixtures up to 52°C. At 90°C, the corrosion rates of alloy C-22 and alloy C-276 in 10% H_2SO_4 + 1% HCl were already extremely high (> 2 mm/a and > 1 mm/a, respectively), while alloy 59 still exhibited very good resistance with a corrosion rate of less than 0.1 mm/a. At a higher hydrochloric acid concentration of 4%, all materials were no longer resistant with corrosion rates > 1mm/a at 79°C. The decrease in corrosion at higher sulfuric and hydrochloric acid concentrations, i.e. at 80% H_2SO_4 + 7% HCl, is remarkable. This can be explained with the increasingly reducing character of sulfuric acid, the good behavior of alloy C-276 cannot be explained in any other way. Compared to the other two nickel alloys, this alloy has a relatively low chromium content – high chromium contents only improve the resistance under oxidizing conditions.

The corrosion behavior of some C series nickel alloys was tested in a mixture of several inorganic acids that may occur in the outlet duct of a FGD plant (see Table 72) [237].

Firstly, with the exception of alloy C-276, the materials still show acceptable resistance in solution 1, which is due to the low amount of added hydrochloric acid. If this is increased to 2.5% HCl, then alloy 686 still is the most favorable compared to the three other NiCrFe alloys – however, with a corrosion rate of 0.58 mm/a it certainly cannot be classified as resistant. All alloys failed in solution 3: the increased HCl concentration and the higher temperature promoted corrosion. See also Table 73.

Alloy	Corrosion rates, mm/a		
	solution 1	solution 2	solution 3
alloy 686	0.36	0.58	6.96
alloy C-276	1.37	0.71	6.05
alloy C-22	0.31	1.02	7.09
alloy 59	0.21	1.20	7.82

Solution 1: 60% H_2SO_4 + 0.5% HCl + 0.1% HF + 0.1% HNO_3 at 85°C
Solution 2: 60% H_2SO_4 + 2.5% HCl + 0.2% HF + 0.5% fly ash at 80°C
Solution 3: 70% H_2SO_4 + 2.5% HCl + 0.2% HF at 105°C

Table 72: Corrosion rates in a simulated solution which may occur in the outlet duct of a FGD plant. Values after 7 days [237]

In boiling mixtures of 10% H_2SO_4 + 1–5% HCl, the C alloys are not resistant and have corrosion rates >> 1 mm/a; while in 5% H_2SO_4 + 0.1% HCl, the values lie at approximately 0.6 mm/a [238].

Material	DIN-Mat. No.	Corrosion rate, mm/a			
		80% H_2SO_4 80°C	2% HCl boiling	10% H_2SO_4 + 2% HCl boiling	10% H_2SO_4 + 5% HCl 80°C
INCONEL® alloy 686[a)]	2.4606	0.10	0.15	3.35	0.86
UNS N10276	c.f. 2.4819	0.58	1.09	3.51	–
INCONEL® alloy 622	c.f. 2.4602	1.32	1.32	7.09	2.08

a) Average after exposure for two weeks

Table 73: Corrosion rate in acids and their mixtures, after an exposure time of one week [237]

Nitric and hydrochloric acids

In mixtures of nitric and hydrochloric acids, the influence of HNO_3 on HCl can be similarly positive as that observed with sulfuric acid; however, only in relatively dilute HCl, as shown in Figure 48.

While all materials are not resistant in 4% hydrochloric acid, additions of nitric acid above a concentration of approx. 5% led to resistance, with the exception of alloy C-276 (DIN-Mat. No. 2.4819). For higher additions of nitric acid (> 25%), the corrosion rates markedly increased again and only the materials alloyed with sufficiently high amounts of chromium remained resistant. In mixed acids based on 9% HNO_3 and HCl up to approx. 25 %, pitting corrosion must be expected as well as high uniform corrosion above 10% HCl (Figure 49).

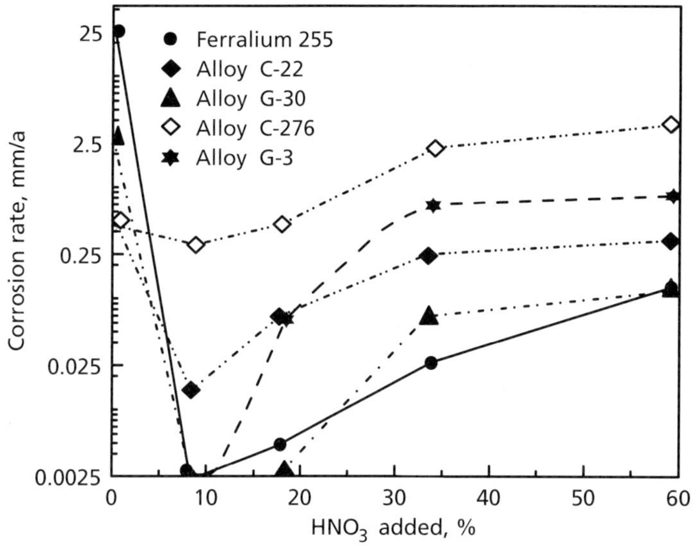

Figure 48: Influence of HNO_3 on the corrosion of various alloys in a 4% HCl solution at 80°C [220]

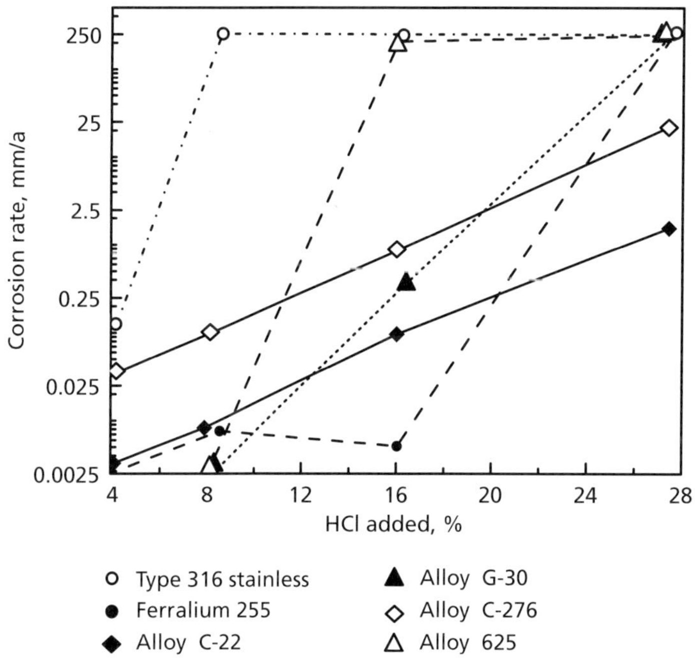

Figure 49: Influence of HCl additions on the corrosion rates of various alloys in 9% HNO_3 at 52°C [220]

According to these results, for an addition of up to approx. 10% HCl, all nickel materials and also the duplex steel Ferralium® are resistant, but not a 18Cr10NiMo steel. On addition of up to 20% HCl, only alloy C-22 (DIN-Mat. No. 2.4602) and Ferralium® are resistant [221].

In a mixture of boiling 5% HNO_3 + 1% HCl, both alloy C-22 (0.01 mm/a) and alloy 625 (0.025 mm/a) are very resistant, while alloy C-276 (0.2 mm/a) and alloy C-4 (0.3 mm/a) are just sufficiently resistant (Table 57) [227].

In 20% HNO_3 + 5% HCl at 104°C, alloy 625, alloy C-276 and alloy C-22 have absolutely no resistance with corrosion rates > 20 mm/a [239]. Although the high-chromium nickel alloy ALLCORR® (UNS N06110; 56Ni, 31Cr, 10Mo, 2W) showed the best behavior, it is still not resistant with a corrosion rate of 3 mm/a. The four nickel alloys cannot be used in 28% HCl + 17% HNO_3 at 109°C due to extremely high corrosion rates.

Nitric and hydrofluoric acids

In mixtures of nitric and hydrofluoric acids, additions of nitric acid initially reduce the corrosion in hydrofluoric acid; however, it increases again above 10% HNO_3. In 20% nitric acid even fractions of 1% HF produce a strong increase in the corrosion of all nickel alloys (Figure 50).

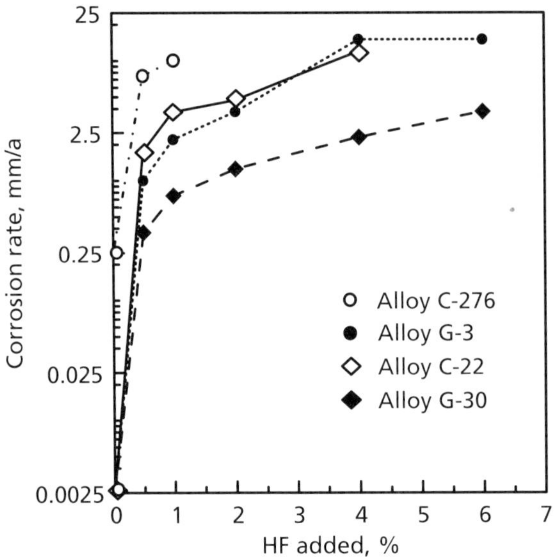

Figure 50: Influence of HF additions on the corrosion of various alloys in 20% HNO_3 at 80°C [220]

This behavior is confirmed by investigations of two high-alloy special stainless steels and two nickel materials that were tested in mixed acids of 12–56% HNO_3 and 0.4–3.5% HF at 90°C (Table 74) [239].

Medium	Temperature, °C	Corrosion rate, mm/a			
Mixed acid		alloy 33	alloy G-30	alloy 28	alloy 690
12% HNO_3 + 0.9% HF	90	0.25	0.28	5.84	0.61
12% HNO_3 + 3.5% HF	90	1.22	1.22	>12	6.40
32% HNO_3 + 0.4% HF	90	0.28	0.51	0.97	1.47
56% HNO_3 + 0.4% HF	90	1.68	2.44	3.43	4.75

Table 74: Behavior of alloys 33, G-30, 28 and 690 in mixed acids [239]

This also explains why these alloys, like the stainless steels, are pickled in mixed acids based on 20% HNO_3 plus 1–3% HF [221].

In mixtures of 5% HNO_3 + 6% HF at 60°C, alloys C-22, C-276, C-4 and 625 are all non-resistant with corrosion rates from 1.5 to 5 mm/a (Table 57) [227].

In a solution of 56% HNO_3+ 0.5% HF at 100°C, which is used to treat nuclear waste, alloy G-30 (1.3 mm/a) and alloy G-3 (3.4 mm/a) can only be used if they are not permanently exposed to the medium [227].

In 10% HNO_3 + 3% HF at 70°C, alloys 625, C-276, C-22 and ALLCORR® have absolutely no resistance [229]. This also applied to mixtures of 15% HNO_3 + 3% HF as well as 20% HNO_3 + 2% HF at 60°C [238].

Table 75 shows the behavior of several NiCrMo alloys and a very high-chromium or high-molybdenum stainless steel in a mixture of 20% HNO_3 and 1–5% HF at 50°C. It is noticeable that only the very high-chromium materials X1NiCrMoCu31-27-4 (DIN-Mat. No. 1.4563) and NiCr30FeMo (DIN-Mat. No. 2.4603) show good resistance for HF additions of 1% (0.25 and 0.2 mm/a, respectively) and still the best relative behavior for additions of 3% HF (0.5 and 0.45 mm/a, respectively) [240]. The area-related rates of mass loss of the material 2.4603 in 2 M HNO_3 at 90°C increase linearly with increasing HF concentration up to 2 M. However, the material is resistant only up to an addition of approx. 0.5 M HF (approx. 0.25 mm/a). In contrast, in mixtures containing 0.25 M HF, a parabolic increase up to 14 M HNO_3 was found; but, in this case too, the material 2.4603 is resistant (0.3 mm/a) only up to approx. 4 M HNO_3 (and 0.25 M HF, 90°C).

The favorable corrosion behavior of material 2.4603 (alloy G-30) in nitric acid-hydrofluoric acid mixtures is supported by positive operating experience with plate heat exchangers and pickling racks [240]. A pickling rack made of 18Cr10Ni steel (DIN-Mat. No. 1.4550) had a service life of only 6 months in a mordant of 10 to 20% HNO_3 and 2–3% HF at 65°C. In contrast, the new rack made of alloy 2.4603 has an expected service life of 10 years on the basis of the above investigations.

A manufacturer has published corrosion rates of the nickel materials alloy G-30, alloy G-3 and alloy 625 (Table 76) [241].

HF concentration %	Corrosion rate, mm/a DIN-Mat. No.						
	1.4404	2.4660	1.4539	1.4563	2.4856	2.4858	2.4603
1	9	1.45	0.55	0.25	0.75	0.7	0.2
3	7.6	3.5	1.25	0.5	1.35	1.65	0.45
5	29	6.25	2.5	0.85	3.5	3.3	0.75

Table 75: Corrosion rates in mixtures of 20% nitric acid and 1–5% hydrofluoric acid at 50°C [240]

Medium	Concentration mass-%	Test temperature °C	Average corrosion rate, mm/a		
			alloy G-30	alloy G-3	alloy 625
acetic acid	99	boiling	0.03	0.02	< 0.03
formic acid	88	boiling	0.05	0.13	0.23
nitric acid	10	boiling	0.01	0.02	0.03
	60	boiling	0.13	0.22	0.41
	65	boiling	0.13	0.28	0.51
nitric acid + 1% HF	20	80	0.79	1.88	3.12
nitric acid + 6% HF	20	80	4.50	13.7	60.9
nitric acid + 1% HF	50	80	4.88	10.67	–
nitric acid + 0.5% HF	56	110	1.19	2.79	–
nitric acid + 0.5% HF + 2,000 ppm Cl^-	56	110	1.27	2.87	–
sulfuric acid + 10% nitric acid	50	boiling	0.41	0.76	–
sulfuric acid	2	boiling	0.20	0.15	0.15
	10	boiling	0.79	0.48	1.17
	20	boiling	1.37	0.76	3.15
	50	107	0.94	0.94	5.66
	80	52	0.30	0.58	0.84
	99	130	1.09	1.88	–
	99	140	1.17	1.45	–

Table 76: Comparative corrosion data in acid mixtures for alloy G-30, G-3 and 625 [241]

Table 76: Continued

Medium	*Concentration mass-%*	*Test temperature °C*	*Average corrosion rate, mm/a*		
			alloy G-30	*alloy G-3*	*alloy 625*
sulfuric acid + 42 g/l $Fe_2(SO_4)_3$ (ASTM-G-28A)	50	boiling	0.18	0.28	0.58
sulfuric acid + 5% nitric acid	70	boiling	3.38	6.10	–
sulfuric acid + 5% nitric acid	60	boiling	1.14	2.13	2.67
sulfuric acid + 8% nitric acid + 4% HF		54	0.01	0.04	–
nitric acid + 8% HCl		80	0.05	0.46	0.15
nitric acid + 11% HCl		80	0.58	23.2	3.20
nitric acid + 3% HCl		80	0.13	0.86	0.51

Table 76: Comparative corrosion data in acid mixtures for alloy G-30, G-3 and 625 [241]

Only materials with corrosion rates markedly lower than 0.25 mm/a can be regarded as resistant. As can be seen from the Table 76, none of the materials are resistant in mixtures of HNO_3 + HF at high temperatures (80–110°C). In a ternary mixture of H_2SO_4 + HNO_3 + HF, alloy G-30 and G-3 have very good resistance, as well as in a mixture of HNO_3 + HCl.

The same manufacturer of semi-finished products gives the average corrosion rates for alloy C-22 in some pure acids and mixed acids (Table 77) [242]. Alloy C-22, which contains a higher amount of chromium than alloy C-276 and C-4 (22% Cr compared to 16% Cr), exhibits the highest resistance in most acidic media, with some exceptions such as very dilute, boiling sulfuric acid as well as the mixed acid 5% HCl + 2% HF at 70°C.

Medium	*Concentration mass-%*	*Testing temperature °C*	*Average corrosion rate, mm/a*			
			alloy C-22	*alloy C-276*	*alloy C-4*	*alloy 625*
hydrochloric acid	1	boiling	0.08	0.33	0.64	0.03
	1.5	boiling	0.36	0.81	1.63	8.97
	2	90	nil	0.03	0.79	nil
	2	boiling	1.55	1.30	2.08	14.15
	2.5	90	< 0.03	0.30	0.86	1.83

Table 77: Comparative corrosion data in acids and mixed acids for nickel alloys of the C series [233]

Table 77: Continued

Medium	Concentration mass-%	Testing temperature °C	Average corrosion rate, mm/a			
			alloy C-22	alloy C-276	alloy C-4	alloy 625
hydrochloric acid	2.5	boiling	3.58	2.16	1.12	15.37
	10	boiling	10.16	7.32	5.79	16.31
hydrochloric acid + 2% HF	5	70	1.50	0.66	0.86	3.12
hydrofluoric acid	2	70	0.23	0.23	0.43	0.51
	5	70	0.36	0.25	0.38	0.41
P_2O_5 + 2,000 ppm Cl^-	38	85	0.03	0.30	–	0.05
P_2O_5 + 0.5% HF	38	85	0.18	1.14	–	0.23
nitric acid	10	boiling	< 0.03	0.18	0.18	< 0.03
	65	boiling	3.40	22.56	5.51	0.53
nitric acid + 6% HF	5	60	1.70	5.26	5.18	1.85
nitric acid + 25% H_2SO_4 + 4% NaCl	5	boiling	0.30	1.63	2.46	18.11
nitric acid + 1% HCl	5	boiling	< 0.03	0.20	0.28	0.03
nitric acid + 2.5% HCl	5	boiling	0.05	0.53	0.66	< 0.03
nitric acid + 15.8% HCl	8.8	52	0.10	0.84	2.90	> 250
sulfuric acid	2	66	nil	0.03	nil	nil
	2	boiling	0.13	0.15	0.15	0.15
	5	79	< 0.03	< 0.03	< 0.03	< 0.03
	5	boiling	0.23	0.30	0.41	0.41
	10	boiling	0.30	0.48	0.64	0.94
	20	66	< 0.03	< 0.03	< 0.03	< 0.03
	20	79	0.03	0.08	0.05	0.33
	20	boiling	0.84	0.99	0.91	2.31

Table 77: Comparative corrosion data in acids and mixed acids for nickel alloys of the C series [233]

Table 77: Continued

Medium	*Concentration mass-%*	*Testing temperature °C*	*Average corrosion rate, mm/a*			
			alloy C-22	*alloy C-276*	*alloy C-4*	*alloy 625*
sulfuric acid	30	66	< 0.03	< 0.03	< 0.03	< 0.03
	30	79	0.08	0.10	0.08	0.69
	30	boiling	1.63	1.40	1.85	5.77
	40	38	< 0.03	< 0.03	< 0.03	< 0.03
	40	66	< 0.03	0.03	0.23	0.03
	40	79	0.23	0.25	0.38	0.89
	50	38	< 0.03	nil	< 0.03	< 0.03
	50	66	0.03	0.10	0.33	0.64
	50	79	0.41	0.30	0.64	1.47
	60	38	< 0.03	< 0.03	0.03	< 0.03
	70	38	nil	nil	0.05	< 0.03
	80	38	nil	< 0.03	< 0.03	< 0.03
sulfuric acid + 0.1% HCl	5	boiling	0.66	0.84	1.24	3.84
sulfuric acid + 0.5% HCl	5	boiling	1.55	1.24	2.31	11.02
sulfuric acid + 1% HCl	10	70	< 0.03	0.28	0.61	3.07
sulfuric acid + 1% HCl	10	90	2.39	1.14	1.68	8.28
sulfuric acid + 1% HCl	10	boiling	5.72	2.95	4.88	22.07
sulfuric acid + 2% HF	10	boiling	0.74	0.56	0.66	1.40
sulfuric acid + 200 ppm Cl^-	25	70	0.28	0.30	0.94	2.79
sulfuric acid + 200 ppm Cl^-	25	boiling	5.46	4.72	4.62	8.26

Table 77: Comparative corrosion data in acids and mixed acids for nickel alloys of the C series [233]

Table 77: Continued

Medium	*Concentration mass-%*	*Testing temperature °C*	*Average corrosion rate, mm/a*			
			alloy C-22	*alloy C-276*	*alloy C-4*	*alloy 625*
sulfuric acid + 1.2% HCl + 1% $FeCl_3$ + 1% $CuCl_2$	11.5	boiling	0.08	1.07	21.26	46.10
sulfuric acid + 1.2% HCl + 1% $FeCl_3$ + 1% $CuCl_2$ (ASTM G28B)	23	boiling	0.20	1.40	54.74	69.11
sulfuric acid + 42 g/l $Fe_2(SO_4)_3$ (ASTM G28A)	50	boiling	1.02	6.35	3.63	0.58

Table 77: Comparative corrosion data in acids and mixed acids for nickel alloys of the C series [233]

Phosphoric and sulfuric acids

During the production of phosphoric acid, a mixture of 30% H_3PO_4 and 3% H_2SO_4 is formed along with small amounts of fluoride ions. Exposure tests in an ageing tank at 66°C gave a corrosion rate of 0.125 mm/a for alloy G (NiCr22Mo6Cu, 2.4618) [243].

Exposure tests with the nickel alloy 825 (NiCr21Mo, DIN-Mat. No. 2.4858) in phosphoric acid production plants that use the wet process gave the corrosion rates presented in Table 78 in the individual plant zones.

Test conditions	*Temperature °C*	*Test duration days*	*Corrosion rate mm/a*
re-used liquid from the washing column: 15% H_3PO_4, 20% H_2SiF_6, 1% H_2SO_4	74–85	16	0.025
solution of: 20% H_3PO_4 + 20% HF	21–29	13	0.036
suspension of: 20% H_3PO_4, 2% H_2SO_4, 1% HF, 40% H_2O and $CaSO_4$	77–93	117	0.018
suspension from the acid transport tank: 37% H_3PO_4, (27% P_2O_5)	66–88	46	0.018

Table 78: Exposure tests of alloy 825 in plants used to produce phosphoric acid [243]

Table 78: Continued

Test conditions	*Temperature °C*	*Test duration days*	*Corrosion rate mm/a*
suspension of: 31.4% H_3PO_4, 1.6% H_2SO_4, 1.5% H_2SiF_6, 0.12% HF, $CaSO_4$	46–60	8.3	< 0.003
suspension from the thickener: 54% H_3PO_4, 1.7% HF, 2% H_2SO_4, 2% $CaSO_4$	52–66	51	0.013
heated acidic gases from the evaporator: 53% H_3PO_4, 1–2% H_2SO_4, 1.5% HF and Na_2SiF_6	121	42	0.152
from the liquid separator: 50–55% H_3PO_4 + HF	107–149	21	0.787
defluorinator: 75–80% H_3PO_4, 1% H_2SO_4 + HF	121–157	8	3.05

Table 78: Exposure tests of alloy 825 in plants used to produce phosphoric acid [243]

During the testing of alloy C-263 (NiCo20Cr20MoTi, DIN-Mat. No. 2.4650) for its polishing behavior, electrochemical measurements were carried out in mixtures of phosphoric, sulfuric, nitric and acetic acids at 35°C [244], Figure 51.

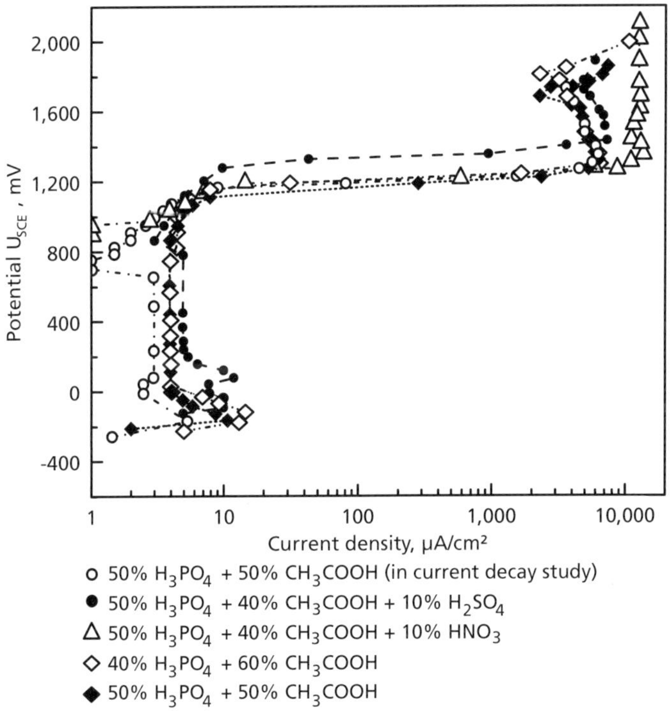

Figure 51: Anodic polarization behavior of Superni® 263 A in binary and ternary mixtures of phosphoric acid-rich solutions with acetic acid and/or sulfuric acid or nitric acid [244]

It can be seen from the curves that the material is spontaneously passivated under all of the given conditions and is thus resistant. Similar behavior can be seen from the comparable curves in 72% H_3PO_4 + 28% H_2SO_4 as well as in 75% H_3PO_4 + 25% CH_3COOH. In the last two mixed acids, the material is resistant in every case if anodic protection is used.

During the production of phosphoric acid, phosphate ores are treated with concentrated sulfuric acid at about (85–95°C). The phosphoric acid, which still contains sulfuric acid and calcium sulfate as reaction products, is pumped from the reactor and fed to the evaporator where the phosphoric acid is concentrated to the usual commercial concentration (54% P_2O_5) or to superphosphoric acid (70% P_2O_5). The requirements on the filter pumps and circulation pumps in this part of the plant are extreme on account of both the corrosiveness of the medium and wear due to abrasion. A centrifugal pump made of cast alloy C-276 (corresponds to DIN-Mat. No. 2.4819) showed sufficiently good behavior over a period of 2 years and the price/performance ratio was much more favorable compared to other materials used. However, because of the approx. 30% solids contained in the process solution, considerable abrasion was found. This was countered in newer pump constructions by halving the rotation speed (to 1700 rpm) [245].

The nickel alloy G-30 (DIN-Mat. No. 2.4603) is used in technically contaminated phosphoric acid, i.e. with sulfuric and hydrofluoric acids, because of its excellent corrosion resistance which is essentially based on its extremely high chromium content [227]. Table 79 lists applications of alloy G-30 in industrial practice, whereby in addition to its use in phosphoric acid production and processing, its uses in further acids and mixed acids are also given.

Application	***Acid/mixed acid***	***Component***
phosphoric acid production	54% P_2O_5 (75% H_3PO_4)	bolts, nuts, pumps
phosphoric acid production	28% P_2O_5 + H_2SO_4 + HF	stirrer blades, filling ports
phosphoric acid production	70% P_2O_5 (96% H_3PO_4)	heating pipes and evaporator pipes
phosphoric acid production	54–70% P_2O_5	cooling pipes
phosphoric acid production	58–64% P_2O_5	rods, sheets, pipes
stainless steel – pickling	HNO_3 + HF	pickling baskets
surface treatment of metals	10% H_2SO_4, 70°C	anodising tank
treatment of nuclear waste	HNO_3 + HF +H_2SO_4	scrubber column, heater
manufacture of special plastic	2% HCl + Fe^{3+}, 50°C	hammer works
paper and pulp manufacture	H_2SO_4 + chloride + sulfite	thickener, metallic linings

Table 79: Industrial applications of Hastelloy® G-30 (c.f. DIN-Mat. No. 2.4603) [227]

The nickel alloy Nicrofer® 5923 hMo (DIN-Mat. No. 2.4605) has also been tested in contaminated phosphoric acid because of its very high content of molybdenum and chromium (16Mo and 23Cr), so that good resistance can be expected, even under the harshest conditions in the production and processing of phosphoric acid [224]. Both the base material as well as welded samples were tested in several process solutions used in practice. The material exhibited very good to good resistance in the individual media. However, the limits of application lie at approx. 100°C, see Table 80. Large amounts of chloride also considerably harshen the corrosion conditions. It is important that the corrosion rates of the welded samples are never higher than those of the unwelded material. This is valid not only for inert gas welding (e.g. TIG) but also for manual arc welding (MAW). Therefore, for practical usage, it is guaranteed that the region of the weld seam of a welded construction (vessels, pipelines) shows equivalent corrosion behavior to the base material and that preferred attack of this area is not expected.

Medium	*Temperature* °C	*Test duration* d	*Corrosion rate* mm/a	
			base material	*welded*
42% H_3PO_4 (30.4% P_2O_5) + 2.4% H_2SO_4 + 2.3% H_2SiF_6 + 1.0% Fe_2O_3, + 1,000 ppm Cl^-	80	21	0.02	0.02
72% H_3PO_4 (approx. 52% P_2O_5) + 4.5% H_2SO_4 + 0.9% H_2SiF_6 + 1.5% Fe_2O_3 + 400 ppm Cl^-	80	21	0.02	0.02
	120	21	0.99	0.89
52% P_2O_5	120	21	0.13	0.12
54% P_2O_5	120	21	0.15	0.12
54% P_2O_5 + 2,000 ppm Cl^-	120	7	2.24	2.03
54% P_2O_5 + 2,000 ppm Cl^-	100	21	0.47	0.41

Table 80: Behavior of Nicrofer® 5923 hMo (alloy 59, DIN-Mat. No. 2.4605) in contaminated phosphoric acid [224].
Base material and welded samples

In formic acid with additions of inorganic acids, i.e. in 10% HCOOH + 1% HCl + 0.1% HF at 101°C, alloy 625, C-276, C-22 and ALLCORR® (see also above under HNO_3 + HCl) have absolutely no resistance [229].

Nickel-copper alloys

NiCu30Fe (alloy 400, DIN-Mat. No. 2.4360) can be used in dilute acids, but only under reducing conditions, i.e. in the absence of air, oxygen, oxidizing salts or heavy metals and oxidizing acids. In pure hydrofluoric acid, NiCu30Fe is resistant at room temperature and at all concentrations as long as the acid is free of air, even up to elevated temperatures. The extent to which this behavior can be transferred to other mixed acids with reducing properties has not been reported and should be validated by investigations. In any case, NiCu30Fe is not resistant in oxidizing acids and mixed acids.

In mixtures of sulfuric and nitric acids, chromium-free NiCu alloys, such as NiCu30Fe (alloy 40, DIN-Mat. No. 2.4360), are strongly attacked with increasing nitric acid contents [221]. This is based on the fact that they cannot be passivated because of the lack of chromium and the oxidizing effect of nitric acid causes polarization into the anodic dissolution range. Thus, NiCu30Fe has absolutely no resistance in a boiling mixture of 50% H_2SO_4 + 50% HNO_3 as well as in a boiling mixture of 75% H_2SO_4 + 25% HNO_3 [226].

Nickel-molybdenum alloys

Of primary importance is that nickel-molybdenum alloys are very resistant in reducing mixed acids, particularly in hydrochloric and sulfuric acids, even at higher concentrations and temperatures. However, they fail rapidly with relatively low additions of oxidizing acids and salts as well as heavy metal ions.

In mixtures of sulfuric and nitric acids, NiMo alloys, such as NiMo28 (alloy B-2, DIN-Mat. No. 2.4617), are increasingly attacked as the content of nitric acid increases [221]. This is due to the fact that they cannot be passivated because of the lack of chromium and the oxidizing effect on nitric acid polarizes them into the anodic dissolution range. Thus, alloy B-2 has absolutely no resistance in a boiling mixture of 50% H_2SO_4 + 50% HNO_3 as well as in a boiling mixture of 75% H_2SO_4 + 25% HNO_3 [226].

Alloy B-2 is resistant in a mixture of sulfuric and hydrochloric acids under the following conditions: 15% H_2SO_4 at 80°C and addition of HCl up to 16% (see Figure 46) [221].

Chlorimet® 2 is a cast alloy containing 30–33% Mo and max. 3% Fe, bal. nickel. It is primarily used as a pump and valve material [246]. The alloy exhibits excellent resistance in reducing inorganic acids. Thus, in all concentrations of hydrochloric acid, it is resistant up to the boiling point, the same applies to sulfuric acid up to approximately 120°C as well as phosphoric acid of the highest concentration. However, it is important that for this alloy, which behaves similarly to the wrought alloy B-2 (DIN-Mat. No. 2.4617), the corrosion rate in all reducing media is drastically increased up to total instability in the presence of oxidizing impurities or additives, i.e. even aeration or addition of oxygen. Thus, even small amounts of iron(III) chloride lead to destructive local corrosion in the form of pitting and crevice corrosion. This behavior should be transferable to mixtures of the three above-mentioned reducing acids.

The statements made above with regard to the resistance in reducing acids and mixed acids are also valid for the NiMo wrought alloy Nimofer® 6629 (NiMo29Cr, DIN-Mat. No. 2.4600) [247]. The careful addition of the alloying elements iron and chromium greatly delays the precipitation of the detrimental Ni_4Mo phase so that heat treatments of semi-finished products are less of a problem. In addition, the resistance towards stress corrosion cracking and intergranular corrosion is improved compared to alloy B-2.

Other nickel alloys

The high-cobalt CoCrNiMo alloy ULTIMET® (CoCr26Ni9Mo5W, DIN-Mat. No. 2.4681) is an alloy with similarly high corrosion resistance in inorganic acids and mixed acids as the nickel alloys of the C series. In addition it has very high resistance to abrasion similar to the Stellites® on account of its high hardness and strength [248].

Medium	*Concentration mass-%*	*Test temperature °C*	*Average corrosion rate, mm/a*			
			ULTIMET®	*HAYNES® 6B*	*HASTELLOY® C-276*	*Carpenter® 20CB-3*
HCl	1	boiling	0.01	4.32	0.33	2.06
HNO_3	65	boiling	0.15	138.0	21.5	0.20
P_2O_5	54	116	0.20	0.38	0.71	0.91
H_2SO_4	10	boiling	2.51	5.89	0.46	0.56
H_2SO_4 + 1.2% HCl + 1% $FeCl_3$ + 1% $CuCl_2$ (ASTM-G-28B)	11.5	boiling	0.05	73.4	1.40	69.1
H_2SO_4 + 42 g/l $Fe_2(SO_4)_3$ (ASTM-G-28A)	50	boiling	0.20	0.36	6.35	0.25

Table 81: Comparative wet corrosion data for the wrought alloy ULTIMET® [248]

Table 81 and Table 82 give some corrosion data of ULTIMET®, for both the wrought state (deformed, rolled semi-finished products, such as sheets, coils, pipes, wire) as well as for the cast state, compared to HASTELLOY® C-276 (c.f. DIN-Mat. No. 2.4819) and CX-2MW (c.f. to alloy C 22, DIN-Mat. No. 2.4602) and to more highly alloyed special stainless steels. It can be seen, particularly with regard to the behavior in the mixed acid, that the cast state usually exhibits less favorable corrosion behavior. In general, at least for the wrought alloy, the corrosion behavior is much better than that of all other comparable materials and also better than the behavior of the C alloy HASTELLOY® C-276.

Medium	*Concentration mass-%*	*Test temperature °C*	*Average corrosion rate, mm/a*		
			ULTIMET®	*CX-2MW [1)] (c.f. DIN-Mat. No. 2.4602)*	*CF-3M [1)] (c.f. DIN-Mat. No. 1.4404)*
iron chloride	10	boiling	0.03	3.96	531.5
HNO_3	65	boiling	0.20	2.97	0.36
P_2O_5	54	116	0.20	0.33	0.71
H_2SO_4	10	boiling	2.13	0.66	16.1
H_2SO_4 + 1.2% HCl + 1% $FeCl_3$ + 1% $CuCl_2$ (ASTM-G-28B)	11.5	boiling	1.04	0.69	90.4
H_2SO_4 + 42 g/l $Fe_2(SO_4)_3$ (ASTM-G-28A)	50	boiling	0.25	1.80	0.69

1) annealed

Table 82: Comparative wet corrosion data of cast ULTIMET® [248]

Nitric Acid

Author: *K. Hauffe* / Editor: *R. Bender, R. Durham*

Nickel

Nickel and nickel alloys have little resistance to oxidizing acids such as nitric acid [249, 250].

The material consumption rate and the corrosion potential of nickel are shown in Figure 52 as a function of the concentration of nitric acid at room temperature [251]. At a concentration of about 8 mol/l, a sharp decrease in the material consumption rate from about 500 to 15 g/m^2 h occurs, coupled with a rise in the corrosion potential from about 0.2 to 0.85 V_{SHE}. While ammonia is the main product in dilute nitric acid, the reduction in concentrated nitric acid proceeds only to nitrous acid. By polarization in the potential range between + 0.6 and + 1.0 V_{SHE}, it was possible to keep the corrosion of nickel in 1 and 4 mol/l nitric acid at 30°C below 1 g/m^2 h [251, 252].

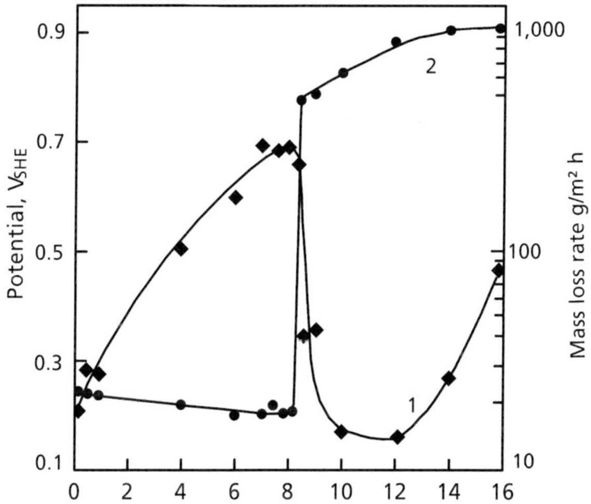

Figure 52: Dependence of (1) the material consumption rate and (2) the potential of nickel in nitric acid at room temperature on the acid concentration [251]

While nickel (99.99%) corrodes with a material consumption rate of 114 g/m^2 h at 25°C in pure 8 mol/l nitric acid, the same acid saturated with nickel nitrate ($Ni(NO_3)_2 \cdot 6H_2O$) causes a material consumption rate of only 7.69 g/m^2 h, although this is still too high for industrial purposes [253]. The inhibitor p-nitroaniline has proved unusable.

The behavior of nickel in nitric acid solutions depends greatly on the acid concentration. At concentrations below 40%, Ni dissolves rapidly, while it is passivated in a stable form above 50%. If chlorides are also present, pitting corrosion can be expected. Passivity is also achieved in dilute acid by oxidizing agents, such as, for example potassium dichromate [254].

The potentiodynamic polarization curves of nickel in 1 mol/l nitric acid with 2% aqueous methanol as a solvent are characteristic of pitting corrosion. Nitrate ions can greatly accelerate dissolution of the metal in the passive state [255].

Thick nickel layers are attacked by nitric acid [256].

To detach a nickel layer from chromium-nickel steels, the layer is dipped in triethanolamine-containing nitric acid with 5 g/l bromide at 60°C for 15 min in order to keep the attack on the steel as low as possible [257].

According to current density/potential curves in an electrolyte consisting of 6 or 0.9 mol/l sodium nitrate + 0.1 mol/l nitric acid flowing at a rate of 10 m/s, anodic dissolution of nickel takes place under industrial conditions (15 to 20 A/cm^2) at a 100% current yield in the transpassive potential range (>> 1 V_{SHE}). Evolution of oxygen is no longer observed under these conditions [258].

Layers of nickel about 500 μm thick precipitated on copper from a bath of $Ni(CF_3\text{-}COO)_2$ halide in methanol at 30°C and 1.5 A/dm^2 corrode in 1.5 mol/l nitric acid + 20 g/l NH_4Cl or 8 g/l potassium bromide at material consumption rates of 1,068 and 143 g/m^2 h, respectively, after short-term tests (10 min) at 25°C. The corrosion rates did not change after additions of $AlCl_3$, NH_4Br and KCl. A commercially available rolled nickel specimen corroded in 1.5 mol/l nitric acid at 50°C at a rate of 116 g/m^2 h [259].

The following pickling solution, which does not attack steel, consisting of 500 ml nitric acid + 5.7 g/l copper sulfate + 3.3 g/l sodium chloride + 3.0 g/l iron chloride, with and without Na_2TeO_3, is used to strip nickel layers applied to stainless steel in cladding devices. The pickling time is 90 to 140 s [260].

31 and 103 min respectively are needed to remove a 12 or 52 μm thick nickel layer on steel using the following pickling solution (100 ml/l HNO_3 + 200 ml/l ethylenediamine + 50 g/l sodium m-nitrobenzenesulfonate; pH = 10) at 65°C. The steel is not attacked during this process [261].

According to current density/potential curves on nickel in a solution containing 5 mol/l sodium nitrate + 1 mol/l nitric acid at 25°C, agitated at a rate of 100 to 1,000 cm/s, the current density starts to rise from < 0.1 to 4 to 15 A/cm^2 during polishing of steel above + 18 V_{SCE} at a running speed of 0.5 V/s. Almost 100% polishing of the surface is achieved at about 5 A/cm^2 and an electrolyte agitation of 100 cm/s [262]. These test results are of importance for electrolytic processing of nickel.

Nickel-chromium alloys

According to Table 83, the nickel-chromium alloy Chromel® (UNS N06010) heat-treated at 1,100°C is severely attacked by nitric acid. The corrosion cannot be reduced sufficiently by the addition of 5.0 g/l oxalic acid alone. Only if hydrochloric acid is present at the same time is a significant reduction in the material consumption rate achieved [263].

Solution, g/l	*Without addition*	*With addition*
	Material consumption rate g/m² h	
50 HNO_3	120	22
100 HNO_3	270	52
100 HNO_3 + 50 HCl	5.3	5
50 HNO_3 + 50 HCl	3	2
200 HNO_3 + 50 HCl	11	9

Table 83: Influence of 5.0 g/l oxalic acid on the corrosion of Chromel® in HNO_3 and HNO_3 + HCl at room temperature [263]

According to Figure 53, the material consumption rate of nickel-chromium alloys with increasing chromium content in 16 to 18% nitric acid + 3 to 4% hydrogen fluoride at 60 to 70°C drops from >> 1 to about 0.15 g/m^2 h with 50% chromium. Specimens which had been heat-treated at 1,100°C and then quenched in water were analysed [264]. According to Table 84, the influence of the alloying additions – with the exception of 0.17% carbon and 1.7% niobium – on the corrosion of the alloy ChN60V in the solution can be ignored. The intercrystalline corrosion depends on the temperature and duration of the heat treatment.

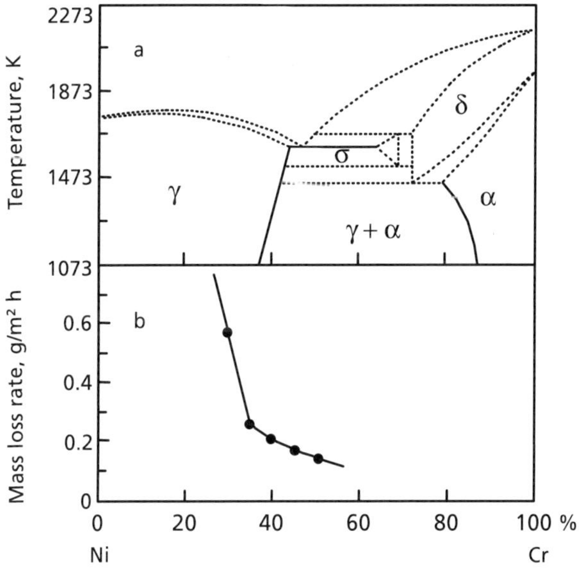

Figure 53: Phase diagram (a) and dependence of the material consumption rate (b) of NiCr alloys on the chromium content in 16–18% HNO_3 + 3–4% HF at 60–70°C after 500 h tests [264]

Alloying elements, %							Material consumption rate g/m² h	IC* of ≥ 15 μm	
C	N	Zr	Ti	Nb	Ce	B		Duration h	Temperature °C
0.01	0.09	–	0.1	–	0.015	–	0.17	20	650
0.01	0.09	0.05	0.1	–	0.015	–	0.17	6	630
0.17	0.09	–	0.1	–	0.015	–	0.4	1	600
0.01	0.09	–	0.35	–	0.015	–	0.25	1	730
0.02	0.09	–	0.1	1.7	0.015	–	0.3	2	700
0.01	0.09	–	0.1	–	0.05	–	0.17	30	730
0.01	0.09	–	0.1	–	0.015	0.009	0.19	20	650

* intercrystalline corrosion

Table 84: Influence of alloying elements on the material consumption rate of ChN60V (quenched) with 40% chromium in 16 –18% HNO_3 + (3–4)% HF at 60 to 70°C, test duration 500 h [264]

According to Figure 54, corrosion of the alloy Ni-40Cr in the above HNO_3/HF solution at 60 to 70°C is altered by addition of 1% tungsten or 3% molybdenum, depending on the temperature and duration of the heat treatment [264].

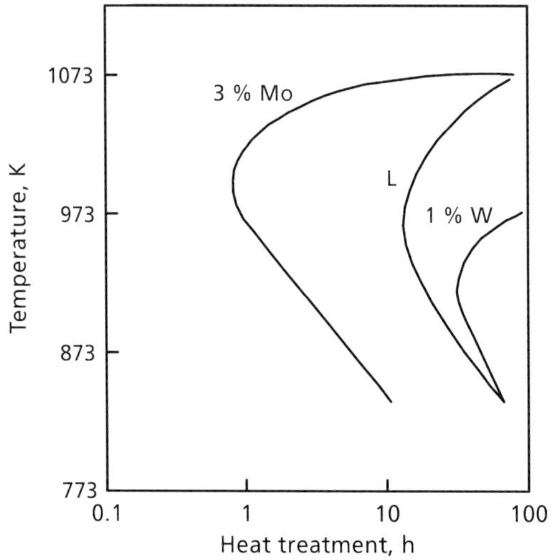

Figure 54: Influence of molybdenum and tungsten on the course of the isocorrosion curve (0.3 g/m² h) of ChN60 in 16–18% HNO_3 + 3–4% HF at 60 to 70°C in the time/temperature diagram of the heat treatment [264]
L = ChN60 without additions

According to Figure 55, above 23% chromium the corrosion rate of the nickel alloys (Ni-xCr-10W-2C) in 10% nitric acid at room temperature reaches values of < 0.2 g/m² h, which are of industrial interest, in 48 h tests [265].

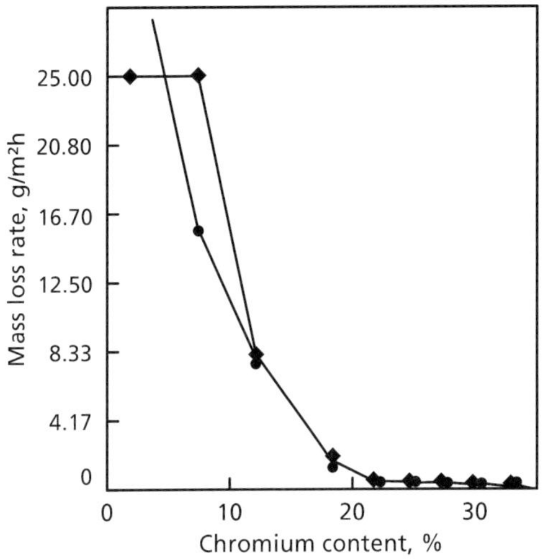

Figure 55: Corrosion of Ni-xCr-10W-2C in the cast state (●) and homogenized at 1,050°C for 180 h in 10% HNO_3 at room temperature (◆) after 48 h tests, as a function of the chromium content [265]

In the modified Huey test (48 h in boiling 25% nitric acid), the alloy I-82 ((%) Ni-0.007C-20.48Cr-3.04Mn-0.75Fe-0.38Ti-0.06Al-0.05Co-0.02Cu-0.13Si-0.007S) corrodes at a material consumption rate of 0.053 g/m^2 h [266].

For effective removal of the scale layer on the nickel-chromium alloy EI-345 ((%) Ni-0.12C-19.22Cr-1.0Fe-0.7Mn-0.2Cu-0.15Al-0.15Ti-0.8Si-0.01P-0.01S), it is immersed in a melt of 75% NaOH + 25% $NaNO_3$ at 400°C for 1 h and then treated in a solution of 10% HNO_3 + 4% HF at 55°C. The material consumption rate of the alloy was determined as about 110 and 100 g/m^2 h, depending on the heat treatment temperature of 1,000 and 1,150°C respectively, with quenching in air. If treatment of the alloy in the melt was carried out at 500°C, the rate was about 40 and 45 g/m^2 h respectively, under otherwise identical conditions [267].

The HNO_3/HF mixtures at which the corrosion rate does not exceed 0.5 mm/a can be seen from the isocorrosion curves (0.5 mm/a) shown in Figure 56 for the nickel-chromium alloy 03ChN60V (Russian type) at various temperatures [268]. At a rotational velocity of the specimen of 500 or 2,500 rpm, the corrosion rate of the alloy in 10 mol/l HNO_3 + 0.1 mol/l HF at 90°C rises from 1.7 to 2 or 2.5 mm/a respectively.

The alloy (%) N-(39-42)Cr-1Fe-1Mn-(0.5-2)W-(0.05-0.15)Mg-(0.01-0.02)Ce-0.15S-0.03C-0.02S-0.02P is said to be particularly resistant to nitric acid containing fluoride; no intercrystalline corrosion occurred even on welded joints [269].

The alloy Ni-(39-41)Cr-1W has proven to be suitable as a tank material for the etching solution consisting of nitric acid and hydrogen fluoride [270]. A solution of 250 g/l HNO_3 + 100 g/l $NH_4F \cdot HF$ is the best etching solution for the alloy 36NChTJu [271].

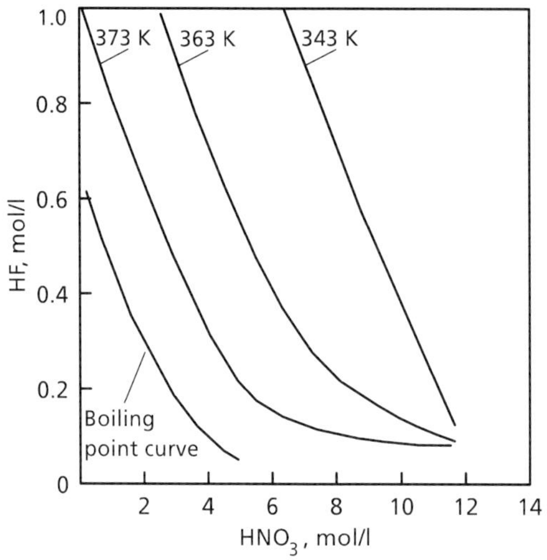

Figure 56: Isocorrosion curves (0.5 mm/a) for 03ChN60V at various temperatures in HNO_3/HF solutions [268]

The rate of dissolution of the nickel-chromium alloy 36NCh (measured as the anodic corrosion current density) at 25°C in 1% nitric acid was increased in the presence of a magnetic field [272].

The main advantage of nickel-chromium alloys in comparison with steels, such as, for example, 12Ch18N10T (SAE 321, 1.4541), is their resistance in nitric acid solutions containing fluoride ions [273].

Regardless of the method of melting of the alloy EI-437B (Nimonic® alloy 80 A, 2.4631; (%) Ni-0.05C-0.27Si-0.16Mn-20.45Cr-0.70Al-0.48Fe-2.69Ti-0.11W-0.006S-0.006P), the rate of etching in 15% nitric acid + 4% hydrogen fluoride at 60°C after annealing at 950 or 1,100°C is about 150 to 160 or 260 g/m^2 h respectively [274]. However, the vacuum-melted alloy shows an etching rate of 380 g/m^2 h after annealing at 1,100°C.

The nickel-chromium alloy (%) Ni-0.02C-34Cr-1.0Fe-0.6Mn-3.7Si with good forgability and malleability is resistant to 65% boiling nitric acid and recommended as the material for boiling Purex-IWW solution containing fluoride ions and $Al(NO_3)_3$ for reprocessing fuel rods [275].

Nickel-chromium-iron alloys

In the modified Huey test (48 h in boiling 25% nitric acid), Inconel® 690 (2.4642) is more resistant than Inconel® 600 by a factor of 100. In contrast to Inconel® 600 (2.4816), no intercrystalline corrosion occurs on this alloy [266]. Some test values taking into account the penetration depth of intercrystalline corrosion are shown in Table 85 [266, 276].

Alloy	*Heat treatment*	*Material consumption rate* $g/m^2\,h$	*Depth of attack** mm
600	(a)	0.883	0.18
	(b)	2.03	0.20
	(c)	1.23	0.18
	(b) + (c)	2.89	0.48
690	(a)	0.0083	–
	(b)	< 0.001	–

* intercrystalline corrosion

Table 85: Corrosion of Inconel® 600 and 690 in boiling 25% HNO_3 (a) in the delivery state, (b) after sensitization at low temperature (LTS: 24 h at 449°C) and (c) after sensitization in an annealing furnace (SA: 24 h at 621°C) [266, 276]

The corrosion rate of Inconel® 600 in boiling 25% nitric acid is increased from 0.12 to 8.8 or 2.1 mm/a by subsequent heat treatment, e.g. 4 h at 593°C or 15 h at 704°C respectively. However, the intercrystalline fractures drop from 62.2 to 18.6% by this procedure [277].

According to Figure 57, the material consumption rate of alloy 600 ((%) Ni-0.05C-16.32Cr-8.68Fe-0.24Mn-0.26Ti-0.04Nb-0.007S-0.009P-0.015N) in boiling 25% nitric acid reaches values of 0.35 g/m^2 h after only 20 h by heat treatment at 700°C. After treatment of the specimen at 600°C, a period of 100 h was not sufficient to bring the material consumption rate to below 50 g/m^2 h [278]. As a result of the high intercrystalline corrosion, the partly oxidized specimens could be broken by hand.

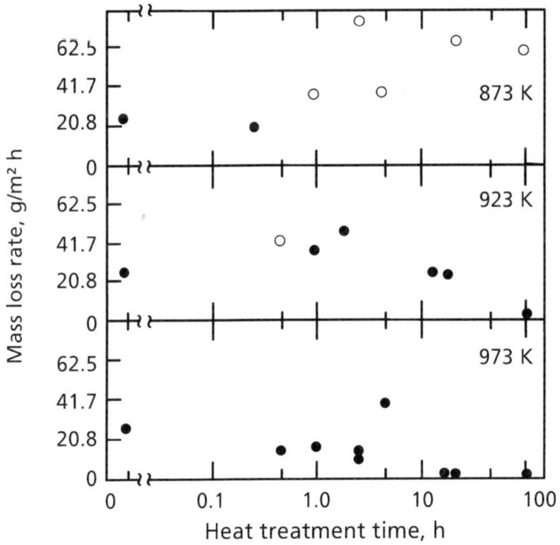

Figure 57: Material consumption rate of Inconel® 600 in boiling 25% HNO_3 as a function of the heat treatment time at 600, 650 and 700°C. The open circles indicate specimens which could be broken by hand as a result of intercrystalline corrosion [278]

The corrosion behavior of various hard nickel-based alloys in hot 30% nitric acid at 50°C is shown in Table 86 [279]. According to this table, the corrosion resistance of the alloys increases with increasing chromium content, decreasing molybdenum content and the addition of tungsten.

Alloy	Chemical composition, %*								Material consumption rate $g/m^2\,h$
	C	Cr	Mo	Fe	W	Nb	Mn	Si	
Akrit® Ni 40	1.44	24.85	13.09	22.9	0.02	0.42	0.11	0.95	0.68
Alloy A	1.31	24.13	8.20	23.1	4.30	0.45	0.24	1.35	0.22
Alloy B	1.37	30.06	6.37	3.02	4.68	–	0.03	1.42	0.06

* balance Ni

Table 86: Composition and corrosion behavior of some Nickel alloys in 30% HNO_3 at 50°C [279]

There is a direct relationship between the magnetic susceptibility, the Curie point and the corrosion rate of alloy 600 in 40% nitric acid via the chromium content of the alloy. The Curie point thus drops as the chromium content increases, from 407°C at 2% to 93°C at 15% chromium. The susceptibility and corrosion of the alloy decrease accordingly as the chromium content increases [280].

While the corrosion rate of the nickel-chromium-iron alloy (0.04-0.07C and 7–10% Fe) in boiling 65% nitric acid was still 2.3 mm/a at 24% chromium, it was reduced to 0.12 mm/a by a chromium content of 30% [281].

Table 87 shows the dependence of intercrystalline corrosion on Inconel® 600 ((%) Ni-0.030C-15.7Cr-7.63Fe-0.33Mn-0.21Ti-0.12Al-0.02Cu-0.005P-0.004S) in boiling 10 mol/l nitric acid on the heat treatment [282].

Heat treatment		Corrosion rate mm/a	$d\sigma_B/dt$ MPa/h
Temp., °C	Duration, h		
350	200	4.4	–
	2,000	17.4	–
600	0.2	2.6	5
	1.0	20.1	23
	10	84	97
	500	10.4	12
700	0.2	34.8	40
	1.0	82	95
	10	30	35
	500	13	15

Table 87: Dependence of the intercrystalline corrosion and rate of decrease of the tensile strength on the heat treatment of the alloy Inconel® 600 in boiling 10 mol/l HNO_3 [282]

Of nine heat–treated alloys of Inconel® 600 with different compositions, the following showed the lowest material consumption rates in the Huey test: (%) Ni-0.016C-15.4Cr-7.0Fe-0.20Mn-0.008Mg-0.37Al-0.33Ti-0.02Cu-0.11Si-0.014S-0.012P-0.008N and (%) Ni-0.016C-15.5Cr-7.1Fe-0.19Mn-<0.005Mg-0.38Al-0.33Ti-0.01Cu-0.11Si-0.0002S-0.032P-0.007N. The material consumption rate of both alloys, which were annealed at 1,120°C for 1 h and then quenched in water, in boiling 65% nitric acid was 0.70 and 0.52 g/m^2 h, with corresponding depths of intercrystalline corrosion of 0.15 and 0.0 mm [283]. In boiling nitric acid, the corrosion rate of Inconel® 600 increases as the carbon and sulfur content increase.

In Figure 58, the curves (1) (solution–annealed at 1,093°C) and (2) (solution–annealed at 1,093°C and 25% cold–worked) enclose corrosion regions of the alloy Inconel® 600 (2.4816) in which the corrosion rate in boiling 70% nitric acid is > 30 mm/a. In the regions outside the curves, the corrosion rate of the alloy is < 3 mm/a [284].

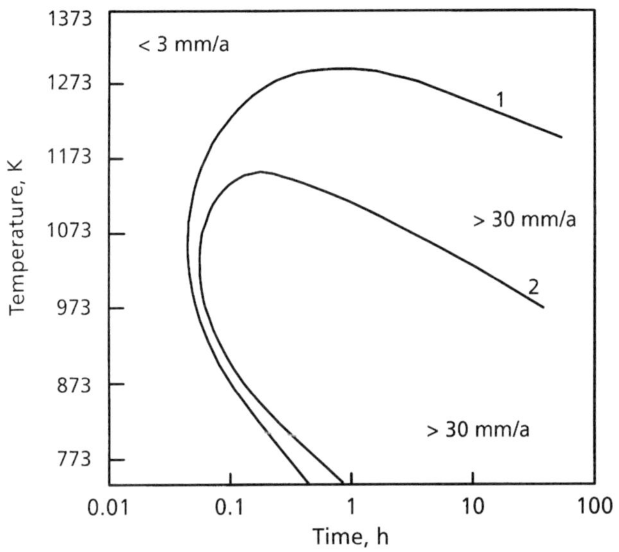

Figure 58: Time/temperature sensitization diagram for Inconel® 600 during corrosion in boiling 70% HNO_3 [284]
(1) solution-annealed at 1,093°C, (2) as (1) with 70% cold working

The corrosion rate of the complex alloy Hastelloy® G (c.f. 2.4816; (%) Ni–0.05C–(21–23.5)Cr–(18–21)Fe–(5.5–7.5)Mo–(1–2)Mn–(1.5–2.5)Cu–(1.75–2.5)(Nb+Ta)–<1.0W–1.0Si–0.030S–0.040P) in boiling 70% nitric acid at 124°C is not low enough, at < 0.5 mm/a [285]. Below 87°C, the corrosion rate is < 0.1 mm/a.

As soon as boiling 65% nitric acid contains 0.1% hydrogen fluoride, the material consumption rate of the alloy 690 solution–annealed at 1,100°C for 0.5 h also reaches values of about 6 to 7 g/m^2 h with 30% chromium [280]. Addition of chromium salts to the nitric acid causes no change in the corrosion.

Inconel® 625 (1.4856) is also mentioned as a suitable material for processing fuel rods from a pressurized water reactor in solutions of 0.5 mol/l nitric acid and hydrofluoric acid, while the NiCrFe alloys are unsuitable because their corrosion rates are too high [286]. Corrosion preferentially occurs in particular at welding joints.

Inconel® 600 is recommended for spring, elastic pipe connections for turbine pipes operated with superheated steam. An alloy structure obtained by solution–annealing at 900°C with subsequent etching in HNO_3/HF solutions leads to no intercrystalline corrosion during long–term turbine operation at 330°C [287].

In boiling 2 and 6 mol/l nitric acid + 0.5 mol/l hydrogen fluoride, Inconel® 690 (2.4816) corroded at a corrosion rate of 1.8 and 3.8 mm/a respectively [288]. In spite of this, it is one of the materials which is used during processing of fuel elements from nuclear reactors in nitric acid solutions containing hydrogen fluoride. In addition to Inconel® 690, the corrosion behavior of some other alloys is summarized in Table 88 [289].

Figure 59 shows that the intercrystalline corrosion of solution–annealed Inconel® 600 (2.4816), determined as the fracture stress after exposure to a solution of 300 ml 70% HNO_3 + 14 g $K_2Cr_2O_7$ + 700 ml water decreases with the contents of phosphorus and silicon in the alloy. However, a higher silicon content is far less hazardous than a higher phosphorus content [290].

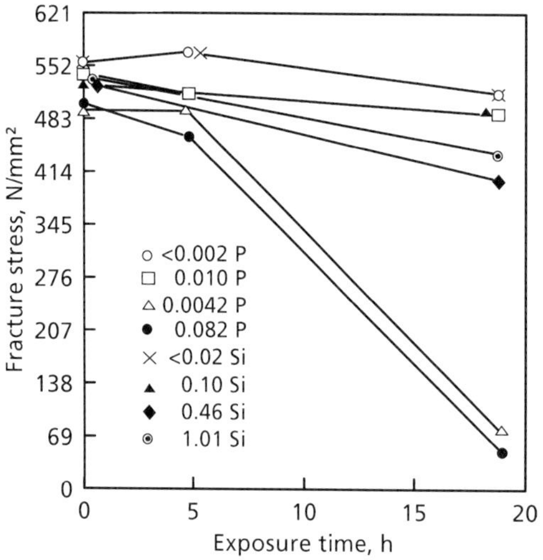

Figure 59: Influence of the exposure time in boiling HNO_3 + Cr^{6+} on the fracture stress of Inconel® 600 with various P and Si contents [290]

An electrolyte of 3% nitric acid + 2% perchloric acid in methanol was used at a polarization potential of + 1.30 V_{SCE} for quantitative extraction of the γ-phase ($Ni_{2.88}Al_{0.37}Ti_{0.51}Cr_{0.12}Fe_{0.12}$) in the NiCrFe alloy ((%) Ni-0.056C-14.68Cr-8.90Fe-2.25Ti-1.12Al-0.02Mn-0.004P-0.005S). The percentage content of the γ-phase was determined as 15.5% after heat treatment at 700°C for 1 hour [291].

Alloy	*Forged*		*Welded*
	HF addition, mol/l		
	0.01	*0.1*	*0.1*
	corrosion rate, mm/a		
Inconel® 690 (2.4642)	0.23	1.60	1.73
Inconel® 671 (52Ni-48Cr)	0.20	1.04	1.14
Hastelloy® G (c.f. 2.4618, Ni-22Cr-18Fe-6Mo)	0.46	3.51	5.59
Inconel® 625 (1.4856, Ni-21Cr-2.5Fe-9Mo)	0.46	3.73	4.32

Table 88: Corrosion of some NiCrFe alloys in 10 mol/l HNO_3 with HF additions at 95°C, test duration 120 h [289]

Nickel-chromium-molybdenum alloys

The good corrosion resistance of Hastelloy® C-4 alloy ((%) Ni-0.015C-(14-18)Cr-(14-17)Mo-<3Fe-<1Mn-<0.7Ti<2.8Co-0.08Si-<0.040P-<0.03S; c.f. 2.4610) in nitric acid can be seen from the isocorrosion curves in the temperature/concentration graph in Figure 60 [285]. The corrosion rate, e.g. in 10% nitric acid at 100°C, is thus 0.2 mm/a [292].

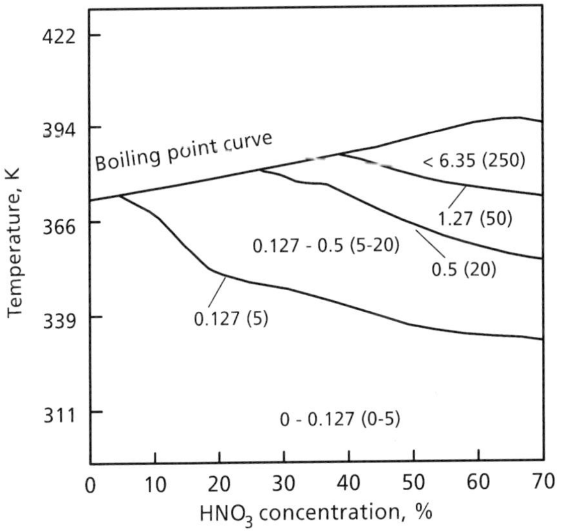

Figure 60: Isocorrosion curves in mm/a of Hastelloy® C-4 in nitric acid in the temperature/concentration graph [285]

A comparable path of the isocorrosion curves is also obtained with Hastelloy® C-276 (c.f. 2.4819, 2.4886) [285].

While the corrosion rate of Hastelloy® C-276 in boiling 10% nitric acid is 0.43 mm/a, the corrosion resistance of the Hastelloy® G-3 (c.f. 2.4619; (%) Ni-0.015C-(21-23.5)Cr-(6.0-8.0)Mo-1.5W-(18-21)Fe-(1.5-2.5)Cu-0.50(Nb+Ta)) of 0.02 mm/a is considerably better [293, 294]. For comparison, the corrosion rate of Hastelloy® C-4 (c.f. 2.4610) under comparable test conditions (10% HNO_3 at 100°C) is 0.20 mm/a. These results have also been confirmed elsewhere [295].

Alloys of the type NiMoCr 55 19 17 have only limited resistance (0.1 to 1 g/m^2 h) in 5 to 65% nitric acid at 20°C and are not resistant at 100°C [296].

The alloy ChN60MB (Russian type; (%) Ni-0.015C-0.25Mn-0.1Si-23.8Cr-16.9Mo-0.18Nb-0.005S-0.007P) corrodes in boiling nitric acid (10, 20, 30 and 40%) with rates of 0.02, 0.055, 0.20 and about 1 mm/a respectively [297].

The corrosion behavior of some nickel-chromium-molybdenum alloys in mixed acids containing nitric acid can be seen in Table 89 [298]. The approximate compositions of the alloys tested in Table 89 are shown in Table 90. Hastelloy® C-22 is accordingly the most resistant to nitric acid electrolytes.

Medium %	*Temperature* °C	*Hastelloy® C-22*	*Hastelloy® C-4*	*Hastelloy® C-276*	*Inconel® 625*
		2.4602	*2.4610*	*2.4819*	*1.4856*
		corrosion rate, mm/a			
5 HNO_3 + 15.8 HCl	Boiling	0.01	0.28	0.20	0.03
8.8 HNO_3 + 15.8 HCl	52	< 0.10	2.90	0.84	> 250
5 HNO_3 + 25 H_2SO_4 + 4 NaCl	boiling	0.03	2.46	1.63	18.1

Table 89: Corrosion of some NiCrMo alloys in mixed acids [298]

Alloy	*Chemical composition, %*					
	Cr	*Mo*	*W*	*Fe*	*Nb*	*C*
Hastelloy® C-22	22	13	3	3	–	≤ 0.010
Hastelloy® C-276	16	16	4	5	–	≤ 0.010
Hastelloy® C-4	16	16	–	≤ 3	–	≤ 0.010
Inconel® 625	21.5	9	–	≤ 5	3.5	≤ 0.10

Table 90: Chemical compositions of the NiCrMo alloys shown in Table 89 [298]

Inconel® 625 ((%) Ni-<0.1C-(20-23)Cr-(8-10)Mo-<1Co-<5Fe-(3.15-4.15)Nb/Ta-<0.04Al-<0.5Si) corrodes in 10 mol/l nitric acid + 0.01 mol/l and 0.1 mol/l hydrogen fluoride at a corrosion rate of 0.46 and 3.7 mm/a respectively [289]. That of the welded alloy of about 4.3 mm/a is even higher.

Hastelloy® C-276 can be used for processing the spent fuel elements from pressurized water reactors in solutions of HNO_3 + HF + dissolved substances. However, the welding joints of the alloy are preferentially attacked. Hastelloy® C-4 would perhaps be preferable [286]. With a total concentration of the process solution, containing Zr salts, of 2 mol/l, including 0.5 mol/l HNO_3, a corrosion rate on the alloy C 276 of 3 mm/a results after exposure at the boiling point for 168 h. At a total concentration of the process solution of 4 mol/l, the corrosion rate approximately doubles.

The corrosion behavior of the high–chromium NiCrMo alloy 47ChNM-2 in nitric acid/hydrogen fluoride solutions can be seen from the 100 h tests in Table 91 [299].

Concentration, mol/l		*Temperature °C*	*Corrosion rate mm/a*
HNO_3	*HF*		
1.5	0.02	50	0.021
1.5	1.0	50	0.046
1.5	0.02	94	0.012
1.5	1.0	94	0.372
4.0	0.2	50	0.011
4.0	1.0	50	0.083
4.0	0.2	94	0.285
6.0	0.2	70	0.103
6.0	1.0	70	0.286
10.0	0.6	82	0.656
14.0	1.0	50	0.457
14.0	1.0	80	2.00
14.0	1.0	94	4.35

Table 91: Corrosion rates of the NiCrMo alloy 47ChNM–2 in HNO_3 with additions of HF [299]

The alloy Ni–15Cr–15Mo used as a sintered compact is resistant to nitric acid containing hydrofluoric acid at room temperature [300].

The high–chromium alloy ChN58W (Russian type EP-795) corrodes in a solution of 18% nitric acid + 4% hydrogen fluoride at 60°C at a material consumption rate of about 0.1 g/m^2 h [273].

A nickel-chromium-tungsten-molybdenum alloy EP-567 (Russian type; (%) Ni-0.03C-(14.5-16.5)Cr-(15-17)Mo-(3.0-4.5)W-1.5Fe-1.0Mn-<0.12Si-0.02S-0.025P) shows neither intercrystalline nor pitting corrosion after 96 h in a solution of 30% H_2SO_4 + 10% HNO_3 + 60% water at 80°C if the silicon content is < 0.12%. The corrosion rate is < 0.1 mm/a [301].

The nickel-chromium-molybdenum-niobium alloy JS-625 shows not only high strength but also good corrosion resistance, also in contaminated nitric acid [302].

Nickel-copper alloys

Nickel-copper alloys are not recommended as materials in the presence of oxidizing acids, such as, nitric acid, because their corrosion rates are too high [303–305].

For rapid and reliable differentiation between Monel® 400 (2.4360; (%) Ni-0.13C-33.7Cu-1.20Fe-0.98Mn-0.20Si-0.01S) and Monel® K 500 (2.4375; (%) Ni-0.16C-29.86Cu-2.81Al-0.52Ti-1.19Fe-0.62Mn-0.20Si-0.005S), the U.S. Navy utilizes rapid dissolution in 50% nitric acid at room temperature. The titanium salt which occurs in the solution is detected by a red-violet coloration with chromotropic acid [306].

Nickel-molybdenum alloys

The nickel-molybdenum alloys Ni28Mo (for example, Hastelloy® B-2, 2.4617; (%) Ni-<0.02C-(26-30)Mo-<2Fe-<1Mn-<1Co) and Ni30Mo (e.g. Hastelloy® B, 2.4810; (%) Ni-0.05C-(26-30)Mo-(4-7)Fe-<1Cr-1.0Mn, with up to 2.5% Co and 0.6% V, Cu as permissible alloy components) are not resistant in nitric acid [307].

Hastelloy® B-2 corrodes in boiling 10% nitric acid with a corrosion rate of 483 mm/a [308]. Hastelloy® B also suffers catastrophic corrosion in nitric acid even at room temperature [285, 296, 300, 308].

As can be seen from the comparison of the relative values of the material consumption of the nickel-molybdenum alloys with phosphorus and boron additions in Table 92, only corrosion of the alloy Ni-17Mo-0.3B in 7.5 mol/l nitric acid at 25°C is significantly lower than that of pure nickel-phosphorus Ni-11.0P [309].

Alloy	*HNO_3 concentration, mol/l*	
	7.5	*15*
	*relative dissolution rate**	
Ni-11.0P	1	1
Ni-11.0Mo-0.2P	2	2
Ni-7.3Mo-3.3P	20	55
Ni-17Mo-0.3B	0.07	0.9
Ni-1.5Mo-6.0B	2.25	90

* determined in mg/min

Table 92: Relative dissolution rate of some NiMo alloys with respect to the corrosion rate of Ni-11.0P in nitric acid at 25°C, test duration 5 min [309]

Other nickel alloys

The nickel-cobalt-chromium-molybdenum alloy MP35N (2.4999; 35Ni-35Co-20Cr-10Mo) corrodes in boiling 65% nitric acid with a corrosion rate of 0.93 mm/a [310].

The nickel alloy MP35N corrodes in 10 and 65% nitric acid at 50°C at a corrosion rate of 0.0025 and 0.0125 mm/a. Hastelloy® F (Ni-22Cr-19Fe-6Mo-3Co-2Mn-1Si) is comparably good [311].

The nickel alloy 617 (UNS N06617, 2.4663; (%) Ni-0.07C-22Cr-12.5Co-9Mo-1Al), which is an excellent material for gas turbines because of its high strength and thermal stability, is also resistant to 20 and 70% nitric acid at room temperature with a corrosion rate of 0.025 and 0.5 mm/a respectively [312].

The nickel-chromium-iron-molybdenum-cobalt alloy ((%) Ni-(0.05-0.15)C-22Cr-18Fe-9Mo-2.5Co-1.0W-1.0Mn-1.0Si) corrodes in boiling 5 and 25% nitric acid at rates of 0.14 and 0.27 mm/a respectively [656].

The alloy IN-504 with the preferred composition (%) Ni-9.5Si-2.85Ti-3.0Mo-2.5Cu (structure: β-phase with Ni_3Si precipitates) corrodes in 65% nitric acid at 160°C at a rate of 1.0 mm/a [313].

According to Table 93, corrosion of the nickel-phosphorus alloys in 1 mol/l nitric acid at 20°C decreases as the phosphorus content increases and reaches a value of < 0.1 g/m^2 h at 16% phosphorus [314] . Only uniform corrosion occurs, without fractures caused by intercrystalline corrosion.

Phosphorus content %	***Material consumption rate*** g/m^2 h
–	15.8
4	9.54
6.6	4.43
10.0	3.60
15.0	0.55
16.0	< 0.1

Table 93: Average material consumption rates of NiP alloys in 1 mol/l HNO_3 at 20°C, test duration 120 d [314]

The hardenable spring alloy Beryvac® 520, a nickel-beryllium alloy NiBe2, is not resistant in 10 and 65% nitric acid even at room temperature. Its corrosion rate in 10% nitric acid is > 10 mm/a [315].

Nickel-phosphorus coatings containing 14% phosphorus are severely attacked or destroyed in 78% nitric acid even at room temperature (8 d tests) [316].

According to Figure 61, the relatively high chromium content in the alloys Ni-25Cr-20W-2C-xSi and Ni-30Cr-20W-2C-xSi has a decisive influence on the extent of corrosion in 20% nitric acid containing 4% hydrogen fluoride with an increasing silicon content at room temperature [265].

In the chemical surface treatment of the nickel-iron alloy Permalloy® 52N (Ni-38.4Fe and Ni-49.5Fe) it is pickled in a solution of H_2SO_4:HCl:H_2O of 2:1:2 percent by volume at 70°C for 2 min, a layer of 2.5 μm being removed. This is followed by polishing in a solution of HNO_3:CH_3COOH:HCl of 30:70:1 percent by volume for 6–9 s at the same temperature, a layer of 3.39 μm being removed [317].

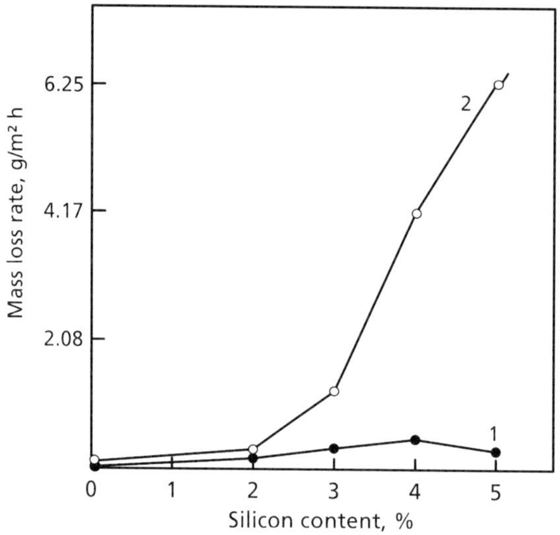

Figure 61: Corrosion of the nickel alloys (1) Ni-30Cr-20W-2C-xSi and (2) Ni-25Cr-20W-2C-xSi in 20% HNO_3 + 4% HF at room temperature as a function of the silicon content, test duration 48 h [265]

The intermetallic γ-phase (Ni_3Al) precipitated in a nickel-chromium-aluminum alloy can be rendered visible by pickling in an electrolyte of 43% nitric acid + 12% phosphoric acid + 45% sulfuric acid at room temperature at a potential of 5.5 V in about 20 s using two platinum needle electrodes [318].

Because of the dissolution of the γ-phase and the resistance of the γ-phase (carbides) of the nickel superalloy ((%) Ni-0.19C-8.5Cr-4.4Al-3.7Ti-4.8Mo-5.2W-8.4Co-2.0Nb) in a solution of 5 g $(NH_4)_2SO_4$ + 15 ml conc. HNO_3 + 35 g citric acid + 1,000 ml water at 15–20°C and anodic polarization with 1.02-1.06 V_{SHE} and 30 mA/cm^2, this solution is used for deep electrochemical etching of the alloy [319].

The following solution is suitable for etching the nickel alloy (%) Ni-0.18C-10Cr-15Co-5.7Al-5Ti-3Mo-0.9V-0.05Cu-0.05Zr-0.015B-0.015P-0.015S: (percent by volume) 40% HNO_3 (65%) + 50% H_2SO_4 (97%) + 10% H_3PO_4 (85%) at a direct voltage of 4–5 V overlapped by an alternating voltage [320].

In addition to the customary etching agent 5 ml HNO_3 + 2 ml HF + 93 ml H_2O and 6 ml HNO_3 + 14 ml HF + 60 ml H_2O for contrasting nickel alloys of high heat resistance, such as, for example, Inconel® 100 ((%) Ni-14.75Co-9.68Cr-5.55Al-3.01Mo-4.73Ti-0.77V-0.17Fe-0.060Zr-0.05Mn-0.05Cu-0.04Si) and Inconel® 939 ((%) Ni-19.11Co-22.67Cr-3.70Ti-2.04W-1.37Al-1.05Nb-0.45Mo-0.070Fe-0.10Zr-0.05Cu-0.03 Mn-0.05Si), by visualizing the dendritic structure, the following are also suitable: 50 ml HNO_3 + 50 ml HCl + 50 ml glycerol and 41 ml HNO_3 + 47 ml H_2SO_4 + 12 ml H_3PO_4 (the latter at a polarization potential of 3 V) and 3 ml HNO_3 + 10 ml HCl + 100 ml ethanol at 50–60°C for a period of 15 s [321, 322].

Phosphoric Acid

Author: *L. Hasenberg* / Editor: *R. Bender, R. Durham*

Nickel

Nickel has good resistance in non-oxidizing acids at room temperature and just above room temperature [323].

It is resistant in pure phosphoric acid at room temperature, but is attacked if oxidizing substances, such as, for example Fe^{3+} ions, are present. Corrosion rates of between 1 and 3.1 mm/a are thus achieved at 20°C in 8.4% H_3PO_4 if it is aerated [324].

As a base metal, nickel provides resistance to corrosion for numerous alloys in several acids, including H_3PO_4 [325, 326].

According to [327], the resistance to corrosion of nickel in aerated phosphoric acid is lower, compared with non-aerated phosphoric acid. The corrosion rates of nickel at room temperature rise as the concentration of acid increases. In aerated H_3PO_4 at 38°C (this is probably pure acid), the corrosion rates in the concentration range of 0–100% are above 1.27 mm/a. This would also apply to air contaminated acid, as already indicated above. Phosphoric acid is used for the chemical removal of nickel, without information on the concentration of the acid. The maximum corrosion rates of nickel between room temperature and 100°C are reported as being 0.01–0.06 mm/min [328].

The material penetration rate of nickel in 100% phosphoric acid at temperatures above 100°C is given in [329], in comparison with other metals. According to [329], the corrosion rates at 150°C are about 5 mm/a, and they increase sharply at higher temperatures.

The good resistance of nickel-based alloys in contaminated phosphoric acid (30–32% P_2O_5) with max. 0.1% Cl^- and up to 3.5% F^- at 75°C under erosion corrosion conditions is mentioned in [330], without further information on the alloy composition. Nickel based alloys have better resistance to corrosion than high-alloy steels with 20% Cr and 25% Ni [330].

Table 94 contains an overview of corrosion rates from various published references.

In summary, the resistance of nickel in phosphoric acid is determined more by the temperature than by the concentration of the acid, while impurities and ingress of oxygen also reduce its resistance. While nickel has good to adequate resistance in all concentrations of pure phosphoric acid at room temperature, this resistance decreases rapidly at higher temperatures, and especially above 50°C. Corrosive attack is then usually localized. Nickel is therefore not particularly suitable as a material for equipment construction for phosphoric acid service. There are more resistant materials and other nickel alloys for this type of service.

Nickel-chromium alloys

This group of Ni-Cr materials is not commonly used as a material for equipment construction for exposure to phosphoric acid, especially under continuous exposure because other materials with better resistance are found among the nickel alloys. Nevertheless, Ni-Cr alloys show very good resistance in (probably pure) phosphoric acid at relatively low temperatures over the entire acid concentration range [331].

H_3PO_4 %		Temperature °C	Test duration d	Corrosion rate mm/a	Literature
1.5		90	51	0.33	[332]
3–14[1) 2)]		66	–	0.09	[333]
5–15[1)]		65	60	0.13	[333, 334]
10		RT	1	0.015	[333, 334]
10		boiling	1 h	4.0	[333]
25		30	–	0.12–1.2	[335]
3–80		60	–	< 0.12	[335]
40		27	5	0.025	[333, 334, 336, 337]
57		RT	27	0.013	[333, 334]
57		104	2	5.6	[333]
70–80		100	–	6.6	[333]
78–85		RT	–	0.008	[333, 334, 336, 337]
78–85		50	–	0.11	[333, 334, 336, 337]
78–85		104	–	3.9	[333, 336, 337]
78[1)]		60	60	1.27	[333]
78[1)]		85	60	6.4	[333]
90[1)]		85	60	3.9	[333, 338, 339]
100		150	4 h	70.0	[329, Figure 11]
100		250	4 h	98.0	[329, Figure 11]
100		300	4 h	700	[329, Figure 11]
105	(free P_2O_5)	RT	–	0.003	[333, 334]
105	(free P_2O_5)	50	–	0.023	[333, 334]
105	(free P_2O_5)	104	–	0.51	[333, 334]
115	(free P_2O_5)	60	2	0.33	[332, 333, 336, 337]
115	(free P_2O_5)	120	2	1.250	[333, 334, 336, 337]
115	(free P_2O_5)	250	2	37.00	[333, 336, 337]

1) plant experiments
2) contains fluorine

Table 94: Corrosion behavior of nickel 200 (2.4060, UNS N02200, Ni 99,6) in phosphoric acid (no information on the acid, probably pure) at various concentrations and temperatures

Nickel-chromium alloys having a chromium content of 20% or more are resistant to pure phosphoric acid up to 80°C. The corrosion rates at all acid concentrations are ≤ 0.1 mm/a, and this applies in particular to low concentrations and temperatures. Better resistance of Ni-Cr alloys is achieved by alloying with molybdenum [334–341].

Since phosphoric acid is present in different concentrations and states of purity, for example during its production, the action on nickel-chromium alloys is of particular interest, especially under long term exposure to corrosion, where higher-alloys and therefore expensive materials have to be used. Studies on alloy Corronel® alloy 230 containing at least 55% Ni, 36% Cr, max. 5% Fe, max. 1.0% Cu and max. 0.08% C have shown that in 75% H_3PO_4 at 112°C, the presence of Cl^- and F^- ions greatly increase the corrosion rates, as can be seen from Figure 62 and Figure 63.

In comparison with the other Ni alloys contained in Figure 62 (> 0.5 mm/a), Corronel® 230 shows the best resistance at a Cl^- content of 0.01% (100 ppm), with a corrosion rate of about 0.05 mm/a. However, it shows the highest increase in corrosion rate as the Cl^- content increases, and is therefore not used in practice under such conditions.

The low resistance of Corronel® 230 against the action of chlorides in comparison with other nickel alloys can be explained by the relatively low pitting index (about 36), which is a measure of the resistance of materials to pitting corrosion. In fact, enhanced localized corrosion of Corronel® 230 also occurred under exposure to chloride.

Impurities containing F^- in phosphoric acid are in practice about ten times higher than those containing Cl^-. Figure 63 shows the high susceptibility of Corronel® 230 toward F^- impurities in 75% H_3PO_4 at 85°C in comparison with molybdenum-containing nickel alloys (4, 5, 6, and 7 in Figure 63). This demonstrates that Ni-Cr Corronel® 230 is unsuitable under Figure 63 conditions, while the other alloys (excluding the steels) undergo barely any change in the corrosion rates. [341]

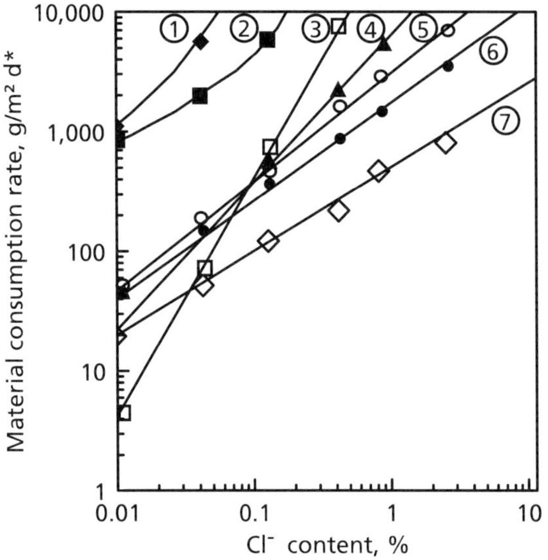

Figure 62: Influence of the chloride content on the corrosion behavior of CrNi steels, CrNiMo steels and nickel alloys in agitated 75% pure phosphoric acid at 85°C [340, 341]
* times 0.365, divided by the density ≈ mm/a

Material	Density	Material	Density
① SAE 304 (1.4301)	7.9	⑤ Hastelloy® G (2.4618)	8.36
② SAE 316 (1.4401)	8.0	⑥ Inconel® 625 (2.4856)	8.63
③ Corronel® 230 (63 Ni, 35 Cr)	8.15	⑦ Hastelloy® C (2.4819)	9.13
④ Incoloy® 825 (2.4858)	8.18		

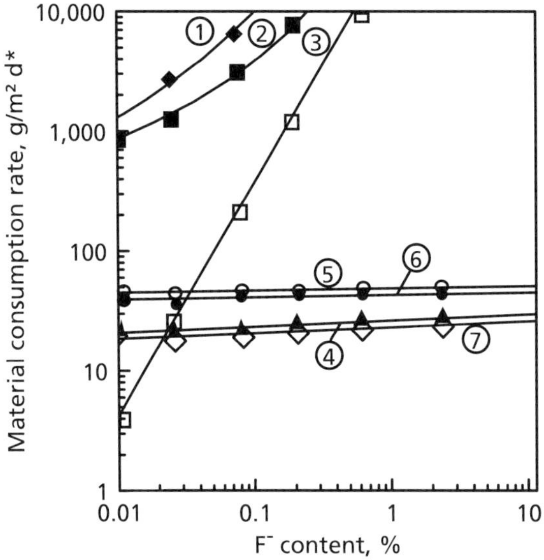

Figure 63: Influence of the fluoride content on the corrosion behavior of CrNi steels, CrNiMo steels and nickel alloys (see Figure 62) in agitated 75% pure phosphoric acid at 85°C [340]
* times 0.365, divided by the density ≈ mm/a

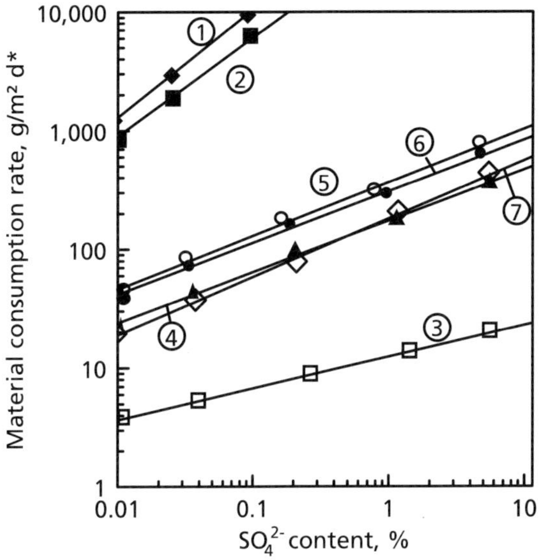

Figure 64: Influence of the SO_4^{2-} content on the corrosion behavior of CrNi steels, CrNiMo steels and nickel alloys (see Figure 62) in agitated 75% pure phosphoric acid at 85°C [340]
* times 0.365, divided by the density ≈ mm/a

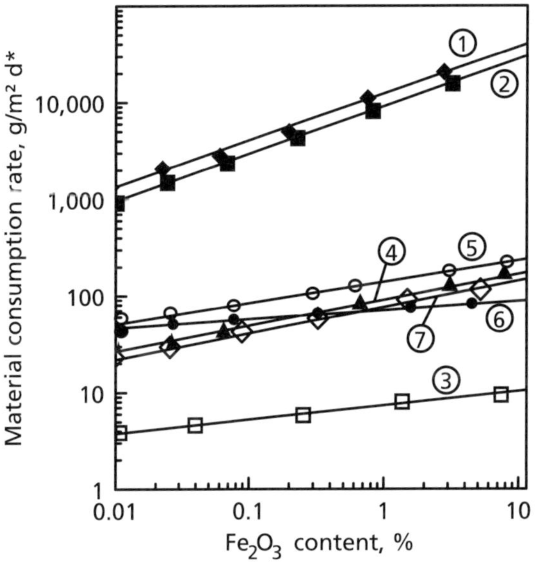

Figure 65: Influence of the Fe^{3+} content on the corrosion behavior of CrNi steels, CrNiMo steels and nickel alloys (see Figure 62) in agitated 75% pure phosphoric acid at 85°C [340]
* times 0.365, divided by the density ≈ mm/a

Additions of sulfuric acid to 75% phosphoric acid at 85°C lead as expected to increased corrosion rates as the concentration of H_2SO_4 increases. As can be seen from Figure 64, Corronel® 230 is superior to the molybdenum-containing alloys under these conditions with corrosion rates of less than 0.5 mm/a (4–10 g/m^2 d) up to a SO_4^{2-} content of 1%. A corrosion behavior virtually identical to that in Figure 64 is found with the addition of S^{2-} [340].

Additions of Fe_2O_3 cause only a slight increase in corrosion in 75% phosphoric acid at 85°C as can be seen from Figure 65. In practice, as expected, there will not be only one impurity present in the acid stream. The mixture of several impurities is expected for the typical plant case. Furthermore, the aggressiveness of fluorides is reduced by additions of SiO_2 and Al_2O_3, as that of sulfuric acid is reduced by metallic cations, such as Ca^{2+}, Mg^{2+} and Al^{3+} [340, 341].

Nickel-chromium-iron alloys

Nickel-chromium-iron alloys are approximately comparable with the nickel-chromium alloys in their corrosion behavior when exposed to pure phosphoric acid.

The corrosion resistance in aerated pure phosphoric acid is lower than in the non-aerated acid. In non-aerated phosphoric acid, the corrosion rates at 121°C up to a concentration of 40% are reported as being < 0.05 mm/a. In aerated phosphoric acid up to 50%, the corrosion rates are at a maximum of 0.5 mm/a at 93°C. Table 95 contains a summary of the corrosion rates for an alloy with 76% Ni, 16% Cr and 7% Fe as the base composition [327].

Phosphoric acid which has been produced by the wet process and is therefore contaminated is more important than pure phosphoric acid.

In comparison with high-alloy steels, nickel alloys are more expensive and may be used only in those cases where their corrosion resistance exceeds that of the stainless steels. This would apply to highly concentrated acids containing impurities and at relatively high temperatures. Chlorides and fluorides may be mentioned as corrosion promotors in phosphoric acid.

As Table 95 shows, nickel-chromium-iron alloys have very good resistance to pure phosphoric acid up to concentrations of 80% and at 93°C [327]. This is provided that the acid is not aerated.

The alloy in Table 95 approximately corresponds to 2.4816 (Inconel® 600). According to [332], a corrosion rate of 0.0051 mm/a is reported for this alloy in pure 25% H_3PO_4 (air-free and non-agitated) at 95°C.

Inconel® 600 (2.4816, NiCr15Fe) showed no sensitivity to intergranular stress corrosion cracking under simulated conditions in conditioned secondary water from nuclear power steam generators in a 0.42 mol/l H_3PO_4 solution at 350°C under a pressure of 17 MPa. After a test duration of 3,000 h, the crack depths were less than 10 μm (0.03 mm/a) and the general corrosion rate was 0.09 mm/a. Under these conditions, the alloy is evaluated as being resistant, and it is also evaluated generally as having better resistance than Incoloy® 800 (1.4876) [342].

H_3PO_4 %	Temperature °C	Corrosion rate, mm/a	
		Non aerated	Aerated
10	149		> 1.27[1)]
≤ 40	121	≤ 0.05	
≤ 50	93	< 0.05	< 0.51
50	121	< 0.51	
60	121	< 0.51	
70	93	< 0.51	
70	121	0.51–1.27	
80	93	< 0.51	
80	149	1.27[1)]	
90	93	< 0.51	
70–90	149	1.27[1)]	

1) including >> 1.27 mm/a

Table 95: Corrosion behavior of NiCrFe alloys (76 Ni, 16 Cr, 7 Fe) in pure phosphoric acid [327]

As has already been described for the Ni-Cr alloys, reduced resistance to predominantly chloride-containing phosphoric acid is also to be expected for these alloys containing iron. The susceptibility of such alloys to chloride-containing phosphoric acid is mainly characterized by localized attack (pitting corrosion). This chloride promoted attack is minimized by added Mo content in the alloys. In fact, the molybdenum-free alloy Inconel® alloy 690 (2.4642, NiCr29Fe, 58% Ni, 27–31% Cr, 7–11% Fe) has a poorer corrosion behavior than Mo-containing Hastelloy G-30 under long term exposure to corrosion in phosphoric acid, as can be seen from Table 96.

According to Table 96, Inconel® alloy 690 (NiCr29Fe) cannot be evaluated as resistant to corrosion in commercially available phosphoric acid. The high susceptibility to attack of 690 to contamination by chloride, in comparison with Hastelloy® G-30 (2.4603, NiCr30FeMo) is evident.

H_3PO_4 %	Temperature °C	Inconel® alloy 690[1)]	Hastelloy® G-30[2)]
		Corrosion rate, mm/a	
78	116	0.76	0.10
78	149	11.35	0.71
58 + 2,000 ppm chlorides as NaCl	85	3.10	0.01
63.5 + 2,000 ppm chlorides as NaCl	85	10.01	0.43

1) 2.4642 (UNS N06690)
2) 2.4603 (UNS N06030, Ni, 30% Cr, 15% Fe, 5% Mo, 2.5% W, 1.5% Cu)

Table 96: Corrosion behavior of nickel alloys (NiCrFe/Inconel® 690; NiCrFeMo/Hastelloy® G-30) in commercially available phosphoric acid without and with a chloride content, test duration 96 h (mean value of in each case 2 specimens) [343]

Nickel-chromium-molybdenum alloys

Table 97 contains a summary of nickel-chromium-molybdenum alloys widely used in chemical plants construction.

Nickel-chromium-molybdenum alloys are distinguished altogether by good to very good corrosion resistance, both in oxidizing and in reducing acids. Very good resistance to pitting corrosion and crevice corrosion, which varies among the particular alloys due to the different chemical composition, is a characteristic of Ni-Cr-Mo alloys.

In the evolution of alloy UNS N08031 Plus – VDM Alloy 31 Plus ® – from alloy UNS N08031 one of the main aims was to keep the corrosion behavior at the level of alloy UNS N08031 or improve it. According to the corrosion behavior in technical phosphoric acid solutions this goal has been accomplished. Corrosion tests were performed in three phosphoric acid solutions being representative for phosphate rock wet digestion in the Central Prayon Process [344]. The chemical compositions of the technical phosphoric acid solutions for testing of alloy UNS N08031 Plus are shown in Table 98. There are three types of acid solutions called attack acid, conversion acid and concentration acid. Certain amounts of chlorides in the form of NaCl have been added to take into account the more severe service conditions as they may occur when higher chloride containing phosphate ores have to be processed. The results of testing are shown in Table 99. The corrosion resistance of alloy UNS N08031 Plus under the test conditions applied is comparable to the corrosion resistance of alloy UNS N08031 (see [344] for details).

Work by Alves et al. on alloy 31 has shown it to be resistant to phosphoric acid solutions up to about 120°C [345–348]. In 1% phosphoric acid full suitability could be observed for Alloy 31 up to at least 300°C.

Material trade name	*DIN designation*	*DIN-Mat.No.*	*Chemical composition, %*						*PREN* 2)
			Cr	*Fe*	*Mo*	*W*	*Cu*	*Ni*	
Incoloy® 825	NiCr21Mo	2.4858	19.5–23.5	≥ 20	2.5–3.5	–	1.5–3.0	bal. 38–46	27.8
Hastelloy® G	NiCr22Mo6Cu	2.4618	21.0–23.5	18–21	5.5–7.5	≤ 1.0	1.5–2.5	bal.	39.2
Hastelloy® G-3	NiCr22Mo7Cu	2.4619	21.0–23.5	18–21	6.0–8.0	≤ 1.5	1.5–2.5	bal.	40.8
Hastelloy® G-30	NiCr30FeMo	2.4603	28.0–31.5	13–17	4.0–6.0	1.5–4.0	1.0–2.4	bal.	41.2
Inconel® 625	NiCr22Mo9Nb	2.4856	20.0–23.0	≤ 3.0	8.0–10.0	–	≤ 0.5	bal.	46.4
Hastelloy® C-4	NiMo16Cr16Ti	2.4610	14.0–18.0	≤ 3.0	14.0–18.0	–	≤ 0.5	bal.	60.2
Hastelloy® C-22	NiCr21Mo14W	2.4602	20.0–22.5	2.0–6.0	12.5–14.5	2.5–3.5	–	bal.	61.3
Hastelloy® C-276	NiMo16Cr15W	2.4819	14.5–16.5	4.0–7.0	15.0–17.0	3.0–4.5	≤ 0.5	bal.	64.0
Hastelloy® C 3)			15.5	5.0	16.0	4.0	–	bal.	68.3 3)

1) different data possible from the various producers
2) PREN (*Pitting* Resistance Equivalent Number) = 1 x %Cr – 3.3 x %Mo (calculated from minimum values)
3) oldest obsolete designation, alloy has been improved

Table 97: Summary of the most important NiCrFeMo and NiCrMo alloys used in construction of chemical equipment

Type of acid	*P_2O_5, wt%*	*SO_3, wt%*	*Cl^-, ppm*	*F^-, wt%*	*SiO_2, wt%*	*Density g/ml*
Attack	31.8	1.35	112	2.00	1.38	1.371
Conversion	28.5	5.29	99	1.83	1.22	1.369
Concentration	42.8	1.22	116	1.51	0.85	1.497

Table 98: Chemical composition of technical phosphoric acid solutions for testing of the corrosion behavior of alloy UNS N08031 Plus [344]

Type of acid	*Corrosion Rate (mm/a) for UNS N08031 Plus*
Attack acid + 1,000 ppm Cl^- at 90°C, liquid phase, samples fixed to rotor blades and rotated	0.02
Conversion acid + 700 ppm Cl^- at 105°C, liquid phase, samples fixed to rotor blades and rotated	0.11
Concentration acid + 400 ppm Cl^- at 130°C, liquid phase, stirred	0.25
Concentration acid without addition of Cl^- at 130°C, liquid phase, stirred	0.25

Table 99: Corrosion loss of alloy UNS N08031 Plus in technical phosphoric acid solutions with the corresponding slurry, mm/a. Total testing time: 3 weeks [344]

A simple Ni-Cr-Mo alloy (UNS N06035) has been found to meet the technical requirements for "wet process" phosphoric acid evaporators [349]. Not only did N06035 show significantly higher resistance to this acid than currently available Fe-Ni and Ni-Fe alloys (around 0.1 mm/a in the tested range for N06035), in the P_2O_5 concentration range 36 to 54% it also possesses higher resistance to chloride-induced localized attack.

As already pointed out in Table 99, the term Hastelloy® C is a designation for a type of alloy which has been modified several times.

The original alloy Hastelloy® C with a C content of about 0.08%, which is high under current practices, had to be solution-annealed after welding because of precipitates, which presented problems especially for large components. The next step was the development of Hastelloy® C-276 (2.4819) with a very low C and Si content. The formation of Mo- and W-rich phases also may not be prevented in this alloy during high temperature exposures for prolonged time. Consequently, it tends to form precipitates after welding in the same way, and thus may become susceptible to intergranular corrosion in the heat-affected zone of the weld seam.

A further step led to the development of the alloy Hastelloy® C-4 (2.4610). Its excellent thermal stability is achieved by reducing the contents of W, Fe, Co and V. The susceptibility of the weld seam region to precipitation is therefore largely eliminated. Nevertheless, C-4 may have the tendency to suffer pitting corrosion and crevice corrosion and, in some cases, higher corrosion rates in comparison with Hastelloy® C-276. For this reason, Hastelloy® C-22 (2.4602) was developed, to improve the not completely satisfactory corrosion properties of the previous alloys, but without reducing the advantages of thermal stability of C-4 and the resistance to pitting corrosion and crevice corrosion of C-276.

Figure 66 shows a comparison of the thermally induced precipitation properties of the Hastelloy® alloys C, C-276 and C-4 [350]. As can be seen from this graph, carbides, for example, precipitate relatively rapidly in the alloys C and C-276, within less than 6 min at temperatures above 800°C while precipitation starts in alloy C-4 after one hour at 800°C. This pattern is of interest when selecting such an alloy, especially for processing large wall thicknesses by welding. The corrosion properties of Hastelloy® C-276 (2.4819), Hastelloy® C-4 (2.4610) and Inconel® 625 (2.4856) are optimized in NiCrFeMo and above all NiCrMo alloys show very good to good corrosion behavior in phosphoric acid in the entire concentration range up to the boiling point [323, 327, 334, 335, 338, 341, 343–362]

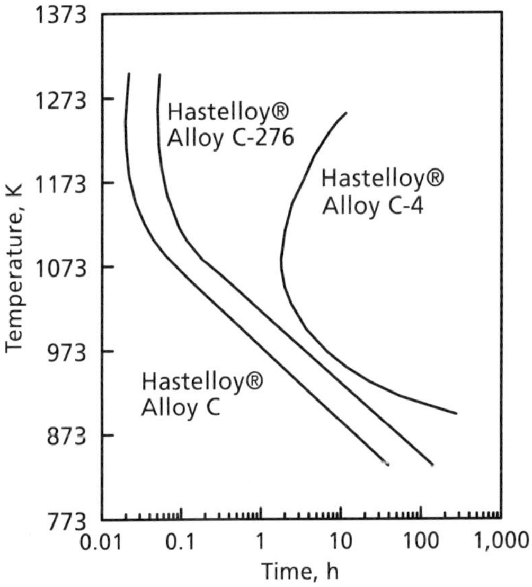

Figure 66: Influence of time and temperature on the start of precipitations in NiCrMo alloys (see Table 97) [350]

Nickel-chromium-iron alloys with a low molybdenum content have better corrosion behavior in aerated pure phosphoric acid in comparison with the nickel-chromium-iron alloy such as Inconel® 600 (2.4816). While corrosion rates of up to 0.51 mm/a are to be expected in aerated pure phosphoric acid at up to 50% and 93°C, they are reported as being not more than 0.051 mm/a for an NiCrFeMo alloy (probably Incoloy® 825 (2.4858)) at up to 60% acid and at 66°C. Above 66°C the corrosion rates correspond to the values of Inconel® 600 [327]. Figures 67–74 are isocorrosion curves to show an overview of the corrosion behavior of several NiCrMo alloys in pure phosphoric acid.

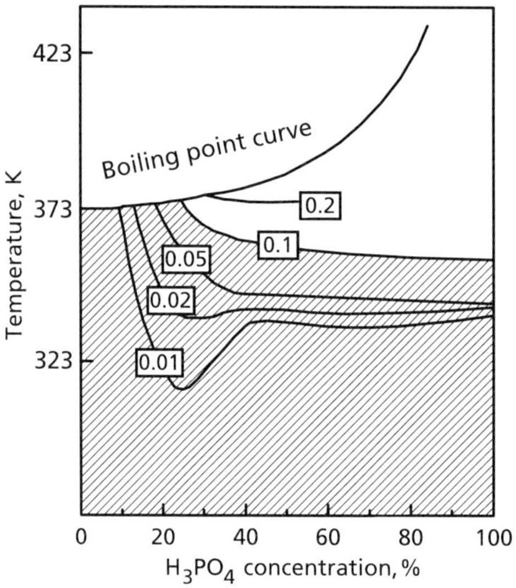

Figure 67: Corrosion behavior of Hastelloy® F (0.1% C, 44% Ni, 30% Fe, 22% Cr, 6.5% Mo, 2% Mn, 1% Si) in pure phosphoric acid (isocorrosion curves in mm/a) [363]

0.01 mm/a (0.4 mpy)	0.05 mm/a (2.0 mpy)	0.2 mm/a (7.9 mpy)
0.02 mm/a (0.8 mpy)	0.1 mm/a (3.9 mpy)	

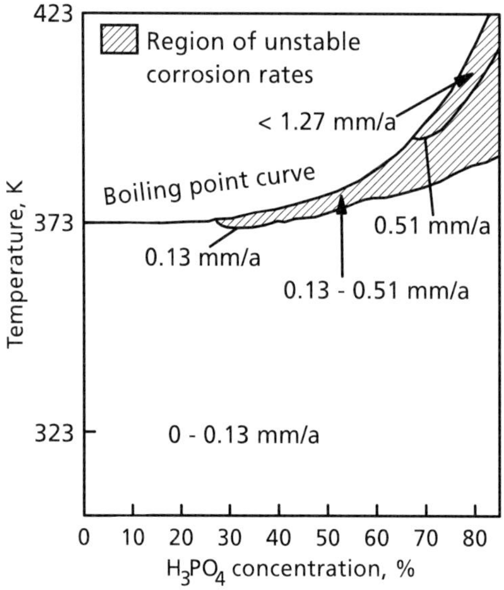

Figure 68: Corrosion behavior of Hastelloy® G (see Table 97) in pure phosphoric acid [359]

0–0.13 mm/a (0–5 mpy)	0.13–0.51 mm/a (5–20 mpy)	< 1.27 mm/a (< 50 mpy)
0.13 mm/a (5 mpy)	0.51 mm/a (20 mpy)	

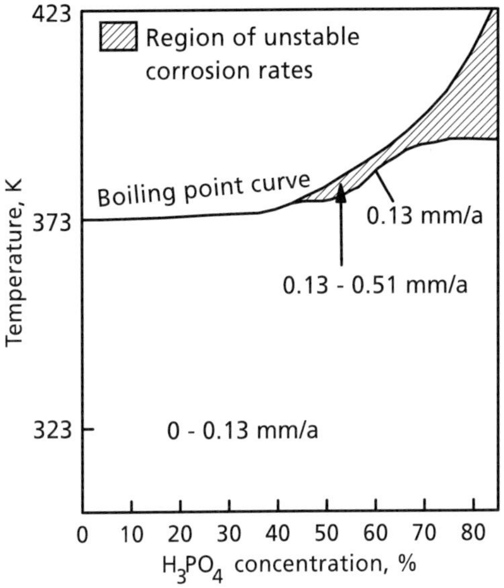

Figure 69: Corrosion behavior of Hastelloy® G-3 (see Table 97) in pure phosphoric acid [364]

0–0.13 mm/a (0–5 mpy) 0.13 mm/a (5 mpy) 0.13–0.51 mm/a (5–20 mpy)

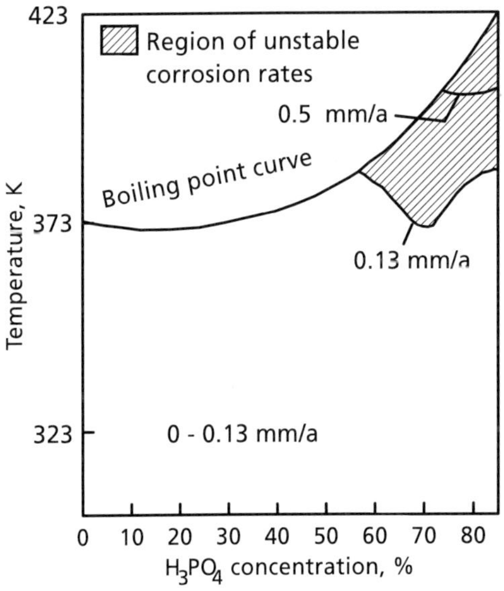

Figure 70: Corrosion behavior of Hastelloy® G-30 (see Table 97) in pure phosphoric acid [364]

0–0.13 mm/a (0–5 mpy) 0.13 mm/a (5 mpy) 0.5 mm/a (20 mpy)

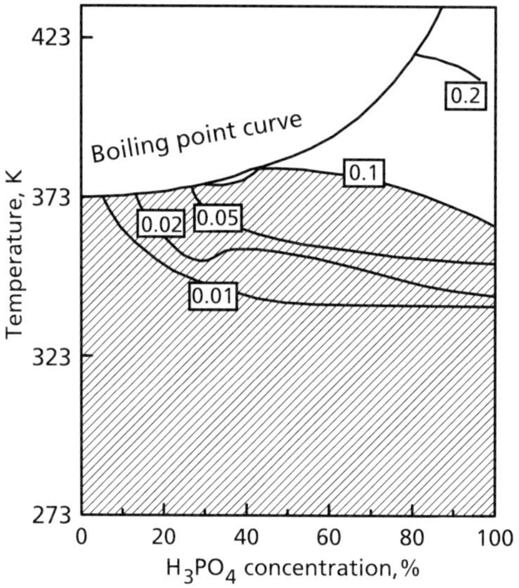

Figure 71: Corrosion behavior of Hastelloy® C (see Table 97) in pure phosphoric acid (isocorrosion curves in mm/a) [363]

0.01 mm/a (0.4 mpy)	0.05 mm/a (2.0 mpy)	0.2 mm/a (7.9 mpy)
0.02 mm/a (0.8 mpy)	0.1 mm/a (3.9 mpy)	

Figures 67–74 show the good corrosion resistance of the alloys in pure, that is to say less corrosive phosphoric acid. Pure H_3PO_4 presents problems only at high concentrations and temperatures. There are differences in the corrosion behavior among the individual materials, especially when contaminated with various impurities. Figure 75 shows the difference in corrosion behavior in pure boiling phosphoric acid as a function of its concentration.

Figure 75 shows that the alloy Hastelloy® B (NiMo) is superior to the NiCrMo alloys. According to Figure 76, NiMo is also "more resistant" than the NiCrMo alloys in phosphoric acid of various concentrations at temperatures of 190°C [365].

In the acid concentration range from 10 to 80%, the corrosion rates of NiMo are between about 0.45 and 1.4 mm/a, while in the case of NiCrMo alloys the corrosion rates are above 1.0 mm/a in 10% acid and 5.0 mm/a in 50% phosphoric acid (Figure 76).

Stainless steels or CrNiMo alloys are more severely attacked by contaminated phosphoric acid than by pure acid. Figure 62–65 show an overview of the effect of impurities on stainless alloys in 75% H_3PO_4 at 85°C. The high-alloyed materials are mainly used under the very corrosive conditions.

Contents of a few hundred ppm (> 0.01%) Cl^- show a noticeable increase in the material corrosion rate, as can be seen from Figure 62. Steep-gradient lines result in the logarithmic plot of the chloride concentration effect on corrosion for the nickel materials; however high-alloy steels also show completely unacceptable corrosion rates as a consequence of the presence of pitting corrosion. According to [341], in practice in phosphoric acid plants, as is the case with wet phosphoric acid, Cl^- contents of about 2% are never to be expected (see also Table 96) [341, 366].

As with the high-alloy steels, the resistance of Ni-Cr-Mo alloys increases with the Cr content and at the same time as the molybdenum content increases (as the pitting index increases), as shown by the sequence of increasing resistance of 2.4858 (Incoloy® 825), 2.4618 (Hastelloy® G), 2.4856 (Inconel® 625) and 2.4819 (Hastelloy® C-276, see also Table 97).

Fluoride contents, which are about 10 times higher than those of the chlorides in crude phosphoric acid or during its production, have hardly any influence on NiCrMo alloys in the range up to 3% F^-, as can be seen from Figure 63. There is a similar resistance sequence here as with the chlorides in Figure 62 [340, 341]. The aggressiveness of the fluorides is reduced in practice by the addition of SiO_2 and Al_2O_3 contained in the phosphate rock, due to the formation of complexes such as SiF_6^{2-} or AlF_6^{3-}. They can also be added for the reduction in the corrosion rate in the crude acid [340, 341].

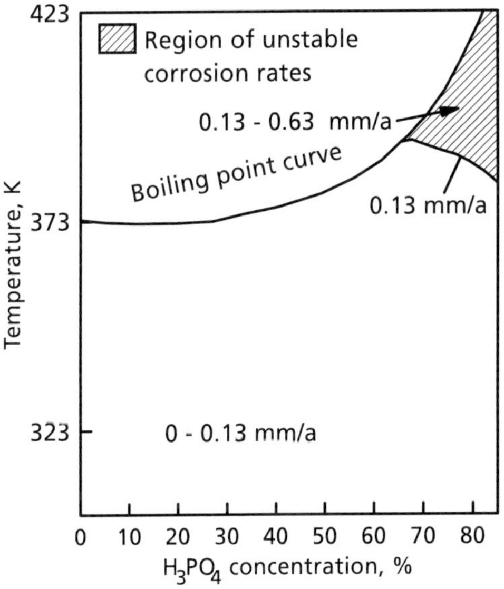

Figure 72: Corrosion behavior of Hastelloy® C-276 (see Table 97) in pure phosphoric acid [359, 360, 364]

0–0.13 mm/a (0–5 mpy) 0.13 mm/a (5 mpy) 0.13– 0.63 mm/a (5–25 mpy)

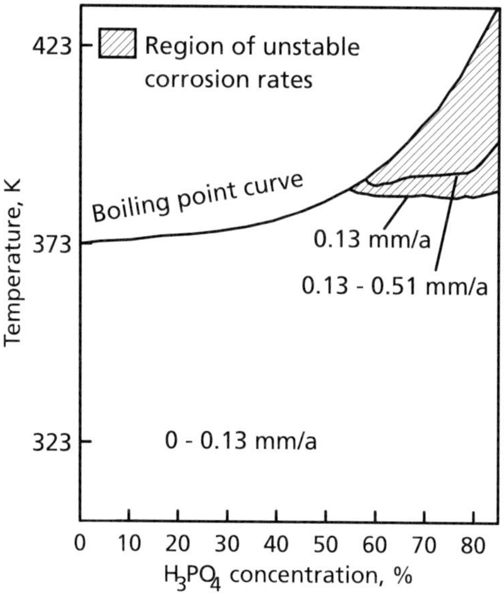

Figure 73: Corrosion behavior of Hastelloy® C-4 (see Table 97) in pure phosphoric acid [350]

0–0.13 mm/a (0–5 mpy) 0.13 mm/a (5 mpy) 0.51 mm/a (20 mpy)

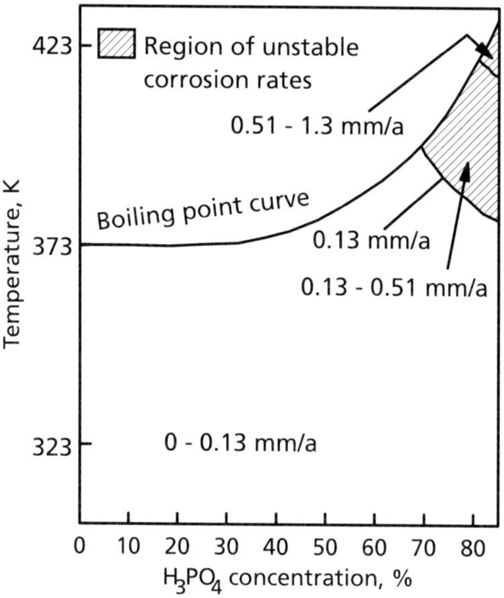

Figure 74: Corrosion behavior of Hastelloy® C-22 (see Table 97) in pure phosphoric acid [364]

0–0.13 mm/a (0–5 mpy) 0.13 mm/a (5 mpy)
0.13–0.51 mm/a (5–20 mpy) 0.51–1.3 mm/a (20–51 mpy)

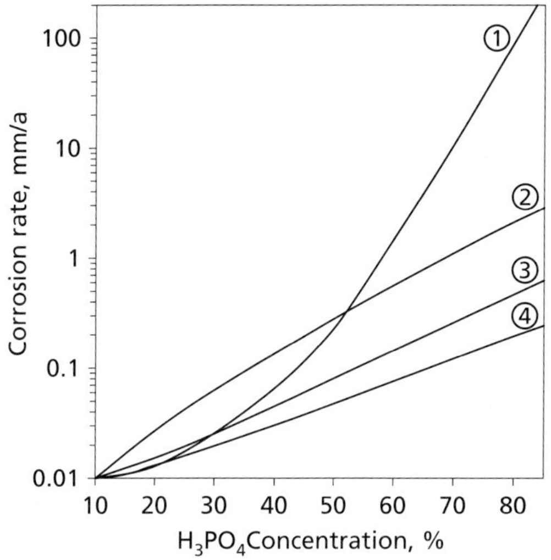

Figure 75: Corrosion behavior of Ni-containing materials in pure boiling phosphoric acid, test duration 3 × 24 h [335, 336, 365]

① SAE 316 (1.4401)
② 45 Ni, 6 Mo, 22 Cr, 1 Mn, 0.5 Si, bal. Fe
③ 16 Mo, 15 Cr, 4 W, 1 Mn, 0.5 Si, 1 Fe, bal. Ni
④ 28 Mo, 1 Mn, 0.5 Si, 5 Fe, bal. Ni

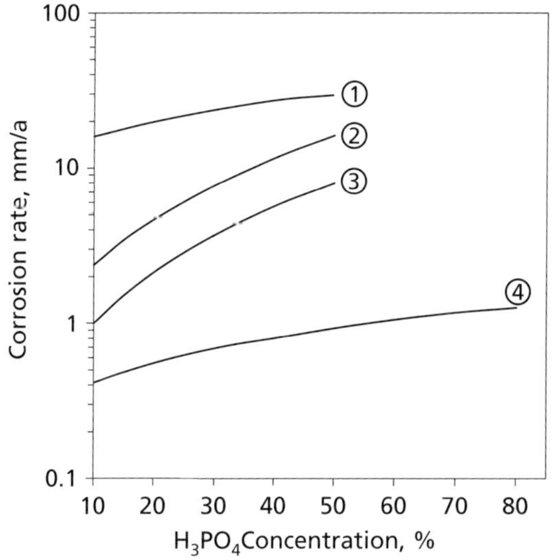

Figure 76: Corrosion behavior of Ni-containing materials in pure phosphoric acid at 190°C (test duration 65 h, specimens were in a closed tube) [335, 336, 365]

① SAE 316 (1.4401)
② 45 Ni, 6 Mo, 22 Cr, 1 Mn, 0.5 Si, bal. Fe
③ 16 Mo, 15 Cr, 4 W, 1 Mn, 0.5 Si, 1 Fe, bal. Ni
④ 28 Mo, 1 Mn, 0.5 Si, 5 Fe, bal. Ni

In synthetic "Florida acid", the high-alloy steel 1.4563 (UNS N08028) is far superior to the nickel alloy 2.4858 (UNS N08825, NiCr21Mo). Here also, the pitting index is a likely explanation. At 41.8, N08028 is far greater than that of the alloy 2.4858 at 27.8.

Like chlorides, sulfuric acid in phosphoric acid also causes an increase in aggressiveness. This can be seen from Figure 64 for 75% H_3PO_4 at 85°C. The increase in corrosion rates approximately corresponds to that in Figure 62. H_2SO_4 concentrations of up to 4% are possible during production of wet phosphoric acid [340, 341].

In the range from 2 to 4% H_2SO_4, the corrosion rates can increase by a factor of 4 to 10 in comparison with H_2SO_4-free phosphoric acid. The corrosiveness of the sulfuric acid is reduced if metallic cations such as Ca^{2+}, Mg^{2+} and Al^{3+} are present, due to the formation of insoluble sulfates [340, 341].

The graph in Figure 64 may mislead the selection of materials in practice if only sulfuric acid were to be regarded as an impurity. If several impurities are present, including sulfuric acid, the corrosiveness must be regarded in more complex terms.

S^{2-} ions in phosphoric acid also have an adverse influence on the corrosion resistance of high-alloy steels and nickel alloys. They generally lead to an increase in corrosion rates, such as, for example, in 75% phosphoric acid at 85°C. Sulfur as sulfide prevents or interferes with the formation of a firmly adhering compact and therefore protective oxide film, which is a prerequisite for corrosion resistance of passivated materials under corrosive effects in the particular medium [340, 341].

Fe^{3+} ions in phosphoric acid (75% at 85°C) can also increase its aggressiveness, as can be seen from Figure 65. Nevertheless, their action on nickel alloys is considerably less than that on 18Cr10Ni2.5Mo steels.

A tabular summary of corrosion rates for the NiCrMo alloys in pure phosphoric acid does not seem appropriate at this point. According to Figures 67–75, corrosion rates will appear considerably below the isocorrosion curves as the concentration and the temperature fall, as can also be seen from the literature references [367–372]. Under the conditions mentioned, at a threshold curve of 0.1 mm/a, the corrosion rates are accordingly considerably below 0.1 mm/a.

On the other hand, the corrosion rates of various alloys in wet phosphoric acid (no further information, probably "Florida acid") as a function of their composition are of particular interest, as can be seen from Tables 100–103 [362].

As already described for the high-alloy steels, chromium contents, and additionally copper contents increase the resistance of alloys to phosphoric acid (Table 100). Molybdenum shows a particularly strong action (see pitting index) and tungsten also leads to an improvement in corrosion resistance (Table 101 and Table 102).

The difficulty of a comparison of corrosion rates in phosphoric acid solutions due to impurities is reported in [362]. Table 103 shows the highest corrosion rates for various alloys under extreme exposure.

In a simulated mixture typical of phosphoric acid production, as expected the NiCrMo alloys 2.4856 (UNS N06625) and 2.4619 (N06985) show adequate resistance, but are more resistant in comparison with steel 1.4563 (N08028).

Material trade name [1)]	*DIN-Mat. No.*	*Corrosion rate mm/a*
Hastelloy® C-276 (16 Cr, 0 Cu)	2.4819	1.12
Inconel® alloy 625 (22 Cr, 0 Cu)	2.4856	0.46
Hastelloy® G alloy (22 Cr, 2 Cu)	2.4618	0.41
Inconel® alloy 690 (30 Cr, 0 Cu)	2.4642	0.13
Hastelloy® G-30 (30 Cr, 1.5 Cu)	2.4603	0.10

1) see Table 97

Table 100: Influence of chromium and copper in nickel alloys on the corrosion behavior in wet process phosphoric acid (probably Florida acid) (46% P_2O_5) at 116°C [362]

Material trade name [1)]	*DIN-Mat. No.*	*Corrosion rate mm/a*
Inconel® alloy 690 (30 Cr, 0 Mo)	2.4642	11.35
Hastelloy® G-30 (30 Cr, 4 Mo)	2.4603	1.55

1) see Table 97

Table 101: Influence of molybdenum in nickel-chromium alloys on the corrosion behavior in wet process phosphoric acid (probably Florida acid) (52% P_2O_5) at 149°C [362]

Material trade name [1)]	*DIN-Mat. No.*	*Corrosion rate mm/a*
Hastelloy® G-30A (30 Cr, 4 Mo, 0 W)	–	4.19
Hastelloy® G-30 (30 Cr, 4 Mo, 2 W)	2.4603	0.97

1) see Table 97

Table 102: Influence of tungsten in NiCrMo alloys on the corrosion behavior in wet process phosphoric acid (probably Florida acid) (54% P_2O_5) at 149°C [362]

The relatively large number of NiCrMo alloys which can be used in contaminated phosphoric acid under extreme exposure to corrosion (and they differ in this respect) can make the selection difficult for the design engineer. This selection should be facilitated with the aid of Table 104. It corresponds to the sequence of the increasing pitting index, also as regards to their resistance.

Material trade name [1)]	*DIN-Mat. No.*	*Phosphoric acid (P_2O_5), %*		
		54%, 116°C	*54%, 149°C*	*70%, 204°C*
		Corrosion rate, mm/a		
Sanicro® 28	1.4563	1.60	82.19	1.35
Incoloy® 825	2.4858	14.1	33.81	0.46
Hastelloy® G	2.4618	2.39	43.18	0.41
Hastelloy® G-3	2.4619	1.04	31.65	0.71
Inconel® alloy 625	2.4856	0.91	20.40	0.56
Hastelloy® G-30	2.4603	0.18	8.05	0.25

1) see Table 97

Table 103: Summary of the highest corrosion rates (mm/a) on high-alloy materials in wet process phosphoric acid (probably Florida acid) after tests on tempered specimens (delivery stage) [362]

Material trade name	*Material designation*	*DIN-Mat. No.*	*Comments*
Incoloy® 825	NiCr21Mo	2.4858	standard material H_2SO_4-containing H_3PO_4
Nicrofer® 4221 hMo	NiCr21Mo6Cu	2.4641	improved resistance to pitting corrosion and crevice corrosion in comparison with 2.4858
Hastelloy® G	NiCr22Mo6Cu	2.4618	similar to 2.4641, more resistant to sensitization because of Nb addition, and therefore better resistance to intergranular corrosion in chloride-containing H_3PO_4
Hastelloy® G-3	NiCr22Mo7Cu	2.4619	newer type based on 2.4858, use in Cl^-- and F^--containing H_3PO_4
Inconel® 625	NiCr22Mo9Nb	2.4856	very good resistance to pitting corrosion, crevice corrosion and therefore also intergranular corrosion (c.f. 2.4618)
Hastelloy® C-22	NiCr21Mo14W	2.4602	use under strongly oxidizing conditions in phosphoric acid with relatively heavy contamination by halogens, hot H_3PO_4

Table 104: Suggestions for materials selection in contaminated phosphoric acid of high aggressiveness (increasing temperature, increasing concentration) [357, 373] (see also Table 97)

Test conditions	*Temperature °C*	*Test duration d*	*Corrosion rate mm/a*
Reflux from the vapor washer of the evaporator, containing 15% H_3PO_4, 20% H_2SiF_6, 1% H_2SO_4	73–86	16	0.025
Sludge in the stirring tank, mixture of 20% H_3PO_4, 2% H_2SO_4, 1% HF, 40% water and $CaSO_4$	77–93	117	0.02
28% wet-decomposed phosphoric acid (20% P_2O_5), 20–22% H_2SO_4, about 1–1.5% fluoride, probably as H_2SiF_6. Moderate aeration, only natural agitation.	82–110	42	0.03
Sludge with 31.4% H_3PO_4, 1.6% H_2SO_4, 1.5% H_2SiF_6, 0.12% HF and $CaSO_4$ in the filter tank	46–60	8.3	< 0.01
Sludge with 37% H_3PO_4 (27% P_2O_5) in the acid converter. Agitation 1 m/s.	66–88	46	0.02
Concentrator: evaporated acid with 54% H_3PO_4, 1.7% HF, 2% H_2SO_4, 2% $CaSO_4$	52–60	51	0.01
55% wet-decomposed phosphoric acid (40% P_2O_5) with 3% H_2SO_4 and suspended $CaSO_4$, sludge.	110	4.7	1.48
Water- and fluorine-containing gases are separated off at the surface of attack. No aeration.			
99% wet-decomposed phosphoric acid (72% P_2O_5), 3.7% H_2SO_4 (3% SO_3), 0.5% fluoride. Moderate aeration. Agitation only vigorous enough for the foam to be dissolved.	116–316	20.8	0.37
Waste gases from the evaporator, H_3PO_4-containing, H_2SO_4 vapor, SO_3, HNO_2, SiF_4, water vapor; atomized with water, containing 0.1% H_3PO_4, 0.06% H_2SO_4, 0.1% bonded fluoride. Vigorous aeration, agitation, constantly sprayed with water.	10–180	20.8	0.32
Gases containing HF, silicon fluoride SO_2, furthermore H_3PO_4 (72% P_2O_5), 3.7% H_2SO_4 (3% SO_3). Vigorous aeration, agitation by fast-flowing gas.	16–343	20.8	0.05

Table 105: Corrosion behavior of Incoloy® 825 (2.4858) under the conditions of phosphoric acid production by the wet decomposition process [374]

At very high levels of impurities and a high temperature and concentration, corrosion testing is recommended under the actual plant conditions. See also the above-mentioned tabular summary in Tables 103–105, which gives comparisons under operating conditions.

Due to the very aggressive conditions during evaporation of wet acid (45–75%), the alloys 2.4856 (Inconel® alloy 625) and 2.4602 (Hastelloy® C-22) are the more suitable materials, because they have the prerequisites of high resistance due to their Cr and Mo contents [341].

According to practical experience, the alloy Incoloy® 825 (2.4858) had a component life of about 4 years in an evaporator in "Florida" phosphoric acid (2.5% HF, chloride-free), while the components failed after 18 months in phosphoric acid from Moroccan phosphorus (1% HF, 0.5% NaCl). The material Inconel® alloy 625 (2.4856) used under these conditions was found to be still completely corrosion free after 2 years. Extensive laboratory studies have confirmed that under heat transfer conditions (hot wall) Incoloy® 825 is generally more suitable than Inconel® alloy 625. On the other hand, Inconel® alloy 625 is preferred to be used under oxidizing conditions in the presence of halides [375]. This would also result from a selection of materials in accordance with Table 104.

H_3PO_4 concentration %	Temperature °C	Test duration h	Corrosion rate mm/a	Comments
36	78		< 0.04	2.9% H_2SO_4, 350 ppm chlorides, 30% gypsum, some HF
36	38–44		< 0.02	2.9% H_2SO_4, 350 ppm chlorides, some H_2SiF_6
52.5	45		< 0.02	2.9% H_2SO_4, 400 ppm chlorides, traces of H_2SiF_6
45	25–130		0.71	45% H_2SO_4, 10% H_2O
45	130		4.7	45% H_2SO_4, 10% H_2O
55	105–128		0.84	3.0% H_2SO_4 (2.5% SO_3), $CaSO_4$ (semihydrate suspension), fluorine compounds
55	108		0.23	0.8% HF
75	100	48	0.58	–
75	100	164	0.18	–

Table 106: Corrosion behavior of Hastelloy® G (2.4618) in phosphoric acid solutions (plant experiments) [376]

Table 106: Continued

H_3PO_4 concentration %	*Temperature °C*	*Test duration h*	*Corrosion rate mm/a*	*Comments*
85	55		< 0.01	–
85	75		< 0.02	–
85	100		0.11	–
85	125		0.25	–
98	199–238		0.23	4–6% H_2SO_4, 2.8–3.0% Fe and Al as trioxides, 0.5–1.0% fluorine compounds
93.5	203–210		0.89	wet acid, about 4.3% H_2SO_4, 4.4% Fe and Al
very low	94–100		0	H_3PO_4 mist, contains some chloride
39	77–84		< 0.03	wet acid, 2% H_2SO_4, 1.2% total fluoride content, 30% gypsum
55	80–85		0.05	wet acid, some fluorides and gypsum
41.3	121	110	0.30 (hot wall)	wet acid, 2% fluorine, 4.5% H_2SO_4
41.3	85	110	0.13	wet acid, 2% fluorine, 4.5% H_2SO_4
41.3	121	96	0.25 (hot wall)	wet acid, 2.4% fluorine, 3% H_2SO_4, 1.2% Fe(II) oxide, 1.1% solids
41.3	85	96	0.06	
74.3	154–169	48	3.6 (hot wall)	1.4% fluorine, 4.3% H_2SO_4
77	107–119	48	0.13 (hot wall)	wet acid, 2% H_2SO_4, 1% fluorine, 1.5% Fe and Al as (II)-oxides, 4% solids

Table 106: Corrosion behavior of Hastelloy® G (2.4618) in phosphoric acid solutions (plant experiments) [376]

Table 106: Continued

H_3PO_4 concentration %	*Temperature °C*	*Test duration h*	*Corrosion rate mm/a*	*Comments*
77	88–110	48	0.04	
77	149	24	1.2 (hot wall)	
77	166	48	2.1 (hot wall)	
91.1	171	48	1.5 (hot wall)	small contents of Co, MgO, Fe, Al_2O_3, SiO_2, Na_2O, Cl, H_2SO_4, and fluoride-free
91.1	163	48	1.1 (hot wall)	
94.8	149–177		0.11	0.33% fluorine with other impurities
97.3	49		0	drum storage

Table 106: Corrosion behavior of Hastelloy® G (2.4618) in phosphoric acid solutions (plant experiments) [376]

Test conditions	*Corrosion rate, mm/a*
28% wet-decomposed phosphoric acid (20% P_2O_5), 20–22% sulfuric acid, about 1–1.5% fluoride, probably as fluorosilicic acid, test temperature 80–110°C test duration 42 days. Moderate aeration, only natural agitation.	0.04
39% wet-decomposed phosphoric acid (28% P_2O_5), 2% sulfuric acid, fluorosilicic and hydrofluoric acid in traces, fluoride equivalent about 1.2% in total. Suspended gypsum, 30% of the suspension weight, liquid phase, temperature 75–85°C test duration 96 days. Moderate aeration, vigorous agitation.	0.02
55% phosphoric acid (40% P_2O_5), 3.0% sulfuric acid (2.5% SO_3), suspended calcium sulfate (semihydrate), fluorine compounds, liquid phase, discharge of gases, containing H_2O and SiF_4, foam distribution process, temperature 105–130°C duration 18.7 days. No aeration, agitation 244 mm/s, test site: reaction region of the extraction vessel.	0.6
98% wet-decomposed phosphoric acid (71% P_2O_5), 4–6% sulfuric acid, 2.8–3.0% iron oxide and aluminum oxide as trioxide, 0.5–1.0% fluorine compounds (calculated as fluoride), temperature 200–240°C test duration 70 days. Vigorous aeration, agitation by pumps, in the hot immersed tube of the evaporator.	0.2

Table 107: Corrosion behavior of Inconel® alloy 625 (2.4856) in media containing phosphoric acid under the conditions of wet decomposition [377]

Table 107: Continued

Test conditions	*Corrosion rate, mm/a*
99% wet-decomposed phosphoric acid (72% P_2O_5), 3.7% H_2SO_4 (3% SO_3), 0.5% fluoride, temperature 15–315°C test duration about 20.8 days. Moderate aeration, agitation only vigorous enough to dissolve foam.	0.4
11–13% fluorosilicic acid plus unspecified impurities from phosphoric acid wet decomposition, liquid phase, temperature 60–75°C test duration 49 days. Vigorous aeration, little agitation.	0.01 0.02
Waste gases from the evaporator, containing phosphoric acid, sulfuric acid vapor, sulfur trioxide, nitrous acid, silicon tetrafluoride, water vapor, atomized with water, containing 0.1% phosphoric acid, 0.06% sulfuric acid and 0.1% bonded fluoride, temperature 10–180°C test duration about 20.8 days. Vigorous aeration, agitation, constantly sprayed with water.	0.3
Gases, containing HF, SiF_4, SO_2, also H_3PO_4 (72% P_2O_5), 3.7% H_2SO_4 (3.0% SO_3), temperature 15–340°C test duration about 20.8 days. Vigorous aeration, agitation by fast-flowing gas.	0.05
Vapors over phosphoric acid, 93.5% wet-decomposed acid (67.8% P_2O_5), containing about 4.3% sulfuric acid and 4.4% iron oxide and aluminum oxide, fluoride less than 1.5%, temperature 190–210°C test duration 52 days. Moderate aeration, rapid agitation.	0.8

Table 107: Corrosion behavior of Inconel® alloy 625 (2.4856) in media containing phosphoric acid under the conditions of wet decomposition [377]

A summary of corrosion rates under conditions very close to operating conditions in contaminated phosphoric acid solutions is contained in Table 105 for Incoloy® 825 (2.4858), Table 106 for Hastelloy® G (2.4618) and Table 107 for Inconel® alloy 625 (2.4856) [374, 376, 377].

When adopting these corrosion rates for construction design, it should be remembered that they originate mainly from studies in chloride free phosphoric acid, probably similar to "Florida acid". Higher corrosion rates are to be expected in chloride-containing acid.

Stress corrosion cracking

As shown in Table 104, the NiCrMo alloys are largely insensitive to stress corrosion cracking. Low carbon contents, high chromium and molybdenum contents increase the resistance to pitting corrosion (see pitting index) and as a result a low tendency to form precipitates as a guarantee for low susceptibility to intergranular corrosion are the reason for this. For precipitation properties, see also the time-temperature-precipitation curves in Figure 66.

No stress corrosion cracking or pitting corrosion was found on the alloys Hastelloy® G (2.4618) and Hastelloy® C-276 (2.4819) in the following aqueous phosphoric acid solutions:

- 0.2% H_3PO_4 + 0.8% NaCl at 141°C
- 1.0% H_3PO_4 + 4.0% NaCl (pH 1.7) at 141°C and
- 2.0% H_3PO_4 (boiling = 102°C).

Erosion corrosion

This type of corrosion occurs if the general surface corrosion is also accompanied by material consumption due to abrasion by solids. This condition almost always exists during processing of crude phosphates to give wet acid, and for this reason this form of corrosion is of particular interest in phosphoric acid production.

As can be seen from practical studies and has been documented by examples of general corrosion by phosphoric acid solutions, relatively low-alloy steels exhibit extreme erosion corrosion. In contrast, high-alloy steels, such as, for example, 1.4439 (SAE 317 LMN, X2CrNiMoN17-13-5) and the nickel alloys Incoloy® 825 and Hastelloy® C-276 show very good resistance, although they have lower hardness values.

Corrosion data for Hastelloy® G confirm that, for good to adequate erosion resistance of the materials corresponding general resistance is a fundamental requirement [378, 379]. It should be remembered here that "Florida acid" is a less aggressive acid because of its relatively low chloride content. High chromium and molybdenum contents in the alloys to be used are therefore a prerequisite for resistance to chlorides (resistance to pitting corrosion).

Nickel-copper alloys

Only alloys with high nickel contents, such as the alloy NiCu 30 Fe (2.4360, Monel® 400) are of importance for use under exposure to phosphoric acid. This alloy has relatively good resistance in pure non-aerated phosphoric acid, as can be seen from Figure 77 [363]. The isocorrosion curves originate from experiments in pure phosphoric acid solution with access to air on the surface of the solution, but without specific aeration or protection by an inert gas.

In non-aerated pure phosphoric acid, corrosion rates of less than 0.05 mm/a are reported up to 66°C while in aerated acid under the same conditions the corrosion rates rise above 1.27 mm/a, and are also in the range up to 0.5 mm/a at –35°C [327].

Figure 78 allows a comparison of the resistance of Monel® 400 (2.4360) with the NiCrMo alloys Incoloy® 825 (2.4858) and Inconel® alloy 625 (2.4856) according to [374]. From Figure 78, Monel® 400 proves to be the less resistant material in phosphoric acid. In fact, Monel® 400 is hardly used or is not used, even in phosphoric acid solutions in which it is resistant, because there are cheaper materials for this purpose.

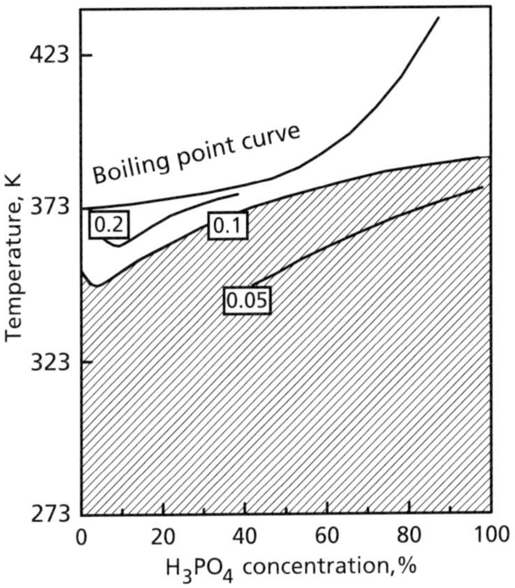

Figure 77: Isocorrosion curves (mm/a) for Monel® 400 (0.15 C, 67 Ni, 30 Cu, 1.4 Fe, 2.4360) in pure phosphoric acid (non-aerated with access of air to the surface of the liquid) [363]

0.05 mm/a (2.0 mpy)
0.1 mm/a (3.9 mpy)
0.2 mm/a (7.9 mpy)

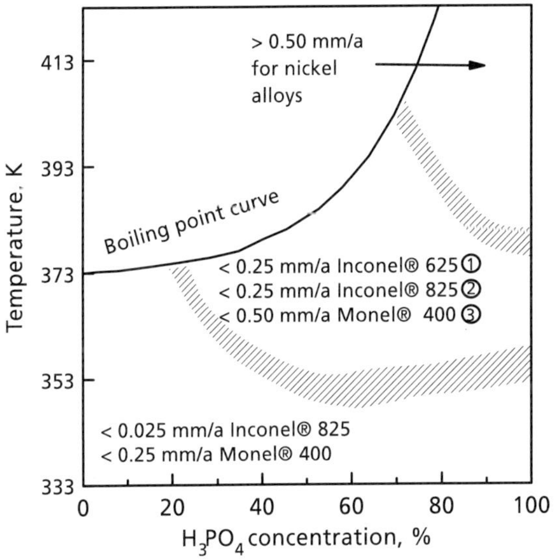

Figure 78: Comparison of the corrosion resistance of Ni alloys (NiCrFeMo, NiCrMo and NiCu) in pure phosphoric acid [374]
① Inconel® alloy 625 (2.4856, NiCr22Mo9Nb)
② Incoloy® 825 (2.4858, NiCr21Mo)
③ Monel® 400 (2.4360, NiCu30Fe)

< 0.025 mm/a (< 1.0 mpy)
< 0.25 mm/a (< 10.0 mpy)
> 0.5 mm/a (> 20 mpy)

Table 108 contains a summary of corrosion rates for Monel® in pure phosphoric acid solutions [53, 54].

According to [355], Monel® is regarded as being less suitable than CrNiMo steels in phosphoric acid solutions, and it is pointed out that vigorous agitation and aeration of the medium are to be avoided with Monel® 400.

The resistance of Monel® 400 and other NiCu alloys in phosphoric acid depends more on the impurities in the acid than the resistance of the other Ni alloys. In particular, oxidizing impurities, such as oxygen and iron compounds (and these are almost always present) intensify the rate of corrosion of NiCu alloys 10-fold or more. Lower phosphoric acid concentrations such as are common during the production of certain drinks cause no corrosion on Monel® 400 [332].

H_3PO_4 concentration %	Temperature °C	Test duration d	Corrosion rate mm/a
3.2	25	49	0.040
3.2	100	0.17	0.12
10	101	2 + 2	0.25
40	27	5	0.025
50	110	2 + 2	0.10
78–85	RT		0.003
78–85	49		0.025
78–85	104		0.23
85[1)]	124	≥ 6	0.25
85[2)]	124	≥ 6	11.0
85	160	1	115
86	98	2 + 2	0.025
117[2)]	60	≥ 6	< 0.25
117[1)]	180	≥ 6	0.084
117	249–255		very high

1) non-aerated
2) aerated

Table 108: Corrosion rates of the Ni alloy Monel® 400 (NiCu 30 Fe; 2.4360) in phosphoric acid (probably pure because of the level of the corrosion rates) according to laboratory studies [336, 337]

The corrosion behavior of Monel® 400 at 30°C was studied for 48 hours in phosphoric acid of various concentrations [380]. Results are presented in Table 109. It can be seen that the corrosion rate gradually increased with increasing phosphoric acid concentration.

H_3PO_4 concentration	Average weight loss, mg	Corrosion rate, mm/a
1 N	8.40	0.0696
5 N	9.50	0.0788
10 N	11.70	0.0971

Table 109: Corrosion behavior of Monel in phosphoric acid [380]

The application ranges of Monel® 400 in thermal phosphoric acid are given in [381]. Corrosion rates of up to 13 mm/a are to be expected in 42% phosphoric acid with 2% H_2SO_4, 2% gypsum and traces of HF at 55–65°C [334].

Nickel-molybdenum alloys

The alloys Hastelloy® B (S-NiMo 30†) and 2.4810 (UNS N10001) and Hastelloy® B-2, as 2.4615 and 2.4617 (NiMo28) as well as Hastelloy® B-3 (UNS N10675, 2.4600 or NiMo29Cr) fall within this group. They are the alloys in the NiMo group which are specially used in the chemical industry.

Ni-Mo alloys have outstanding resistance in non-oxidizing acids, including phosphoric acid, against which a particular resistance is required. Alloy 2.4617 in particular, because of its extremely low contents of carbon, silicon and iron, has a low susceptibility to corrosion of the heat-affected zone alongside the weld seam under exposure to corrosion, and for this reason may require no heat treatment after welding [357, 364].

The exceptional resistance of Ni-Mo alloys in pure concentrated phosphoric acid at high temperatures is shown in Figure 74 and Figure 75 in comparison with other Ni alloys and CrNiMo steel, and in Figure 79. According to [327], corrosion rates of up to 0.5 mm/a are reported in pure non-aerated phosphoric acid up to 70% and 66°C and corrosion rates of up to 0.05 mm/a are reported in 90% pure phosphoric acid up to 93°C. In aerated phosphoric acid up to 50% and 149°C the corrosion rates are less than 0.05 mm/a, and in more highly concentrated pure acid they are up to 0.5 mm/a.

According to [355], the corrosion rates for Hastelloy® B in pure boiling phosphoric acid up to 88% are below 0.15 mm/a. According to [355], Hastelloy® B is more resistant than Hastelloy® C in 88% boiling acid (in this context, see also Figure 74 and Figure 75).

Figures 80–83 give an overview of the usability of Hastelloy® B, B-2 and B-3 with the aid of isocorrosion curves, Figure 81 containing a further classification of the corrosion rates below 0.1 mm/a.

The isocorrosion curves in Figure 81 were determined in an aqueous solution of pure phosphoric acid. In these studies, neither were the solutions specially aerated nor were the specimens protected from the action of oxygen by passing in inert gases [363].

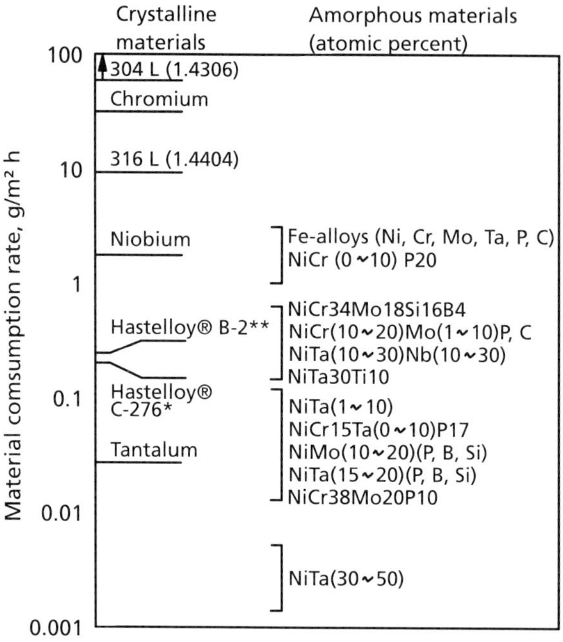

Figure 79: Material consumption rates (times 8.76, divided by the density = mm/a) of amorphous alloys in 87% H_3PO_4 solution at 160°C in comparison with crystalline materials [358]
* 2.4819
** 2.4615, 2.4617

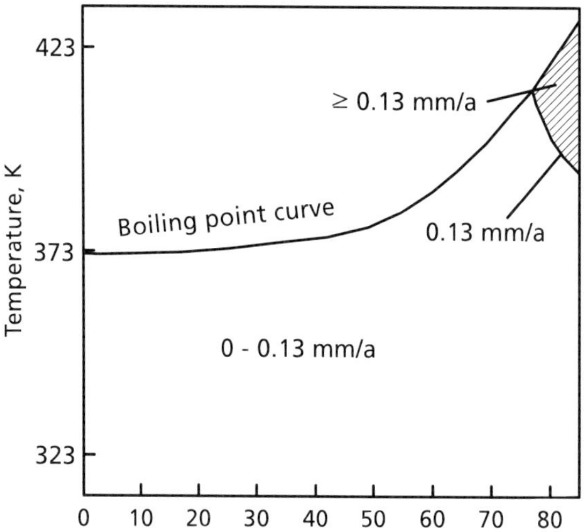

Figure 80: Corrosion ranges for Hastelloy® B (c.f. 2.4800, 2.4810) in pure phosphoric acid (see also Figure 81) [359]

0–0.13 mm/a (0–5 mpy) 0.13 mm/a (5 mpy) ≥ 0.13 mm/a (> 5 mpy)

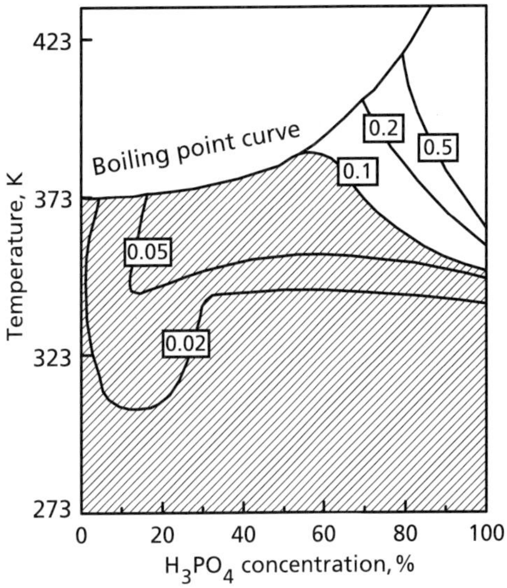

Figure 81: Isocorrosion curves (mm/a) for Hastelloy® B (0.05 C, 61 Ni, 26-30 Mo, 4.7 Fe) in pure phosphoric acid (access of air to the surface of the liquid) [363]

0.02 mm/a (0.8 mpy) 0.1 mm/a (3.9 mpy) 0.5 mm/a (19.7 mpy)
0.05 mm/a (2.0 mpy) 0.2 mm/a (7.9 mpy)

Table 110 gives an overview of the corrosion rates to be expected in pure phosphoric acid.

H_3PO_4 %	Corrosion rate, mm/a		
	20°C	**66°C**	**Boiling**
10	0.008	0.050	0.025
30	0.008	0.020	0.075
50	0.003	0.008	0.075
85	0	0.010	0.71

Table 110: Corrosion of Hastelloy® B (NiMo30) in pure phosphoric acid at various temperatures (mean values from at least five 24 h tests) [336, 337]

Similarly, the corrosion rates of Hastelloy® B-2 in boiling chemically pure phosphoric acid of different concentrations for 120 hours are shown in Table 111 [382].

Concentration, %	Corrosion rate, mm/a
10	0.05
30	0.08
50	0.15
85	0.63

Table 111: Corrosion rates of Hastelloy® B-2 in boiling chemically pure phosphoric acid of different concentrations after 120 hours exposure [382]

Pure phosphoric acid, which is added to foods, is made from oxidizing elemental phosphorus derived from phosphate rock, then adding water. This type of phosphoric acid is much less corrosive than the fertilizer grade, as indicated by the isocorrosion curves for Hastelloy® B-3 in Figure 83 and C-2000 in Figure 84. Indeed, only at high temperatures and concentrations do the Ni-Mo and Ni-Cr-Mo alloys exhibit corrosion rates in excess of 0.1 mm/a in pure phosphoric acid.

Alloy C-2000 (UNS N06200) was also investigated in [383] in various concentrations of phosphoric acid using C-276 for reference. Four 24 hour periods were used for each test. Results of the tests are shown in Table 112. In both concentrations alloy C-2000 exhibited a lower corrosion rate than alloy C-276.

Solution	Temperature	Corrosion rate, mm/a	
		C-2000	C-276
50% H_3PO_4	boiling	0.03	0.18
75% H_3PO_4	boiling	0.84	1.29

Table 112: Corrosion rates of nickel alloys in boiling phosphoric acid [383]

According to [351], use of Hastelloy® B in crude acid is not recommended.

Material		Corrosion rate[1)] mm/a
Illium® R	NiCrMoCu 68 21 5 3	0.16
Illium® G	NiCrMoCu 56 23 6 6	0.20
Hastelloy® G alloy	NiCr22Mo6Cu	0.23
Carpenter® 20 Cb 3	X 5 NiCrMoCuNb 34 20 3 3	0.34
Illium® 98	NiCrMoCu 55 28 8 6	0.40
Incoloy® 825	NiCr21Mo	0.34
Inconel® alloy 625	NiCr22Mo9Nb	0.42
Hastelloy® B	NiMo 30†	0.75
Carpenter® 20 Cb	X 5 NiCrMoCuNb 29 20 3 3	2.60
SAE 317	X3CrNiMo18-12-3	5.10

† old designation
1) Test conditions:
Medium: 55% H_3PO_4 (analytically pure) (= 40% P_2O_5), 0.8% HF
Duration: 2 days
Temperature: 110°C
Agitation: by boiling
Aeration: none

Table 113: Corrosion behavior of nickel-containing materials in phosphoric acid containing hydrofluoric acid [336, 337]

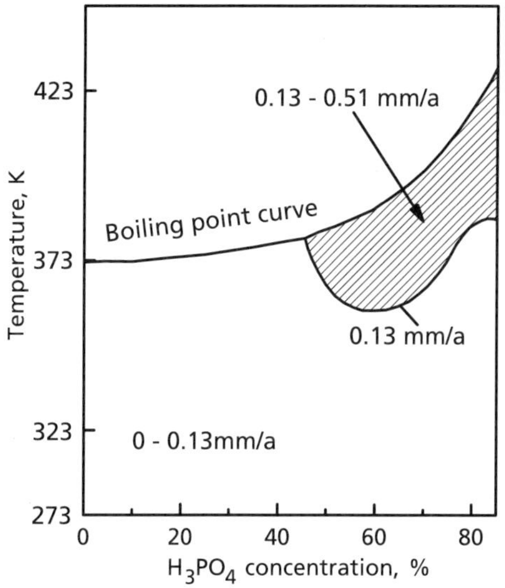

Figure 82: Corrosion ranges for Hastelloy® B-2 (2.4615, 2.4617) in pure phosphoric acid [359, 364, 384]

0–0.13 mm/a (0–5 mpy) 0.13 mm/a (5 mpy) 0.13–0.51 mm/a (5–20 mpy)

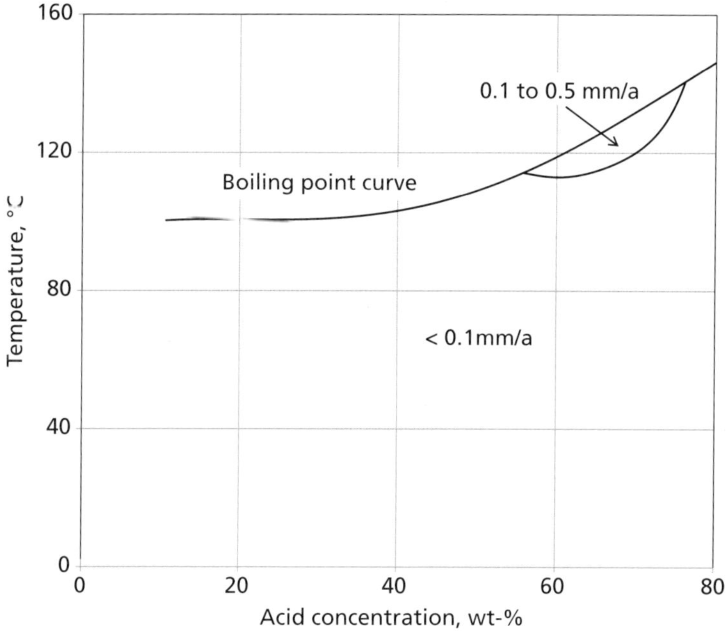

Figure 83: Isocorrosion curve for Hastelloy® B-3 alloy in phosphoric acid (mm/a) [385]

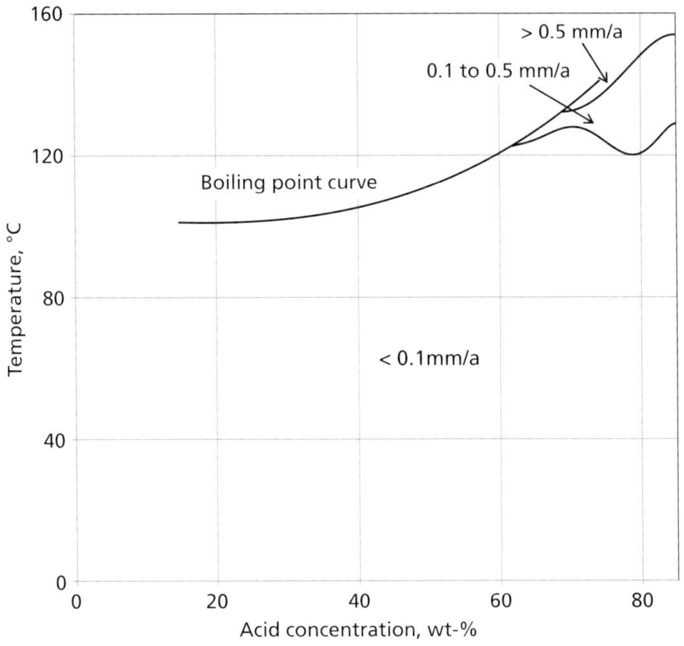

Figure 84: Isocorrosion curve for alloy C-2000 in phosphoric acid (mm/a) [385]

Material		***Corrosion rate***[1] ***mm/a***
Hastelloy® C	G-NiMo 16 Cr†	0.610
Incoloy® 825	NiCr21Mo	0.740
Carpenter® 20 Cb	X 5 NiCrMoCuNb 29 20 3 3	0.840
SAE 316	X5CrNiMo17-12-2	1.40
Monel® 400	NiCu 30 Fe	> 3.50[2)3)]
Hastelloy® B	NiMo 30†	4.90[3)]
Ni-Resist® 1	0.6655, EN-GJLA-XNiCuCr15-6-2	9.30[3)]

† old designation
1) Test conditions:
Medium: moist vapors which occurred during concentration of 50–55% H_3PO_4 (= 36–40% P_2O_5) containing hydrogen fluoride
Test site: specimens in the moisture separator at the top of a drum concentrator
Duration: 21 days
Temperature: 93–169°C average: 107°C
Agitation: moderate
Aeration: moderate
2) one specimen probably destroyed
3) one specimen perforated

Table 114: Corrosion behavior of nickel-containing materials in phosphoric acid in the moisture separator of a drum concentrator with direct heating, according to plant studies [336, 337]

Figure 85 and 86 contain information on the corrosion behavior of Hastelloy® B in crude acids in comparison with other materials. As is also the case with other metallic materials, laboratory studies in pure acid cannot be equated with those at the same concentration and temperature with contaminated acid.

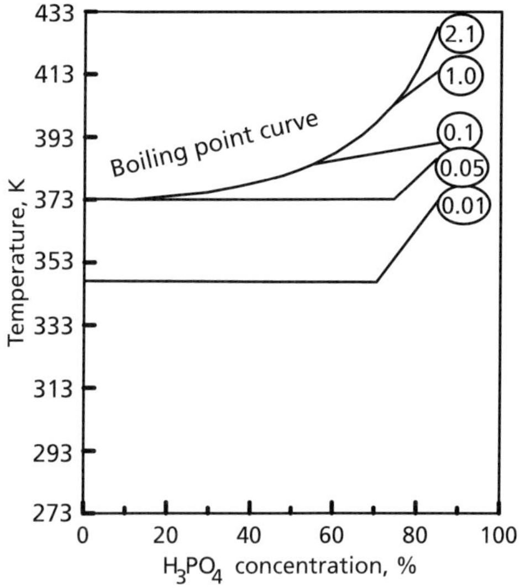

Figure 85: Isocorrosion curves (mm/a) in pure phosphoric acid for 2.4883 (Hastelloy® C alloy, UNS N10002) [386]

0.01 mm/a (0.4 mpy) 0.1 mm/a (4 mpy) 2.1 mm/a (83 mpy)
0.05 mm/a (2.0 mpy) 1.0 mm/a (39.4 mpy)

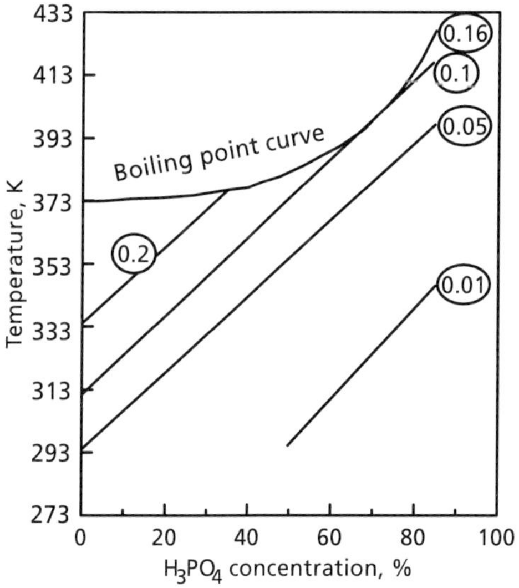

Figure 86: Isocorrosion curves (mm/a) in pure phosphoric acid for 2.4882 (Hastelloy® B alloy, UNS N10001) [386]

0.01 mm/a (0.4 mpy) 0.1 mm/a (4 mpy)
0.05 mm/a (2.0 mpy) 0.16 mm/a (6.3 mpy)

Hastelloy® B is not used if the phosphoric acid contains fluoride [364].

Tables 113–117 contain a summary of corrosion rates under simulated operating conditions [336, 337].

The high susceptibility of Hastelloy® B to fluorides in media containing phosphoric acid under operating conditions can be seen from Tables 113–117. In Table 113, for example, NiCrMo alloys are more suitable under fluoride containing conditions. In the case of phosphoric acid containing hydrofluoric acid in particular, corrosion testing studies should be carried out before possible use under these specific conditions.

Material		***Test 1***[1)]	***Test 2***[1)]	***Test 3***[1)]
			Corrosion rate, mm/a	
Carpenter® 20 Cb	X 5 NiCrMoCuNb 29 20 3 3	2.20	0.56	1.50
Incoloy® 825	2.4858, NiCr21Mo	3.00	0.66	1.40
Illium® G	NiCrMoCu 56 23 6 6			2.80
SAE 317	1.4449, X3CrNiMo18-12-3	3.00		4.10
Hastelloy® C	G-NiMo 16 Cr†	3.30	0.64	1.40
SAE 316	1.4401, X5CrNiMo17-12-2	3.90	1.20	6.10
Hastelloy® B	NiMo 30†	280[2)]	1.40	
Monel® 400	2.4360, NiCu 30 Fe		1.60	
Nickel 200	2.4060, Ni 99,6		1.70	
CA 715	CW354H, CuNi30Mn1Fe		2.40	
Copper	Cu		2.60	

† old designation

1) Test conditions:

	Test 1	Test 2	Test 3
Medium:	75% H_3PO_4 (= 54% P_2O_5), 1% H_2SO_4, HF	75% H_3PO_4 (= 54% P_2O_5), 1% H_2SO_4, HF, SiF_4	75% H_3PO_4 (= 54% P_2O_5), 1% H_2SO_4, HF, SiF_4 (Test 3 was carried out after Tests 1 and 2)
Test site:	specimens immersed in the acid	in the exhaust above the tank	in the gas stream
Duration:	7.8 days	10 days	15 days
Temperature:	121–157°C (average: 150°C)	107–150°C (average: 127°C)	71–182°C (average: 160°C)
Agitation:	vigorous	vigorous	vigorous
Aeration:	none	none	none

2) specimen destroyed

Table 115: Corrosion behavior of nickel-containing materials under conditions for the vapor atomizer and the vapor dome during removal of fluorine compounds [336, 337]

Material	Corrosion rate[1)], mm/a
Incoloy® alloy 804, Incoloy® alloy 825	< 0.003
SAE 304 (1.4301), SAE 309 (1.4828)	< 0.003
SAE 316 (1.4401), SAE 317 (1.4449)	< 0.003
SAE 304 (1.4301), sensibilized	0.008[2)]
SAE 316 (1.4401), sensibilized	< 0.003
Carpenter® 20 Cb	< 0.003
Durimet® 20 (c.f. 1.4500)	< 0.003
NiCrMoCu 68 21 5 3, NiCrMoCu 56 23 6 6	< 0.003
Inconel® 600 (2.4816, NiCr15Fe)	0.740[3)]
Hastelloy® B (c.f. 2.4800)	0.970
Monel® 400 (2.4360, NiCu 30 Fe)	1.25
CA 715 (CW354H, CuNi30Mn1Fe)	1.35
Ni-Resist® 1 (0.6655, EN-GJLA-XNiCuCr15-6-2)	3.30[4)]
Ni-Resist® 2 (0.6660, GGL-NiCr 20 2)	5.80
Nickel 200 (2.4060, Ni 99,6)	> 3.00[5)]

1) Test conditions:
Medium: 0.1% H_2SiF_6, 0.08% H_3PO_4 (= 0.05% P_2O_5), traces of H_2S, pH 2.0
Test site: specimens immersed in waste water used for washing filters
Duration: 76 days
Temperature: 27–80°C average temperature 65°C
Agitation: about 0.6 m/s
Aeration: moderate
2) pitting corrosion: depth to 0.30 mm and intergranular corrosion
3) perforated in the crevice
4) pitting corrosion: depth to 0.58 mm
5) specimen mainly destroyed

Table 116: Corrosion behavior of nickel-containing materials in the waste water of a condenser [336, 337]

Material	Test 1[1)]	Test 2[1)]	Test 3[1)]
	Corrosion rate, mm/a		
Durimet® 20 (c.f. 1.4500)	< 0.003	< 0.003	< 0.003
Carpenter® 20 Cb	< 0.003	< 0.003	< 0.003
Incoloy® alloy 825 (2.4858, NiCr21Mo)			< 0.003
SAE 316 (1.4401)	0.003	0.041	< 0.003
SAE 304 (1.4301)	< 0.003	0.740	< 0.003
Monel® 400 (2.4360, NiCu 30 Fe)	0.230	0.710	0.510
CA 715 (CW354H, CuNi30Mn1Fe)	0.305)	0.970	
Hastelloy® B (c.f. 2.4800)	0.240	0.990	0.940
Inconel® 600 (2.4816, NiCr15Fe)	0.150[2)]	1.02	
Ni-Resist® 1 (0.6655, EN-GJLA-XNiCuCr15-6-2)			1.20
Nickel 200 (2.4060, Ni 99,6)	0.340	1.65	

1) Test conditions:

	Test 1	Test 2	Test 3
Medium:	about 1% fluorides, 0.2% solids, pH about 3	16% O_2, 5% CO_2, small amounts of compounds of sulfur, fluorine, phosphorus and chlorine	16% O_2, 5% CO_2, small amounts of compounds of sulfur, fluorine, phosphorus and chlorine
Test site:	specimens in the sludge of the washer	specimens in the stream of waste gas, exposed to sprayed sludge	specimens in the stream of waste gas 60 cm above the moisture separator
Duration:	47 days	47 days	58 days
Temperature:	average: 60°C highest: 66°C lowest: 22°C	average: 62°C highest: 300°C lowest: 22°C	average: 55°C highest: 60°C lowest: 50°C
Agitation:	vigorous	vigorous	vigorous
Aeration:	vigorous	vigorous	vigorous

2) pitting corrosion: max. depth 0.25 mm, also crevice corrosion

Table 117: Corrosion of nickel-containing materials in a waste gas washer of a furnace for agglomeration of phosphates (plant studies in a pilot and product ion plant) [336, 337]

Other nickel alloys

Also with these other nickel alloy materials a distinction must be made between the behavior in pure and in contaminated phosphoric acid.

Figure 74 shows the corrosion behavior of the alloy G-NiSi10Cu in pure boiling phosphoric acid. Under these conditions, the alloy proves to be virtually equivalent to Hastelloy® C. This even applies at high concentration ranges, in which phosphoric acid is regarded as being very aggressive. Under conditions in practice, however, it is more inferior to Hastelloy® C.

Nickel alloys with added chromium also have, in addition to their good corrosion resistance, the double function of being used as heat-resistant materials, because as well as providing corrosion resistance at room temperature, chromium also improves the resistance to high temperature corrosion, such as scaling.

Inconel® 617 (NiCr23Co12Mo, 2.4663), also has, in addition to a very good suitability for welding as a result of its high Cr, Co and Mo contents, very good resistance to wet corrosion in boiling phosphoric acid (specific data in this context are not available) [387]. It also has good resistance to high temperature corrosion (oxidation resistance).

In the series of hardenable spring alloys having particular properties, the alloy Beryvac® 520 (NiBe2) lies between the known alloys CuBe2 (CW101C, UNS C17200) and CuBe1.7 of high electrical conductivity and the CoNiCr-based alloys with good spring properties and at the same time a high heat resistance. It is chiefly processed to highly stressed catch springs, relay spring elements and relay return springs.

The following corrosion rates are reported for it in (probably pure) phosphoric acid:

- 10 and 89% H_3PO_4 at 20°C at ≤ 0.1 mm/a
- 10% H_3PO_4 at 100°C at 1–10 mm/a and
- 89% H_3PO_4 at 100°C at more than 10 mm/a [387, 388].

One of the most important pulverulent spray materials is composed of 65–85% Ni, 8–20% Cr, 2–5% B, 1–4% Si, 8–10% Fe and 0.3–1.0% carbon. In this alloy, boron and silicon cause "self-flowing" of the spray alloy, and at the same time lower the melting point to about 1,050°C and furthermore have flux-like functions during fusion of these protective coatings.

In phosphoric acid (no precise information) the Ni-Cr-B alloys have good to poor corrosion behavior [389]. This depends on the composition of the alloy and is to be classified in about the same class as that of the NiCr alloys.

Nickel in titanium alloys causes a different corrosion behavior in acid mixtures of 15% H_3PO_4 and 80% H_2SO_4 at room temperature (20°C). The following material consumption rates in 15% H_3PO_4 and 80% H_2SO_4 acid solutions are reported, depending on the Ni content:

0% Ni	1.95 g/m^2 h
4.7% Ni	3.24 g/m^2 h
7.6% Ni	2.3 g/m^2 h
20.0% Ni	~ 1.0 g/m^2 h
55.0% Ni	2.9 g/m^2 h
98.0% Ni	47.6 g/m^2 h

The alloy which is most resistant under these conditions contains 20% Ni. Compared with the other alloys listed, it had comparably good resistance in 80% H_2SO_4, while titanium, at 15.2 g/m^2 h, and 98% nickel with 2% titanium, with a material consumption rate of 643.5 g/m^2 h, were far less resistant [390].

As has already been pointed out several times elsewhere, the use of nickel alloys under exposure to phosphoric acid is justified only under very corrosive conditions, such as at high concentrations and high temperatures. Table 118 shows an overview of the corrosion behavior of amorphous alloys in 87% H_3PO_4 (63% P_2O_5) at 160°C whereas Figure 79 contains a comparison with crystalline materials.

Chemical composition, %[1)]	***Material consumption rate g/m² h***	***Ranking***	***Literature***
NiCr(15-20)P(15-20)	0.016–0.028	3	[358]
NiCr38Mo20P10	0.024/0.040	7	[358, 391]
NiCr15Ta3P17	0.109	10	[391]
NiCr15P17	0.105	9	[391]
NiCr34Mo18Si16B4	> 0.700	17	[358]
NiCr(10-15)Mo10P17	0.200–0.300	14	[358]
NiCr20Mo(1-10)P15C5	0.200–0.300	14	[358]
NiCr20Mo1P17	> 0.700	17	[358]
NiMo20P19	0.022	4	[358]
NiMo10P19	0.027	6	[358]
Ta[2)]	0.024	5	[391]
NiTa30	0.007	2	[358]
NiTa50	0.003	1	[358]
NiTa5P17	0.041	8	[358]
NiTa3P17	0.058/0.152	12	[358, 391]
NiTa30Nb10	0.120	11	[358]
NiTa30Ti10	0.200	13	[358]
NiTa20Zr20	0.360	15	[358]
NiTa10Nb30	0.460	16	[358]

1) atom percent
2) crystalline

Table 118: Overview of the corrosion behavior of amorphous nickel alloys in 87% H_3PO_4 (63% P_2O_5) at 160°C test duration 100 h

Both illustrations show the very good corrosion behavior of NiCr alloys and, above all, NiTa alloys under these conditions. Under conditions in practice, the alloys with 30 and 50 atom-% tantalum can be regarded as being completely resistant. Furthermore, as can be seen from these illustrations, relatively small amounts of titanium, zirconium and niobium greatly impair the material consumption rates, and even lead to an exponential increase [358, 392].

Figures 87 and 88 show examples of electrochemical studies on these alloys.

In Figure 87 increasing tantalum contents in NiTa(x)P17 alloys lead to a passivity of the material in 87% H_3PO_4 solution at 160°C. Under the same conditions, the alloy NiTa3P17 (Figure 88) is in the active state. The free corrosion potential of this alloy is in the active range, and is almost constant over a test duration of 10 min at –0.22 V_{SCE} [392]. Under polarization, these alloys display an active-passive behavior. Additions of tantalum reduce the current density in the passive region of potentials.

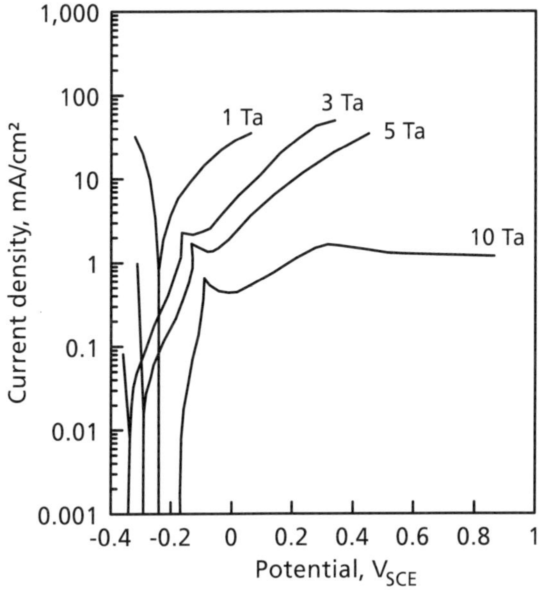

Figure 87: Influence of tantalum on the corrosion behavior (current density/potential curves) of amorphous NiTaP17 alloys (data in atom percent) in 87% H_3PO_4 (63% P_2O_5) at 160°C [358]

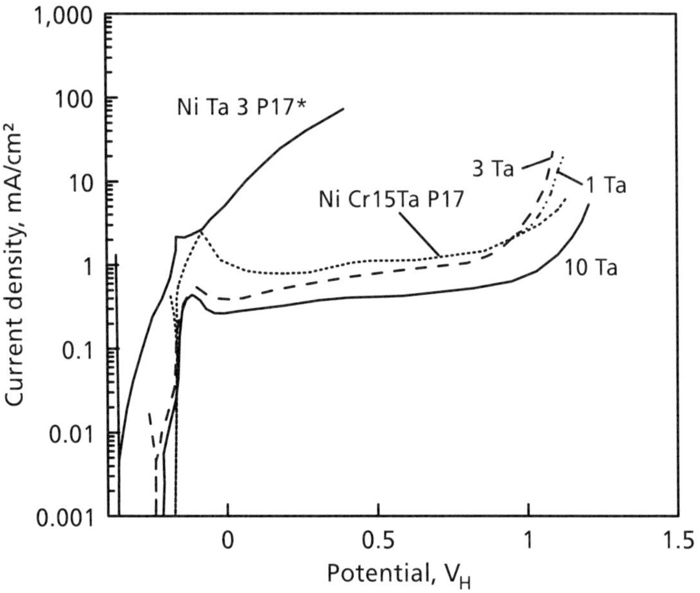

Figure 88: Influence of tantalum on the corrosion behavior (current density/potential curves) of amorphous NiCr15TaP17 alloys (data in atom percent) in 87% H_3PO_4 (63% P_2O_5) at 160°C [391]
* see Figure 87

The alloy NiCr38Mo20P10 is passive under these conditions. The material consumption rate is 0.04 g/m^2h (the alloy constituents are stated in atom percent).

Sulfuric Acid

Author: *R. Durham/L. Hasenberg* / Editor: *R. Bender, R. Durham*

Nickel

Nickel is attacked by solutions of an oxidizing nature, whereas non-oxidizing solutions attack nickel only slightly, if at all [394]. Nickel is sufficiently resistant in up to 80% sulfuric acid at room temperature if air is excluded. The corrosion rates are as 0.125 mm/a or 1.25 mm/a if air is present. Nickel is severely attacked by boiling sulfuric acid [394].

According to [395], nickel is not particularly corrosion-resistant in sulfuric acid. Its resistance is lower than that of Monel® 400 (NiCu 30 Fe) or lead. A protective layer may form on nickel in concentrated sulfuric acid at high temperatures, but cannot be relied upon with certainty. The use of nickel in concentration ranges above 25% H_2SO_4 and at temperatures above room temperature is risky [395]. If possible, the sulfuric acid should be free of air. These conditions also correspond to other studies [396–402]. Figure 89 gives an overall view of the corrosion behavior of nickel in sulfuric acid as a function of the temperature and concentration (compare with Table 119). From this graph, nickel has relatively high resistance in sulfuric acid, so that it can be assumed that the acid is non-aerated. However, there is no indication of this in the literature.

As a supplement to Figure 89, Table 119 shows a summary of the corrosion rates on nickel in sulfuric acid solutions under various states of agitation, temperature and aeration.

Potentiostatic corrosion studies on Ni 99.98 nickel anodes were performed in [403]. As is known, the aggressiveness of halogen ions in sulfuric acid increases in the order: iodide, bromide, chloride. Pitting corrosion is caused mainly by chlorides as a consequence of the destruction of local passive layers. The susceptibility of nickel to pitting corrosion increases as the sulfuric acid concentration increases. Nickel is used when handling sulfuric acid only in non-aerated solutions at low temperatures, and these should also contain inhibiting substances. The sulfuric acid content should not exceed 5%.

One of the few exceptions to this is the evaporation of viscose spinning bath solutions, in which the boiling point is reduced by a vacuum and the presence of organic substances and sulfides has an inhibiting effect. Under these conditions, nickel has proved suitable. Operational corrosion studies in an evaporator for a spinning bath solution showed a corrosion rate on nickel of 0.16 mm/a in a solution of 6% H_2SO_4, 17% Na_2SO_4 and 3% inorganic sulfates at 40°C (test duration 35 d). In a concentrated solution of 18 to 20% H_2SO_4 and 27 to 30% Na_2SO_4 at 54°C in a vacuum at 91.4 kPa, the corrosion rate was only 0.04 mm/a (test duration 120 d). In a solution of 10 to 12% H_2SO_4 and 20 to 22% Na_2SO_4 with small amounts of $ZnSO_4$ the corrosion rate was 0.08 mm/a (test duration 30 d) at 48°C. Further information on corrosion rates in solutions containing H_2SO_4 is contained in [396]. Figure 90 [404] gives a summary of the application ranges of Ni-containing materials in sulfuric acid for pipeline construction.

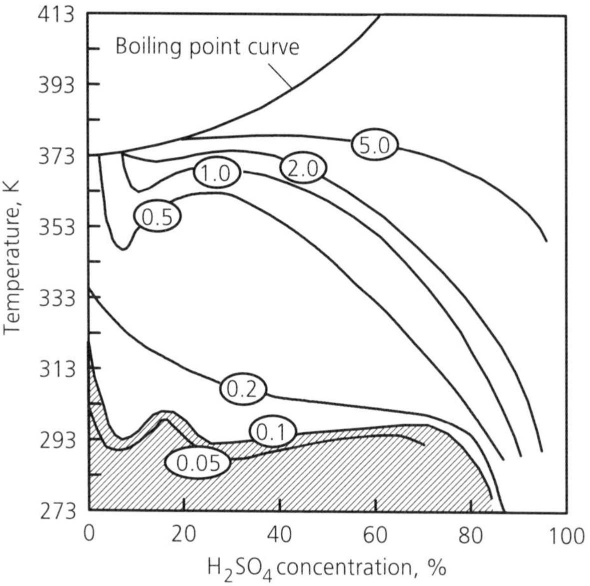

Figure 89: Isocorrosion curves (mm/a) of nickel (99%) in various concentrations of sulfuric acid [405]

H_2SO_4 concentration %	*Temperature °C*	*Test duration h*	*Agitation rate m/s*	*Medium*	
				Non aerated	*Air saturated*
				Corrosion rate, mm/a	
1	30	120	0.079	–	1.24
1	78	20	0.079	–	2.79
2	21	5	none	0.05	–
5	19	100	none	0.06	–
5	30	24	0.081	0.23	1.55
5	60	100	none	0.25	–
5	60	20	0.079	–	2.24
5	71	18	0.081	–	2.62
5	77	120	none	0.53	–
5	78	20	0.079	0.76	5.08
10	21	96	none	0.04	–
10	60	20	0.079	–	2.26

Table 119: Corrosion rates on nickel in H_2SO_4 as a function of the concentration, temperature, state of agitation and aeration (see also Figure 89) [396]

Table 119: Continued

H_2SO_4 concentration %	Temperature °C	Test duration h	Agitation rate m/s	Medium	
				Non aerated	Air saturated
				Corrosion rate, mm/a	
10	77	120	none	0.30	–
10	80	6	none	–	3.05
20	21	5	none	0.10	–
25	82	20	0.133	–	2.11
48	70	–	none	0.46	–
50	30	24	0.081	–	0.41
70	38	24	0.079	0.74	–
93	30	24	0.079	–	0.25
93	65	–	none	3.71	–
95	21	20	none	1.80	–

Table 119: Corrosion rates on nickel in H_2SO_4 as a function of the concentration, temperature, state of agitation and aeration (see also Figure 89) [396]

Pure nickel (not shown in Figure 90) and NiCu 30 Fe (Monel® 400) can be used at room temperature up to a concentration of 80% [405]. The use of nickel in sulfuric acid may be of interest in cases where Cu-, Fe- and Cr-ions are present [406].

Material	Corrosion rate, mm/a
NiCu 30 Fe	4.95
Ni 99.6	2.00
NiCr15Fe	2.75
X5CrNi18-10	10.00
X5CrNiMo17-12-2	4.85
Unalloyed steel	21.3

Experimental conditions:
Medium: 5% HF, 5% H_2SO_4, 10% $FeSO_4$
Duration: 40 h
Temperature: 49°C
Aeration: The specimens were alternately immersed in the bath and removed again.

Table 120: Corrosion behavior of iron materials and nickel materials in a pickling solution for cast iron [407]

Of the materials listed in Table 120 with their resistance in a pickling bath for cast iron, nickel shows the best resistance with a corrosion rate of 2.0 mm/a. The pickling bath consists of 5% H_2SO_4, 5% HF, 10% $FeSO_4$ and water. The specimens were tested for 40 h at a temperature of 49°C by alternate immersion and removal from the bath [407].

– Corrosion behavior in sulfuric acid solutions –

Since in industrial practice mineral acids are used in organic media as reaction partners or catalysts, they may also be present in these solutions as impurities. Figure 91 shows the corrosion behavior of nickel in organic solvents containing 0.05% H_2SO_4. Of the various groups of organic solvents, acetic acid, acetone, ethanol and water were used for comparison. With the exception of acetic acid, the solvents are distinguished by a relatively high dissociating power. Since nickel is subject to corrosion by oxygen and the solubility of oxygen in ethanol is about seven times higher than in water, the higher corrosion rates determined in ethanol containing sulfuric acid can also be explained.

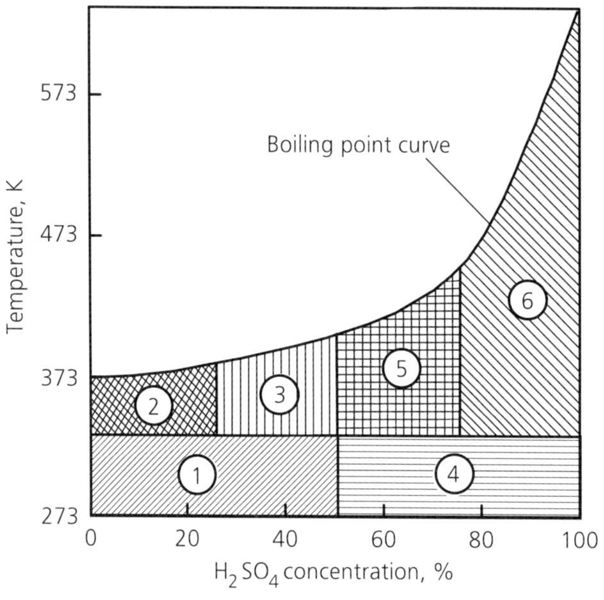

Figure 90: Application ranges of Ni-alloys in sulfuric acid [405]
① NiCu30Fe, NiMo30, NiCr20Mo15, CuNi30, NiCr21Mo
② NiMo30, NiCr20Mo15, NiCr21Mo, NiCu30Fe (non-aerated)
③ NiMo30, NiCr21Mo
④ NiMo30, NiCr20Mo15, NiCr21Mo
⑤ NiMo30, NiCr20Mo15 (below the boiling point)
⑥ NiCr21Mo (non-aerated) > 0.5 mm/a

The purity of the nickel has only an insignificant effect on its corrosion under these conditions. In ethanol the corrosion rate rises sharply initially, but drops off again at the boiling point. This is due to the formation of a surface layer, already referred to above. This surface layer was in fact established on these specimens. The same behavior was found in acetic acid containing sulfuric acid. The excellent corrosion behavior of nickel in acetone containing sulfuric acid was due to the low dissolving power of acetone regarding the corrosion products of nickel. In acetic acid, nickel is corroded without the inhibiting effect of sulfuric acid [408].

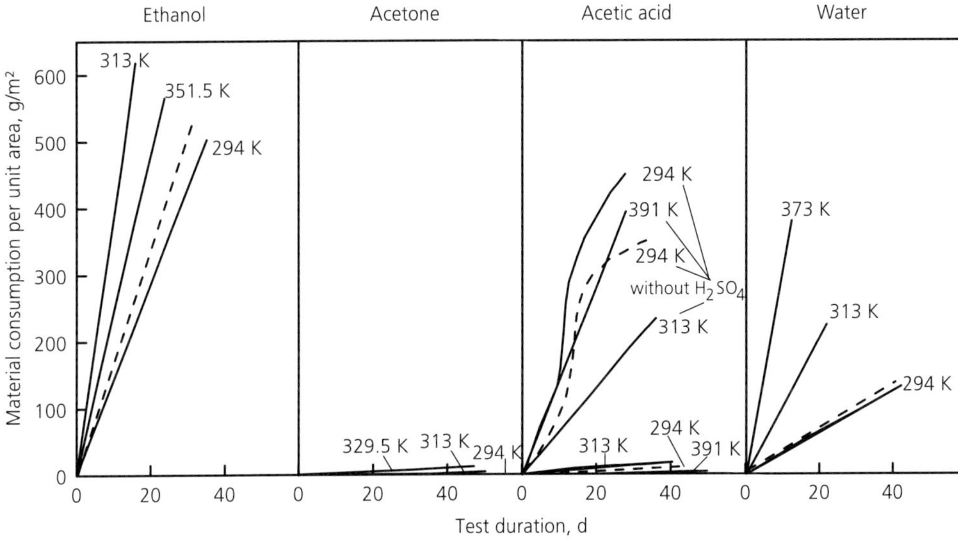

Figure 91: Corrosion/time curves for nickel in sulfuric acid (0.05% H_2SO_4) solutions [408]
—— Ni99.8 %
- - - - Ni99.2 %

– *Stress corrosion cracking* –

Stress corrosion cracking may occur in nickel materials as a consequence of embrittlement of the material due to uptake of atomic hydrogen. The results on specimens exposed to hydrogen are summarized in Figure 92. Figure 92 also shows that nickel may become very brittle under the influence of hydrogen. As the graph shows, short exposure times with atomic hydrogen are sufficient to greatly impair the deformability of the material. The embrittlement decreases with the specimen thickness [409].

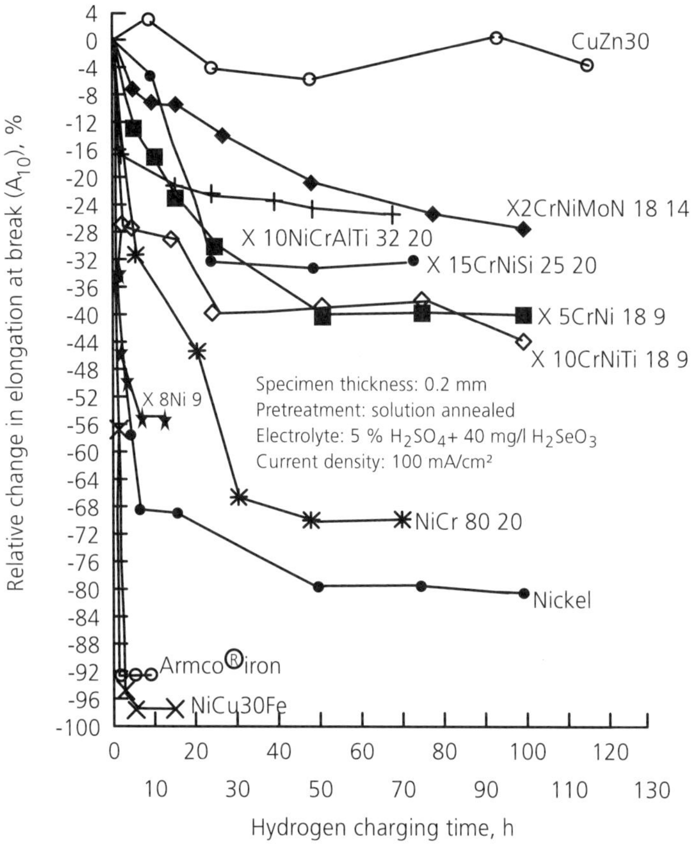

Figure 92: Relative change in the elongation at break as a function of the hydrogen charging time in 5% sulfuric acid [409]

– Influence of a magnetic field on sulfuric acid corrosion –

Nickel specimens (no precise material data) were exposed to a magnetic field of variable strength for a short time in flowing sulfuric acid at 20 ± 2°C. The test duration was 380 h and the action time of the magnetic field was 0.24 min (perpendicular to the direction of flow). The flow rate of the acid was 0.04 m/s. The experimental results are summarized in Table 121. They show that a corrosion maximum occurs at a magnetic field strength of 39,788.5 A/m. The results are based on 4 to 5 exposures using, in each case, 3 to 4 specimens [410]. The corrosion rates were converted from the material consumption rates (g/m^2 h). The corrosion rates are low, so that it can be assumed that they lie within the range of scatter of the measured values in such studies. A comparison with Figure 89 shows a rise in the corrosion rate of nickel in the range 5 to 15% H_2SO_4.

Nickel-chromium alloys

By addition of chromium to nickel, the resistance of the latter to oxidizing media increases. The critical minimum concentration obtained from the current density/potential curves in sulfuric acid is 14% chromium [411]. Nickel-chromium alloys in chemical engineering contain, as a rule, a greater or lesser iron content. These alloys are comparable in their corrosion behavior in sulfuric acid to the NiCr-alloys without iron.

Magnetic field strength A/m	*H_2SO_4 concentration, %*		
	5	*7*	*15*
	Corrosion rate, mm/a		
0.0	0.11	0.09	0.07
5,968.3	0.12	0.10	0.08
7,957.7	0.12	0.12	0.09
28,873.1	0.14	0.13	0.11
31,830.8	0.15	0.14	0.12
39,788.5	0.17	0.15	0.13
47,746.2	0.16	0.14	0.12
55,703.9	0.14	0.12	0.08

Table 121: Corrosion rates on nickel (conversion from g/m^2 h) in flowing sulfuric acid (0.04 m/s) at 20 ± 2°C after the brief action (0.24 min) of a magnetic field of varying strength (test duration 380 h) [410]

NiCr-alloys are practically resistant in cold, non aerated sulfuric acid up to a concentration of 70%, but are generally inferior to the NiCu-alloy NiCu 30 Fe in their corrosion resistance. Under oxidizing conditions, the chromium content renders them considerably more resistant than NiCu 30 Fe. In concentrated sulfuric acid, the oxidizing action of dissolved air is not sufficient to maintain the passivity of NiCr materials in sulfuric acid and solutions thereof. Better results are achieved by addition of Fe(III)- or Cu(II)-ions. It is important for practical purposes that the corrosion rates in dilute, pure sulfuric acid increase on aeration (Table 122). At 30°C, the corrosion rate in 93% H_2SO_4 on aeration of 0.25 mm/a is considerably lower than that without aeration of 6.6 mm/a. The susceptibility of the NiCr-alloys to corrosion in sulfuric acid increases more as the temperature rises than that of the alloy NiCu 30 Fe, which is why these alloys should not be used under such conditions. The corrosion losses also increase with simultaneous aeration [396]. Under reducing conditions, NiCr alloys are comparable to nickel in their resistance to corrosion.

H_2SO_4 concentration %	Temperature, °C	Test duration h	Agitation rate m/s	Medium	
				Non aerated	Air saturated
				Corrosion rate, mm/a	
0.16	100	–	–	0.09	–
1	30	120	0.079	–	1.24
1	78	22	0.079	–	2.80
5	20	100	none	0.06	–
5	30	20	0.079	0.22	–
5	30	23	0.081	–	1.90
5	60	100	none	0.25	–
5	80	20	0.081	0.76	3.81
10	room temperature	24	none	0.11	–
70	30	20	0.079	1.17	–
93	30	20	0.079	6.62	0.25

Table 122: Corrosion behavior of Inconel® (NiCrFe-Alloy) in sulfuric acid solutions under various conditions [396]

In hot 98% sulfuric acid, NiCr-alloys containing about 33% Cr (Lewmet®) exhibit considerably higher resistance than, for example, the NiCrMo-alloys of the type Hastelloy® C [412, 413].

Under the influence of atomic hydrogen, embrittlement of NiCr-alloys may occur and, together with the stresses present in the material, cause stress corrosion cracking. The relative change in the elongation at break as a function of the hydrogen exposure time is shown in Figure 92. The alloy NiCr 80 20 (2.4869, c.f. N06003) in particular shows a marked increase in embrittlement as a result of hydrogen uptake [409, 414].

The reason for studies on the susceptibility of the NiCr-alloy NiCr20TiAl (N07080) to stress corrosion cracking was stress corrosion cracking damage to cap screws M 110 × 6 × 780/830 made of this material [415, 416].

Screws of heat-resistant materials are used in conventional steam turbine construction for fastening the casing halves of steam turbines in the region of higher temperatures. The temperature range is between 350 and 540°C.

As a result of the change to the basic demand being met by nuclear power stations, conventional power stations burning fossil fuel are used to meet the medium and peak demands. Because of the intermittent operation of such plants, formation of condensate and precipitation of impurities in the steam, such as chlorides and sulfates, occurs in the entire turbo-generator during the start-up, stopping and preheating periods. Frequent cold starts and the associated changes in stresses, possible concentration of condensates and conceivable sensitization of the materials may all lead to stress corrosion cracking. Such damage occurred after an operating time of 30,000 h. The features were severe pitting corrosion, and intercrystalline attack starting from these points. For further patterns of damage see [415]. Figure 93 contains a summary of the experimental results.

The alloy Nimonic® 80 A (NiCr20TiAl) tested in the pre-loaded state exhibits sensitivity to stress corrosion cracking in more highly concentrated electrolytes containing SO_3 and SO_4 above a minimum loading of $0.7 \times R_{p0.2}$ and in Cl-containing electrolytes above $R_{p0.2}$ [415, 416].

Nickel-chromium-iron alloys

According to [397], of all the Ni-materials used in sulfuric acid solutions, whether aerated or non-aerated, the alloy NiCrFe 76 16 7 has the poorest corrosion resistance. From this work, corrosion rates of up to 0.5 mm/a can be expected in 10% H_2SO_4 at room temperature, and in excess of 1.27 mm/a at higher concentrations. Corrosion rates of between 0.5 and 1.27 mm/a are quoted for 80% H_2SO_4. In air-free acid, this alloy corresponds to nickel in its corrosion behavior [397].

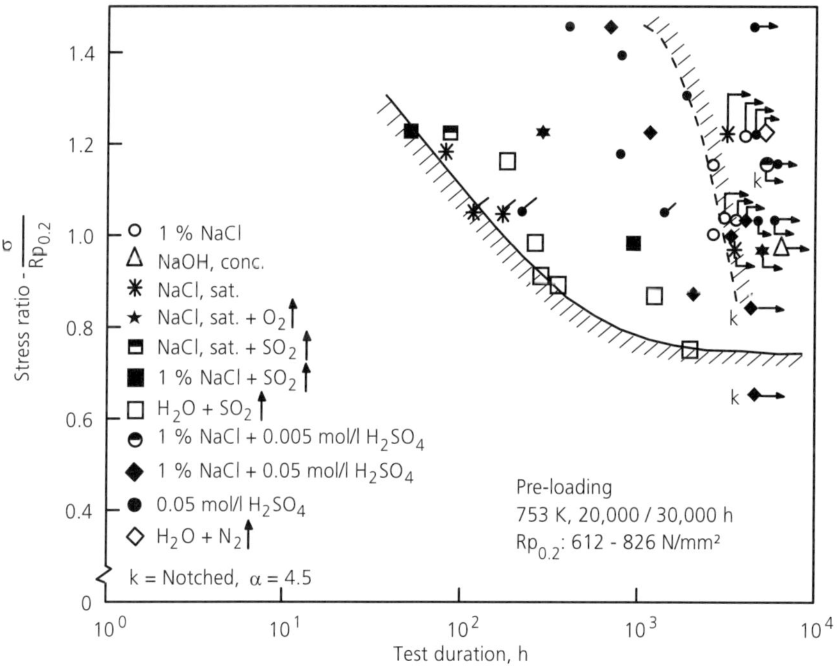

Figure 93: Stress corrosion cracking behavior of the heat-resistant screw alloy NiCr20TiAl in various solutions (simulated turbine condensates) at 90°C (occasionally 80°C) [415, 416]

The corrosion rate of this type of alloy in sulfuric acid increases as the carbon content and sulfur content increase [417]. According to [395], this alloy, with the commercial name Inconel® 600 (2.4816, NiCr15Fe), has only "moderate" corrosion resistance in sulfuric acid. Corrosion rates in excess of 1.22 mm/a are quoted for the concentration range between 15 and 70% H_2SO_4 at room temperature [418]. Inconel® 600 is used in the chemical industry chiefly because of its excellent heat resistance.

Figure 94 gives a summary of the corrosion resistance of Inconel® 600 in sulfuric acid. It is to be assumed from this graph that the acid is non-aerated, and may cause a higher corrosion rate on aeration and does not allow passivation.

In boiling 10% sulfuric acid, a corrosion rate of 5.6 mm/a is to be expected [419], which agrees with Figure 94. NiCrFe alloys are not susceptible to stress corrosion cracking [420].

As already mentioned, these pure NiCrFe-alloys can be passivated by the chromium content. Molybdenum in such alloys increases their resistance by reducing the corrosion rate in the active state, as does copper.

On the basis of this fact, a number of CrNiFe-alloys with Mo- and Cu-additions were developed. These have a broader application range in sulfuric acid (see also Figure 90) and indeed are used more often. These include, for example, the Cu-free alloy Hastelloy® F with a Mo-addition. Its isocorrosion curves are shown in Figure 95. Whereas Inconel® 600 has approximately the same corrosion resistance in sulfuric acid under reducing conditions as nickel, Hastelloy® F has significantly higher resistance (Figure 89, Figure 94, Figure 95). More iron, less molybdenum and the addition of copper are the characteristics of the alloy Incoloy® 825 (2.4858, NiCr21-Mo) containing about 42% Ni, 21% Cr, 30% Fe, 3% Mo and 1.8% Cu. Figure 90 shows the possible uses in sulfuric acid. This alloy group also includes the alloy Ni-o-nel® 825 [421].

In comparison with the CrNiMo-steels a considerable gain in corrosion resistance is evident. The application range in sulfuric acid can be described as being for all concentrations up to 65°C, up to 60% at 80°C and up to 40% at the boiling point [422]. These alloys have good resistance in sulfuric acid under reducing conditions like Monel® 400, and, due to the Cr-content, also under oxidizing conditions.

Inconel® 825 is used for heating coils in media containing sulfuric acid [422–425].

Figure 96 shows an overview of the corrosion rates expected in pure sulfuric acid [426–428].

NiCrFeMo-alloys with Ni-contents of up to 50% are less resistant in sulfuric acid than alloys with a higher Mo-content because of their low molybdenum content, this applies in particular at higher temperatures. They represent approximately the transition from steels to higher-alloy nickel alloys. The alloy Hastelloy® G (c.f. NiCr22Mo6Cu, 2.4618) containing 45% Ni, 22% Cr, 20% Fe, 6.5% Mo and 2.0% Cu, the isocorrosion curves of which are shown in Figure 99, has slightly better corrosion resistance in sulfuric acid, like copper-containing steels.

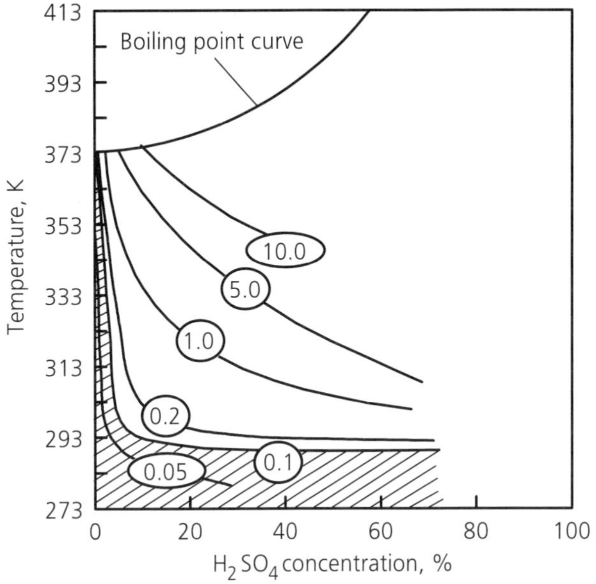

Figure 94: Isocorrosion curves (mm/a) of Inconel® 600 ((%) 76Ni-15Cr-9Fe-0.2C) in sulfuric acid [404]

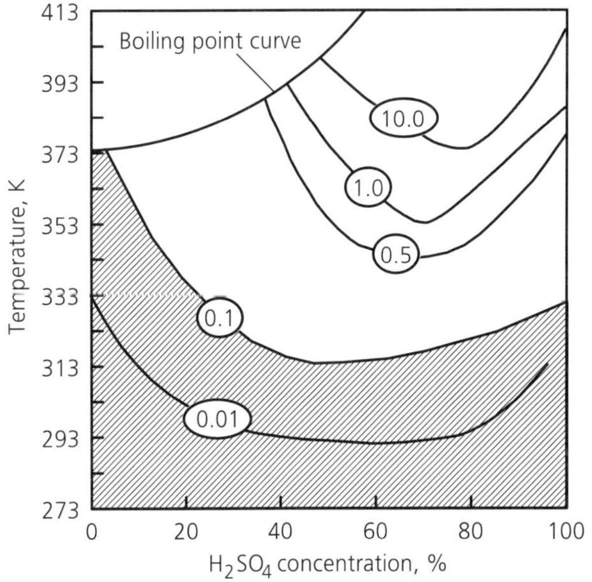

Figure 95: Isocorrosion curves (mm/a) of Hastelloy® F ((%) 44Ni-< 30Fe-22Cr-6.5Mo-2Mn-1Si-0.1C) in sulfuric acid [404]

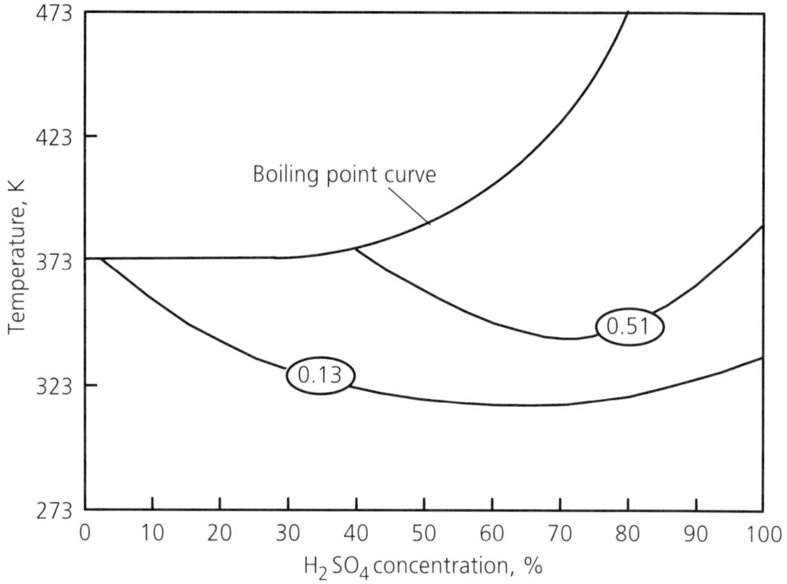

Figure 96: Isocorrosion curves (mm/a) of Incoloy® 825 (2.4858) in sulfuric acid [426, 427, 428]

Incoloy® 825 (NiCr21Mo) is more suitable for screws in sulfuric acid than Inconel® 600 (NiCr15Fe). Corrosion rates of less than 0.12 mm/a, and for NiCr15Fe more than 1.22 mm/a, are quoted for the concentration range 15 to 75% H_2SO_4 at room temperature [418].

The nickel based alloy Carpenter 20 Cb 3® with the addition of niobium avoids some of the earlier alloy failures by avoiding or preventing carbide precipitation. The level of nickel is also highly beneficial to improving the resistance to stress corrosion cracking as well as enhancing the resistance to boiling sulfuric acid solutions, Figure 97 demonstrates this point [429]. However, attack can occur in Carpenter 20 Cb 3 in the region of 65 to 74%. This occurs by preferential attack of certain phases. The addition of oxygen greatly assists in the resistance to this attack. The small amount of Cu in both alloy 825 and 20Cb-3 help in their resistance to corrosion in sulfuric acid.

Alloy 693 (UNS N06693), is a nickel-based with about 30% chromium, 4% iron and 3% aluminum. With its high content of chromium, alloy 693 offers excellent resistance to oxidizing environments. Figure 98 is a comparison of recently developed corrosion rates for alloy 693 compared to other commercially significant alloys in 95 to 98% sulfuric acid at 80 to 130°C [430]. As a result of its high content of chromium and aluminum, alloy 693 exhibits excellent corrosion resistance to hot concentrated sulfuric acid environments relative to other commonly selected nickel alloys such as Alloys 59 and 686.

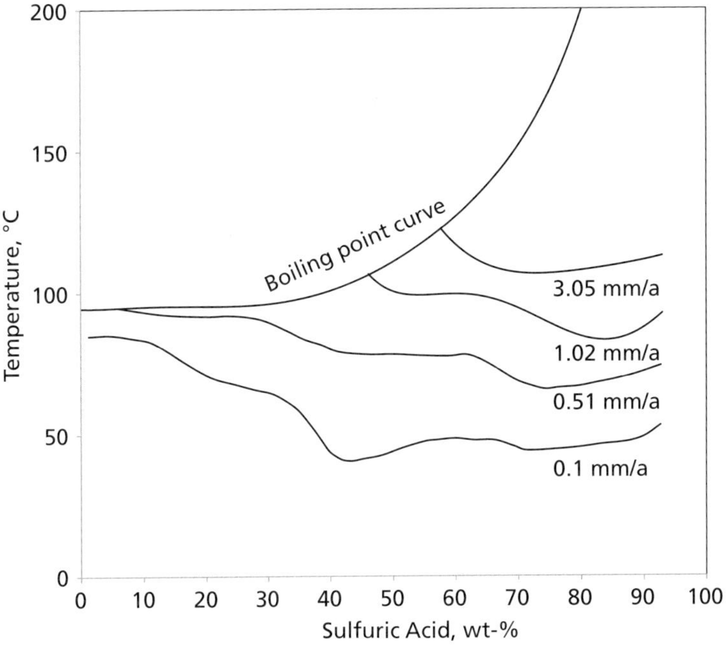

Figure 97: Isocorrosion curve for Carpenter 20 Cb 3 in sulfuric acid [429]

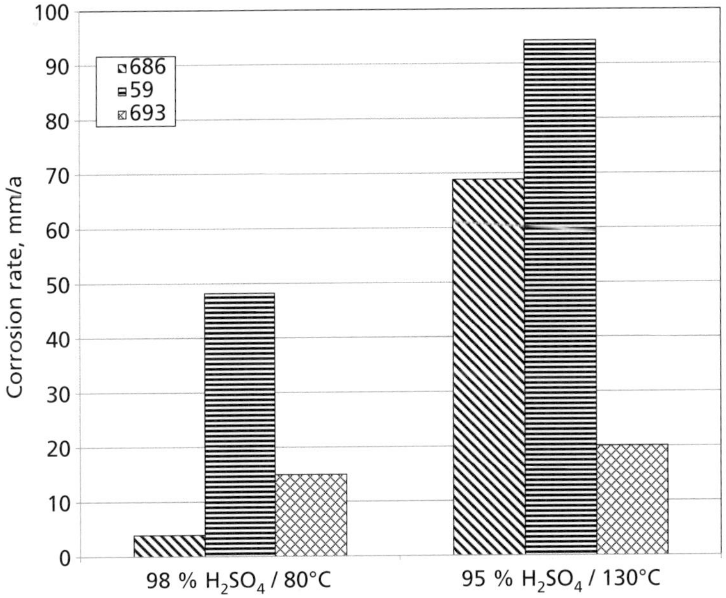

Figure 98: Comparison of corrosion rates for Alloys 686, 59 and 693 in concentrated sulfuric acid [430]

Accelerated corrosion of the cathodes commonly used in anodically protected H_2SO_4 coolers – UNS N06200 (Hastelloy® C-276) and UNS N10665 (Hastelloy® B-2) – can occur under cathodic polarization [431]. Corrosion occurs over a limited range of current densities, which are larger than those typically required to establish and maintain protection of the anodic structure during normal operation. The likelihood of corrosion under cathodic polarization may be reduced by restricting the use of Hastelloy® C-276 to drying acid environments (93 to 94 wt% H_2SO_4) and the use of Hastelloy® B-2 to absorbing acid environments (98 to 99 wt% H_2SO_4).

The corrosion behavior of NiCr21Mo in spent acids can be seen from Table 123 in comparison with the CrNi-steels 1.4439, 1.4539 and 1.4503. NiCr21Mo alloy is inferior to the steels mentioned in spent acids containing between 5 and 20% H_2SO_4 in the temperature range around 50°C, but at 100°C proves to be more corrosion-resistant; in fact the corrosion increases only in 20% H_2SO_4 + 3,000 ppm Cl^-. It is also no longer usable at temperatures of 150 and 202°C. Corrosion rates in excess of 10 mm/a are to be expected at 200°C. As with the steels listed in Table 123, Cu-ions have an inhibiting effect in sulfuric acid, this being the case particularly in 20% sulfuric acid at 100°C. In 50% sulfuric acid, the corrosion rates of 0.26 mm/a are still undoubtedly acceptable under certain conditions.

Solution No.	Composition	DIN-Mat. No.*	Temperature, °C 50 Material consumption rate, $g/m^2\,d$ (Corrosion rate, mm/a)		100		150			
1	5% H_2SO_4 + 500 ppm Cl^-	1.4571	0.03/ 12.0	(<0.01/ 0.55)	109.0	(4.97)	Cl^-		Cl^-	
		1.4439	0.03	(<0.01)	83.5	(3.81)	without	with	without	with
		1.4539	0.07	(<0.01)	6.7	(0.30)	207.3	80.8	(9.5)	(3.)
		1.4503	0.6/1.6	(0.03/ 0.07)	4.2	(0.19)	167.4	80.7	(7.6)	(3.7)
		2.4858	2.6	(0.12)	3.6	(0.15)				

* 1.4571 (X6CrNiMoTi17-12-2)
1.4439 (X2CrNiMoN17-13-5)
1.4539 (X1NiCrMoCu25-20-5)
1.4503 (X3NiCrCuMoTi27-23)
2.4858 (NiCr21Mo)

Table 123: Comparison of the corrosion rates in static dilute H_2SO_4 in air with and without impurities (test duration 21 days) [434]

Table 123: Continued

Solution No.	*Composition*	*DIN-Mat. No.**	*Temperature, °C*					
			50		*100*		*150*	
			Material consumption rate, g/m² d (Corrosion rate, mm/a)					
2	5% H_2SO_4 + 3,000 ppm Cl^-	1.4571	16.3	(0.74)	–	–		
		1.4439	0.03	(< 0.01)	88.5	(4.04)		
		1.4539	3.2	(0.15)	6.8	(0.31)		
		1.4503	3.4	(0.15)	5.8	(0.26)		
		2.4858	3.5	(0.16)	6.6	(0.27)		
3	10% H_2SO_4	1.4571	6.8	(0.30)	154.	(7.02)		
		1.4439	0.03	(< 0.01)	105.3	(4.80)		
		1.4539	0.06	(< 0.01)	11.5	(0.52)		
		1.4503	0.3	(0.01)	9.5	(0.43)		
		2.4858	2.2	(0.10)	3.4	(0.14)		
4	10% H_2SO_4 + 500 ppm Cl^-	1.4571	20.0	(0.91)	–	–		
		1.4439	0.01/5.1	(< 0.01/ 0.23)	–	–		
		1.4539	0.84	(0.04)	7.4	(0.34)		
		1.4503	1.4	(0.06)	2.9	(0.13)		
		2.4858	3.0	(0.14)	2.9	(0.12)		
5	20% H_2SO_4	1.4571	27.9	(1.27)	–	–		
		1.4439	4.4	(0.2)	–	(>> 10)		
		1.4539	0.23	(0.01)	52.7	(2.40)	763.5	(34.8)
		1.4503	0.8	(0.04)	27.5	(1.25)	871.9	(39.8)
		2.4858	2.6	(0.12)	6.5	(0.27)	1,951.9	(89)

* 1.4571 (X6CrNiMoTi17-12-2)
1.4439 (X2CrNiMoN17-13-5)
1.4539 (X1NiCrMoCu25-20-5)
1.4503 (X3NiCrCuMoTi27-23)
2.4858 (NiCr21Mo)

Table 123: Comparison of the corrosion rates in static dilute H_2SO_4 in air with and without impurities (test duration 21 days) [434]

Table 123: Continued

Solution No.	Composition	DIN-Mat. No.*	Temperature, °C					
			50		100		150	
			Material consumption rate, g/m^2 d (Corrosion rate, mm/a)					
6	20% H_2SO_4 + 500 ppm Cl^-	1.4571	90.2	(4.12)	–	–		
		1.4439	0.03/ 10.5	(< 0.01/ 0.47)	–	(>> 10)		
		1.4539	2.0	(0.09)	4.5	(0.21)		
		1.4503	1.6/2.4	(0.07/ 0.11)	4.3	(0.20)		
		2.4858	2.6	(0.12)	3.5	(0.14)		
7	20% H_2SO_4 + 3,000 ppm Cl^-	1.4571	19.8	(0.9)	–	–		
		1.4439	0.03/ 15.0	(< 0.01/ 0.68)	–	–		
		1.4539	3.8	(0.17)	15.7	(0.72)		
		1.4503	3.5	(0.16)	10.2	(0.47)		
		2.4858	2.8	(0.13)	10.4	(0.43)		
8	20% H_2SO_4 + 2,000 ppm Cl^-	1.4571	0.04	(< 0.01)	0.4	(0.018)		
		1.4439	0.02	(< 0.01)	0.5	(0.023)		
		1.4539	0.03	(< 0.01)	0.4	(0.018)		
		1.4503	0.03	(< 0.01)	0.3	(0.012)		
		2.4858	0.4	(0.02)	0.8	(0.033)		

* 1.4571 (X6CrNiMoTi17-12-2)
1.4439 (X2CrNiMoN17-13-5)
1.4539 (X1NiCrMoCu25-20-5)
1.4503 (X3NiCrCuMoTi27-23)
2.4858 (NiCr21Mo)

Table 123: Comparison of the corrosion rates in static dilute H_2SO_4 in air with and without impurities (test duration 21 days) [434]

Table 123: Continued

Solution No.	*Compo-sition*	*DIN-Mat. No.**	*Temperature, °C*					
			50		*100*		*150*	
			Material consumption rate, g/m² d (Corrosion rate, mm/a)					
9	50% H_2SO_4	1.4571		(>> 10)	–	–		
		1.4439	175	(8.0)	–	–		
		1.4539	0.99	(0.04)	38.5	(1.72)		
		1.4503	1.0	(0.05)	10.8	(0.49)		
		2.4858	1.0	(0.04)	5.3	(0.26)		

* 1.4571 (X6CrNiMoTi17-12-2)
1.4439 (X2CrNiMoN17-13-5)
1.4539 (X1NiCrMoCu25-20-5)
1.4503 (X3NiCrCuMoTi27-23)
2.4858 (NiCr21Mo)

Table 123: Comparison of the corrosion rates in static dilute H_2SO_4 in air with and without impurities (test duration 21 days) [434]

In addition to these studies, electrochemical experiments were carried out in these spent acids in comparison with high-alloy steels. The results show that at higher temperatures the alloy NiCr21Mo is superior to the steels investigated. Since under these conditions the repassivation potential of pitting corrosion is considerably more positive than the free corrosion potential, no pitting corrosion is expected [432, 433].

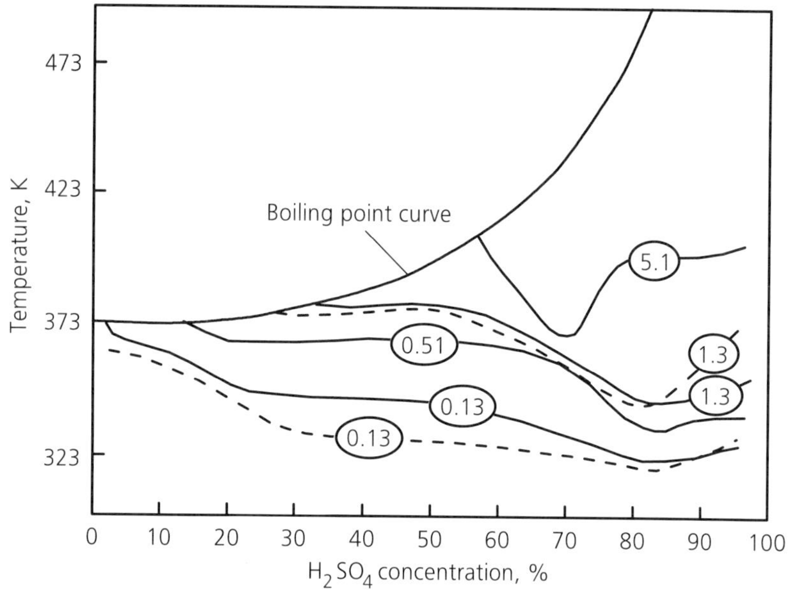

Figure 99: Isocorrosion curves (mm/a) for Hastelloy® G ((%) 45Ni-22Cr-20Fe-6.5Mo-2Cu) in sulfuric acid [434]
——— H_2SO_4
- - - - H_2SO_4 + 200 ppm Cl^- in form of NaCl

Compare also Figure 100 for possible uses of nickel alloys in boiling dilute sulfuric acid. The results originate from accelerated experiments with 4 periods of 24 h. The possible uses of high-alloy materials in hot concentrated sulfuric acid can be seen from Figure 101, in comparison with steels. These graphs show that in 95% hot H_2SO_4 at 100 and 150°C the alloy 2.4858 has the lowest resistance of the materials shown. The alloy 2.4642 (NiCr29Fe, c.f. UNS N06690) proves to have relatively good resistance in the entire temperature range investigated with corrosion rates below 0.35 mm/a [435]. Steels are a more suitable choice of material under these conditions, anodic protection ensuring corrosion resistance. See [435] for the effect of alloying elements under these conditions.

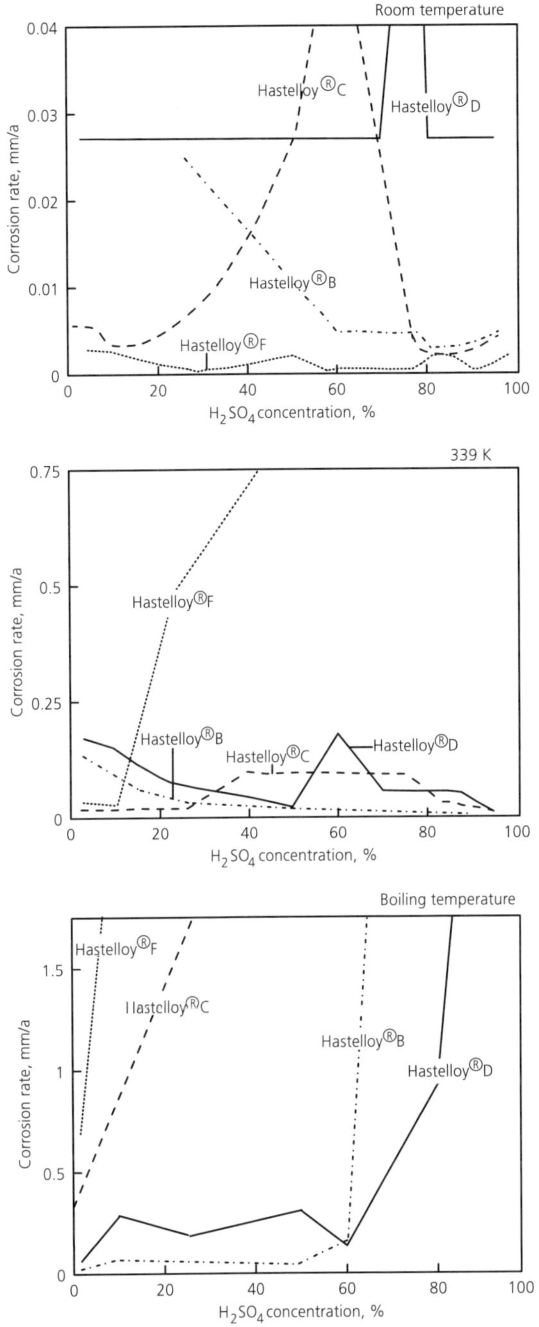

Figure 100: Corrosion behavior of Ni-alloys in boiling (probably analytically pure) sulfuric acid (test duration 4 × 24 h periods) [427, p. 64]
① Inconel® 625 (2.4856)
② Hastelloy® C-22 (2.4602, (%) Ni-22Cr-13Mo-3W-3Fe-max. 0.010C)
③ Hastelloy® C-276 (2.4819)
④ Hastelloy® G-30 (2.4603, (%) Ni-29.5Cr-15Fe-5Mo-<5Co < 0.03C)
⑤ Hastelloy® G-3 (c.f. 2.4619)
⑥ Hastelloy® B-2 (2.4617)

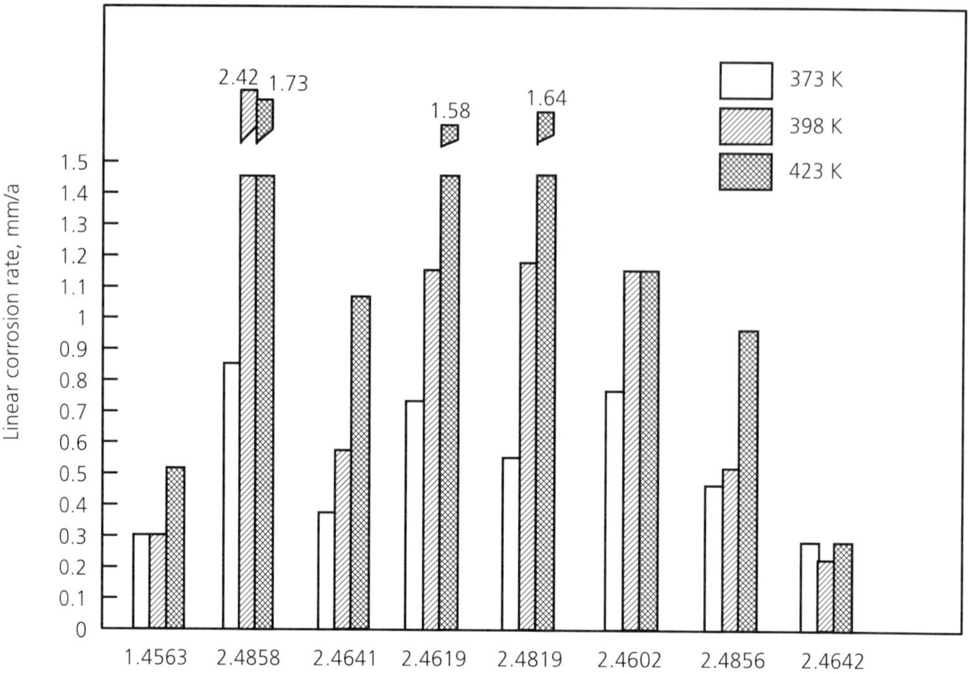

Figure 101: Comparison of the corrosion resistance of nickel alloys with a CrNi-steel in static 95% sulfuric acid at various temperatures (test duration 21 d) [435]

– Pitting corrosion in sulfuric acid solutions –

Table 124 summarizes the pitting corrosion behavior of some metallic materials in a test solution of 0.01 mol/l H_2SO_4 + 0.1% $Fe_2(SO_4)_3$ + 1,000 ppm Cl^-. According to ASTM immersion test procedures. The experimental procedure is documented elsewhere [588].

According to this compilation, the pitting and crevice corrosion temperatures of Incoloy® 825 are about the same or slightly worse than those of the steel SAE 317L [436]. Hastelloy® G exhibits better pitting and crevice corrosion behavior [437]. The use of Inconel® 625 rather than Incoloy® 825 is recommended particularly at high temperatures under oxidizing conditions, where pitting corrosion is increasingly likely [406].

The corrosion resistance of the alloys Hastelloy® G-3 and Hastelloy® G-30 in sulfuric acid is improved by 1% $FeCl_3$, as can be seen from the isocorrosion curves in Figure 102 and Figure 103. The acid causes pitting corrosion in the concentration range up to 50% H_2SO_4 at temperatures above 60°C, and to a considerably greater degree than in the case of the alloy Inconel® 625 (Figure 104), as a consequence of the increasing Mo-content (in this context compare also Figure 105) [427]. The results were obtained in accelerated corrosion tests.

Material	*Pitting corrosion Temperature, °C*	*Crevice corrosion Temperature, °C*
SAE 304L (X2CrNi19-11)	15	10
SAE 316L (X2CrNiMo17-12-2)	30	20
SAE 317L (X2CrNiMo18-15-4)	25	25
Alloy 825 (NiCr21Mo)	25	20
20Cb-3 Alloy (NiCr20CuMo, c.f. 2.4660)	30	20
SAE 317LM (X 2 CrNiMoN 17 13 5)	35	30
3RE60 (X2CrNiMoSi18-5-3)	40	30
SAE 904L (X1NiCrMoCu25-20-5)	55	45
Ferralium® 255 (X2CrNiMoCuN25-6-3)	55	55

Table 124: Pitting and crevice corrosion temperatures, determined in 0.01 mol/l H_2SO_4 + 0.1% $Fe_2(SO_4)_3$ + 1,000 ppm Cl^- [436]

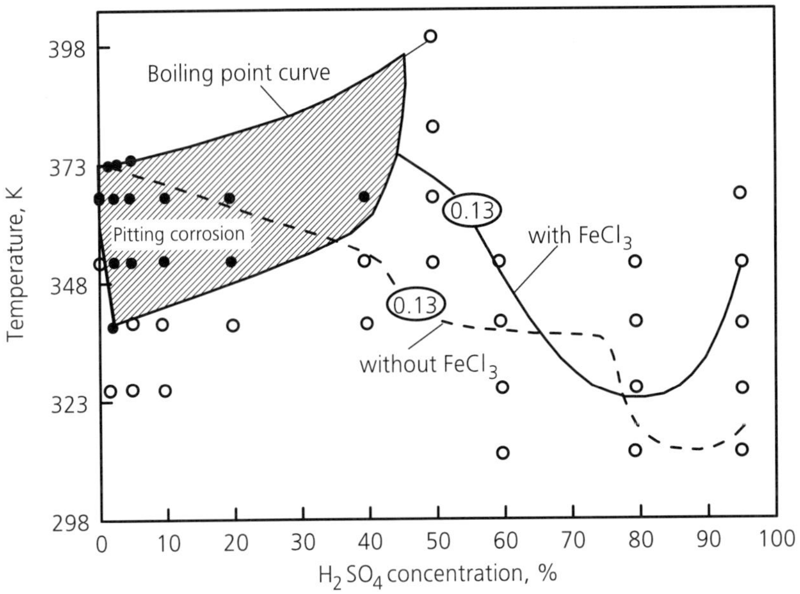

Figure 102: Influence of the addition of 1% $FeCl_3$ to sulfuric acid (analytically pure) on the corrosion behavior of Hastelloy® G-3 (test duration 24 h) [427, p. 64, 458]

● pitting corrosion

○ general corrosion

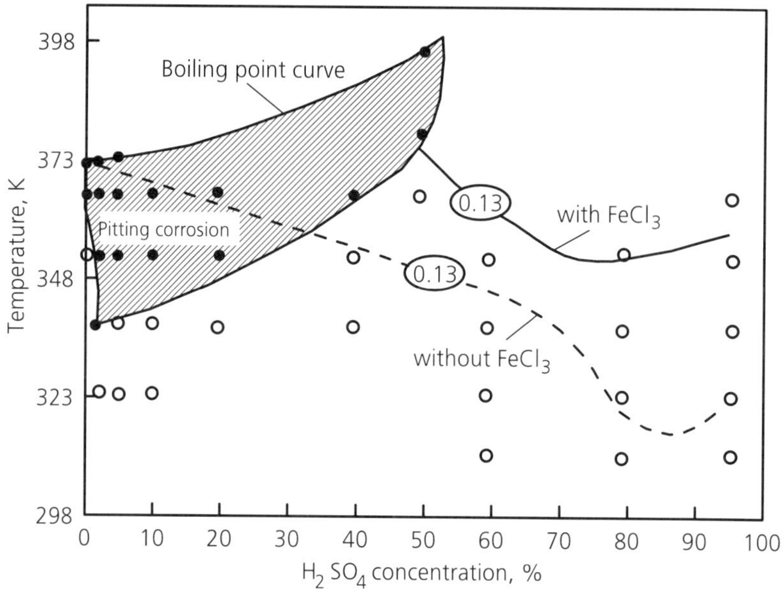

Figure 103: Influence of the addition of 1% $FeCl_3$ to sulfuric acid (analytically pure) on the corrosion behavior of Hastelloy® G-30 (test duration 24 h) [427, p. 65]

● pitting corrosion

○ general corrosion

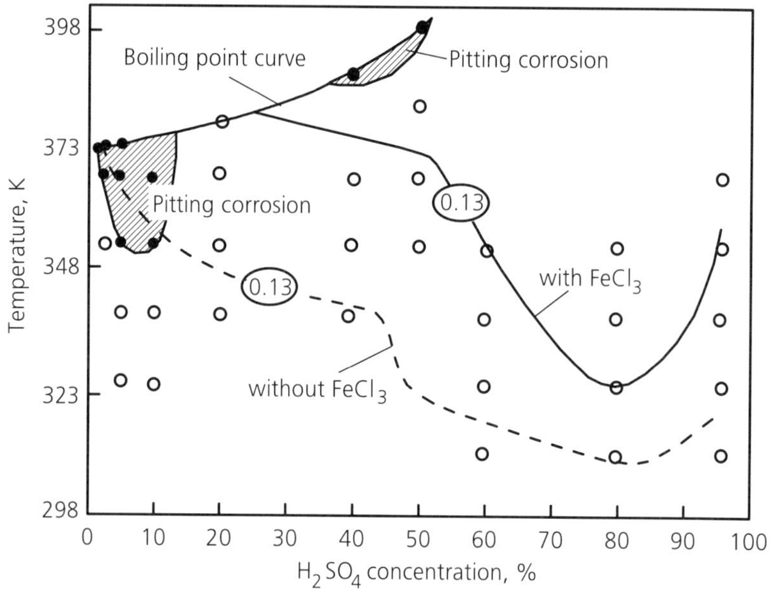

Figure 104: Influence of the addition of 1% $FeCl_3$ to sulfuric acid (analytically pure) on the corrosion behavior of Inconel® 625 (test duration 24 h) [427, p. 65]

● pitting corrosion

○ general corrosion

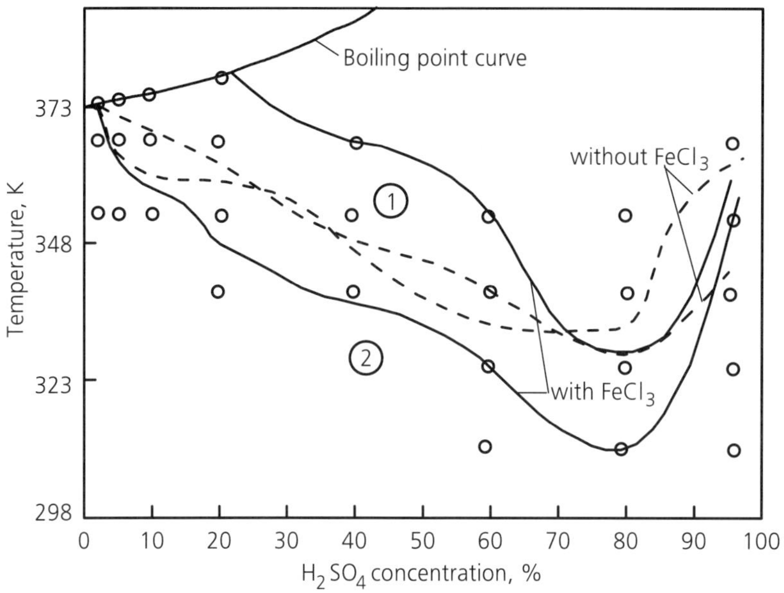

Figure 105: Influence of the addition of 1% $FeCl_3$ to sulfuric acid (analytically pure) on the corrosion behavior of Hastelloy® C-22 and Hastelloy® C-276 (test duration 24 h) [427, p. 65]
○ general corrosion

– Electrochemical corrosion protection processes –

The materials listed in Figure 101 exhibit stable-passive behavior in highly concentrated sulfuric acid at 60°C. As the temperature of the acid increases, this behavior changes once a critical temperature is reached. Variations in potential are observed at and above this temperature, indicating a metastable state, that is to say an active-passive state, which leads to increased corrosion of the materials. The corrosion rates are reduced considerably and practically to zero by applying an anodic protection potential.

Nickel-chromium-molybdenum alloys

This group mainly includes alloys containing more than 50% Ni and about 17% Cr and Mo. Their good corrosion resistance is based on the high Cr-content and the consequently high resistance under oxidizing conditions, and on the high Mo-content, with increased basic resistance as a consequence of lower weight losses in the active range.

NiCrMo alloys can be used in a wide concentration range of sulfuric acid, but from the point of view of the alloy composition, their main use should predominate under oxidizing conditions. They can also be used under reducing conditions due to their Mo-content. In contrast, the Cr-free alloys cannot be passivated and therefore are used chiefly under reducing conditions; although this need not rule out their use under oxidizing conditions at a sufficiently low rate of dissolution [438] .

Figure 106 depicts the isocorrosion lines for the newly developed NiCrMo-alloy UNS N06058 (Ni-21% Cr-19% Mo) in slightly aerated industrial grade sulfuric acid, determined in immersion tests over at least 120 hours [439]. The 0.1 mm/a line is running well above 80°C across nearly the entire concentration range while the 0.5 mm/a line is running clearly above 100°C across the whole concentration range. In further testing in a 90% H_2SO_4 solution at 90 and 105°C the corrosion rate of the alloy was ≤ 0.1 mm/a.

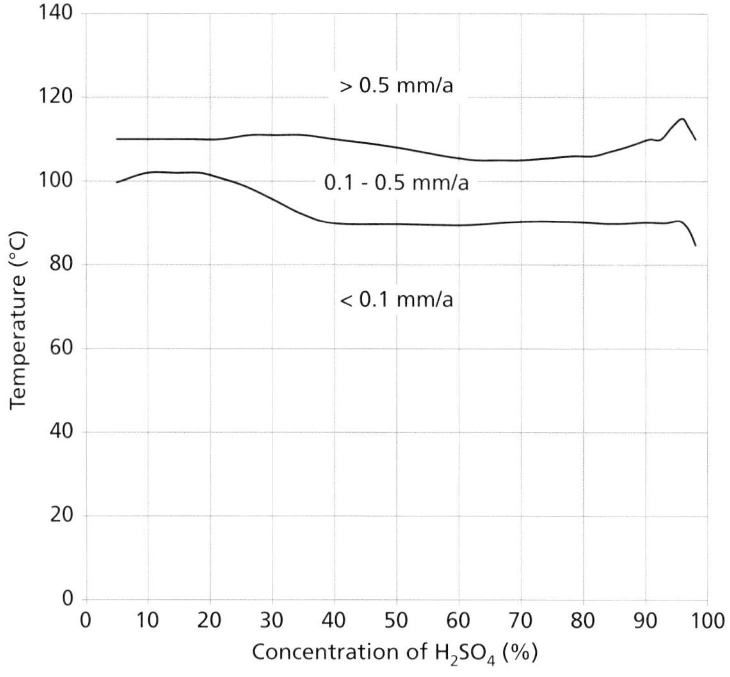

Figure 106: Isocorrosion diagram of alloy UNS N06058 in slightly aerated industrial grade sulfuric acid, determined in immersion tests over at least 120 h [439].

A new wrought nickel alloy which is named Hastelloy® HYBRID-BC1 has been shown to be much more resistant to sulfuric acid than other Ni-Cr-Mo materials [440]. The corrosion behavior of the new alloy in 10 to 90% sulfuric acid at 93°C, relative to the Ni-Cr-Mo materials was shown to be excellent, as seen in Figure 107. In some cases, the differences are greater than an order of magnitude in favor of the new alloy, although N06200 provides lower rates than the new material at concentrations less than 40 wt-%. The fact that N06200 provides lower corrosion rates could be related to the presence of a small amount of Cu in N06200.

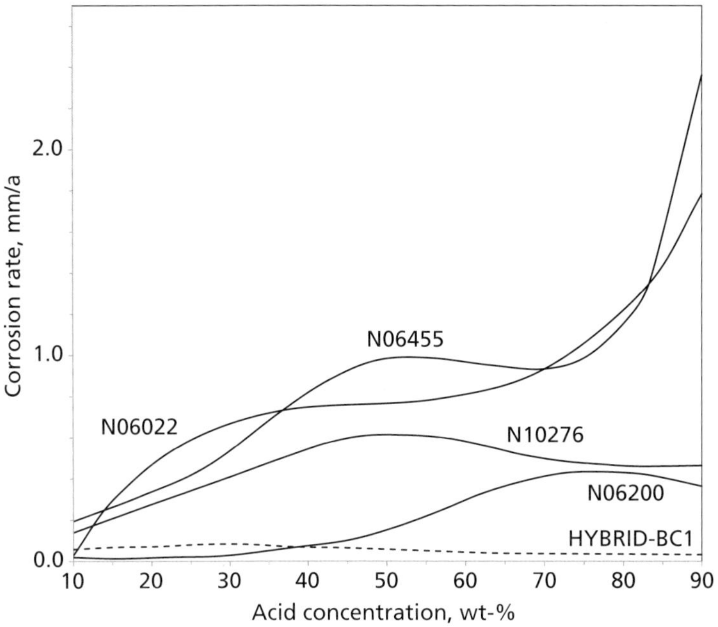

Figure 107: Corrosion rates in sulfuric acid at 93°C [440]

The influence of temperature on the corrosion rates of alloy 31 and alloy 59 is shown in Figure 108. While for alloy 31 small temperature increases can cause a jump in the corrosion rate, for alloy 59 the corrosion rate also increases, but more moderately. It is known that high nickel and molybdenum contents reduce the rate of active current density giving rise to lower corrosion rates, as seen for alloy 59 [441].

UNS N06110 is a Ni-Cr-Mo alloy developed for service in highly corrosive environments. It has a higher chromium content than the numerous other "C-type" alloys that have been developed. The corrosion resistant properties of N06110 were examined in [442]. Experiments were conducted to determine the corrosion rate of N06110 in sulfuric acid solutions of various concentrations from 0 to 96% by weight, at temperatures from 65 to 100°C. The results are shown in Figure 109.

It was reported in [443] that nickel and molybdenum have a beneficial effect on the corrosion resistance at 90–91 weight-% sulfuric acid; the higher the alloying content the lower the corrosion rate regardless of the chromium content. However, when the sulfuric acid concentration exceeded 98% alloys with increased nickel and molybdenum contents had an increased corrosion rate.

Figure 110 shows the corrosion behavior of cast nickel alloys in sulfuric acid solutions of various concentrations and at various temperatures [444]. Substantial differences from the corresponding wrought alloys can be excluded.

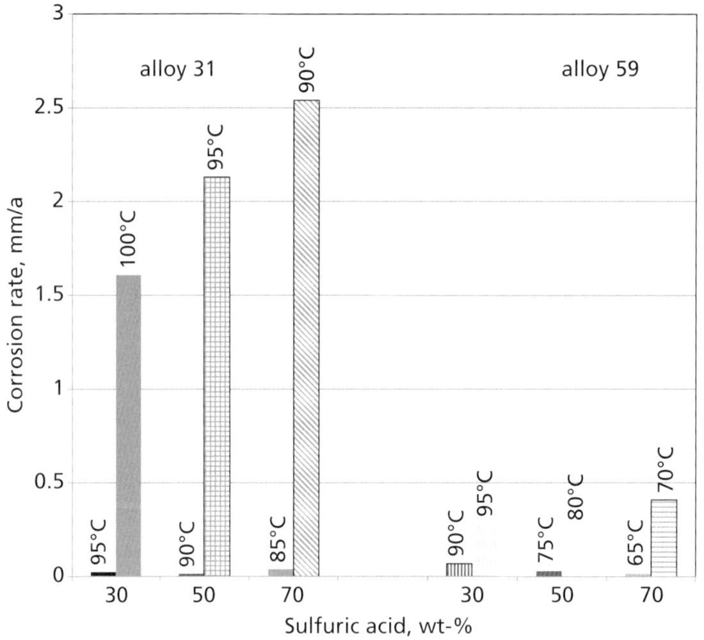

Figure 108: Corrosion rates of alloy 31 and alloy 59 in aerated technical grade sulfuric acid in 24 h immersion test [441]

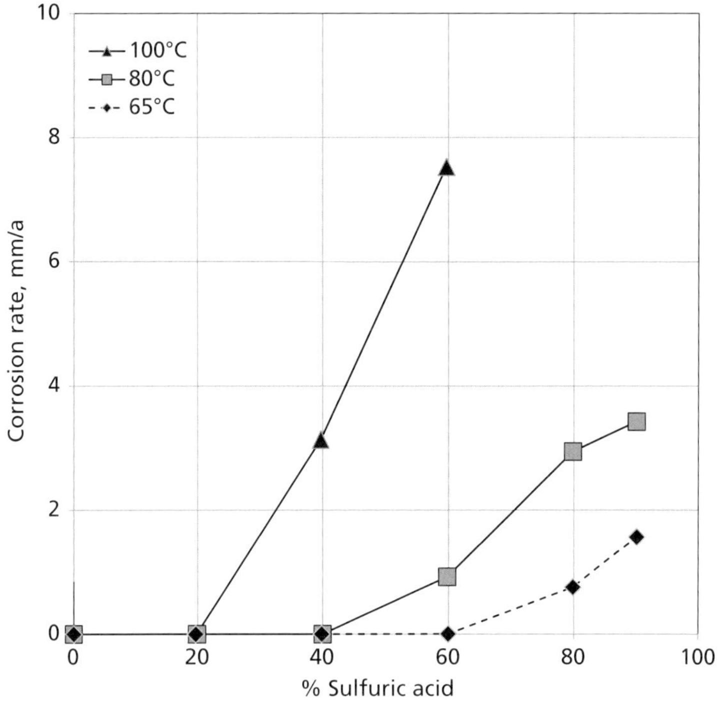

Figure 109: Corrosion of N06110 alloy in sulfuric acid [442]

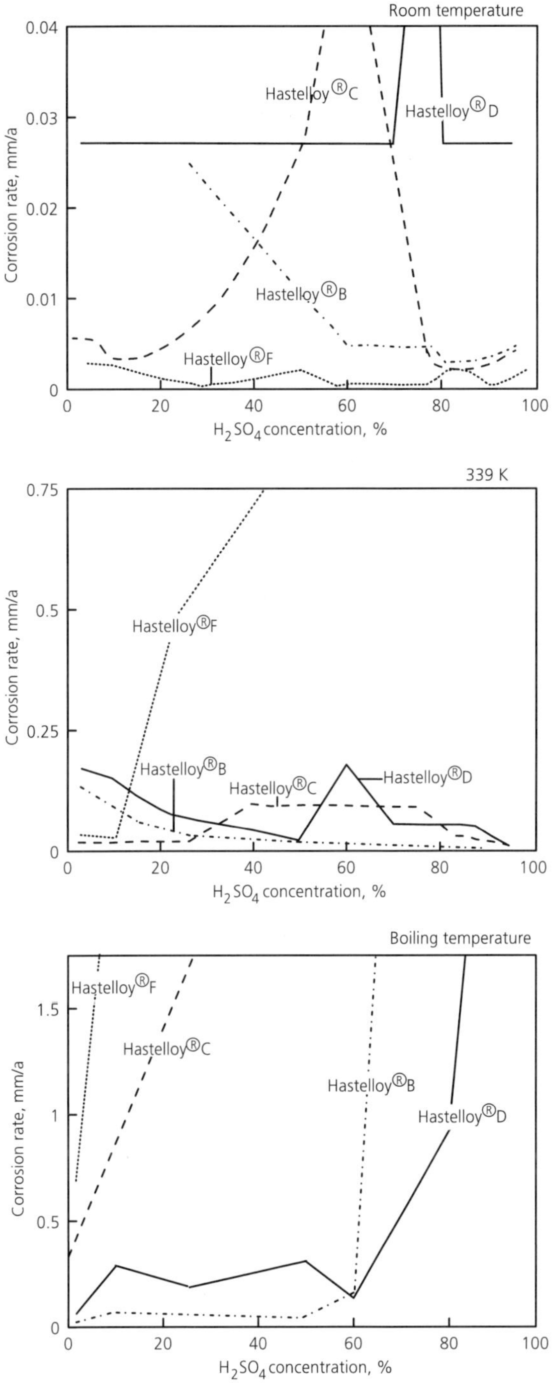

Figure 110: Corrosion behavior of Ni-alloys in sulfuric acid as a function of the temperature and concentration [444]
Hastelloy® B (c.f. G-NiMo 30)
Hastelloy® C (c.f. G-NiMo 16 Cr)
Hastelloy® D (NiSi 10 Cu)
Hastelloy® F (c.f. NiCr 22 Mo)

In this group of NiCrMo-alloys, there is a large number of commercial names with, in some cases, only slight deviations in the alloy composition. The alloy type Hastelloy® C is of particular interest for chemical engineering because it is widely usable. Its further development in terms of chemical resistance has recently aroused particular interest. Some of these alloys and their compositions are listed in Table 125 [445].

Alloy	Chemical composition, %						
	Ni	Cr	Mo	W	Fe	Nb	C
Hastelloy® C-22	balance	22	13	3	3	–	≤ 0.010
Hastelloy® C-276	balance	16	16	4	5	–	≤ 0.010
Hastelloy® C-4	balance	16	16	–	≤ 3	–	≤ 0.010
Inconel® 625	balance	21.5	9	–	≤ 5	3.5	≤ 0.10
Remanit® HC	balance	20	15.5	–	≤ 2	–	≤ 0.015

Table 125: Chemical compositions of some NiCrMo-alloys [445]

Components made of the original alloy Hastelloy® C had to be solution-annealed after welding to avoid intercrystalline corrosion. After the significance of low C- and Si-contents was recognized, the alloy Hastelloy® C-276 and a similar Remanit® HC appeared on the market. They also met the requirements for adequate structural stability. A subsequent step in the development of these alloys was Hastelloy® C-4, which, by reducing the contents of Fe, W, Co and V, has a lower susceptibility in the heat-affected zone of the weld seam than C-276, but a higher tendency to undergo pitting and crevice corrosion and more severe general corrosion than the latter. The development of C-22 provided an alloy which combines the advantages of the other alloys; the thermal structural stability of C-4 and the pitting and crevice corrosion resistance of C-276. Table 126 allows a comparison of the resistances of these alloys in solutions containing sulfuric acid [445].

According to this work, the critical pitting corrosion temperatures in a test solution of (percent by volume) 7 H_2SO_4 + 3 HCl + 1% $FeCl_3$ + 1% $CuCl_2$ increase in the sequence Inconel® 625 ((75°C), Hastelloy® C-4 (90°C), C-276 (110°C) and C-22 (120°C). Above these temperatures, pitting corrosion of the particular alloy is to be expected.

Isocorrosion curves for various alloys are shown below in Figures 111–115. The number of graphs was chosen consciously to demonstrate that the alloys exhibit similar corrosion behaviors in sulfuric acid. No differences are found until higher concentrations and temperatures are reached (Figures 112, 114 and 115). In most cases, the use of Hastelloy® C-22 is appropriate, since this alloy is better qualified from the material point of view [445]. Figure 116 shows the influence of heat treatment on these alloys. As the purity increases, selective corrosive attack, in the form of higher corrosion rates, is greatly suppressed [445–447]. As expected, chloride additions to sulfuric acid reduce the corrosion resistance, especially in the case of alloys which can be passivated (Figure 113).

Medium	Test temperature °C	Hastelloy® C-22	Hastelloy® C-4	Hastelloy® C-276	Inconel® 625
		Corrosion rate, mm/a			
Non-oxidizing					
10% H_2SO_4	boiling temperature	0.28	0.79	0.58	1.17
40% H_2SO_4	79	0.16	0.38	0.25	0.89
10% H_2SO_4 + 1% HCl	70	0.01	0.61	0.28	3.07
25% H_2SO_4 + 200 ppm Cl^-	70	0.28	0.94	0.30	2.79
Oxidizing					
25% H_2SO_4 + 5% HNO_3 + 4% NaCl	boiling temperature	0.03	2.46	1.63	18.1
23% H_2SO_4 + 1.2% HCl + 1% $FeCl_3$ + 1% $CuCl_2$	boiling temperature	0.17	58.3	1.40	97.7

Table 126: Comparison of the corrosion behavior of NiCrMo-alloys in sulfuric acid solutions [445]

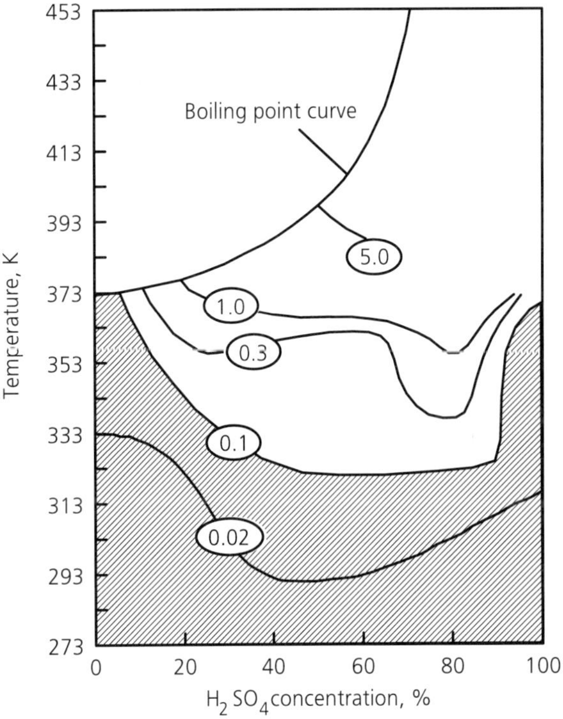

Figure 111: Isocorrosion curves (mm/a) of Hastelloy® C ((%) 54Ni-14–16Cr-15–17Mo-3–4.5W-4–7Fe-0.08C) in sulfuric acid [404, 448]

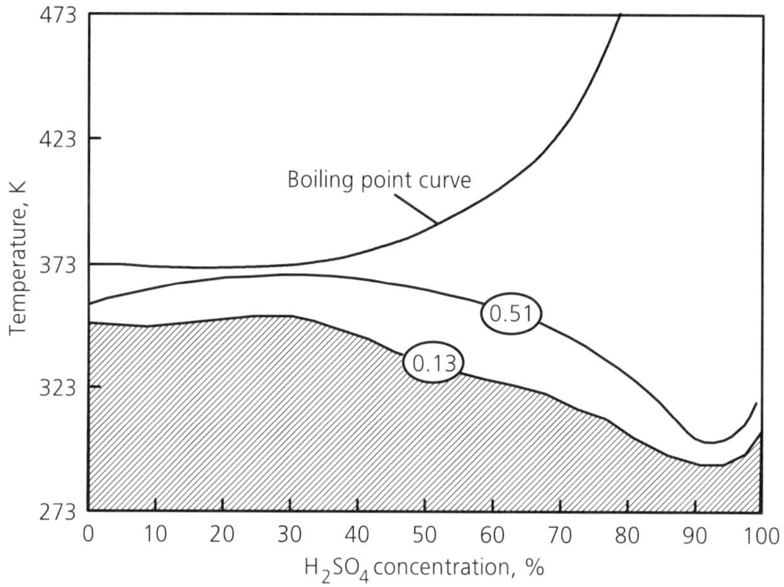

Figure 112: Isocorrosion curves (mm/a) of Inconel® 625 (2.4856) ((%) Ni-21.5Cr-9Mo-5Fe-3.5Nb(Cb)-max.0.10C) in sulfuric acid [425–427, 449]

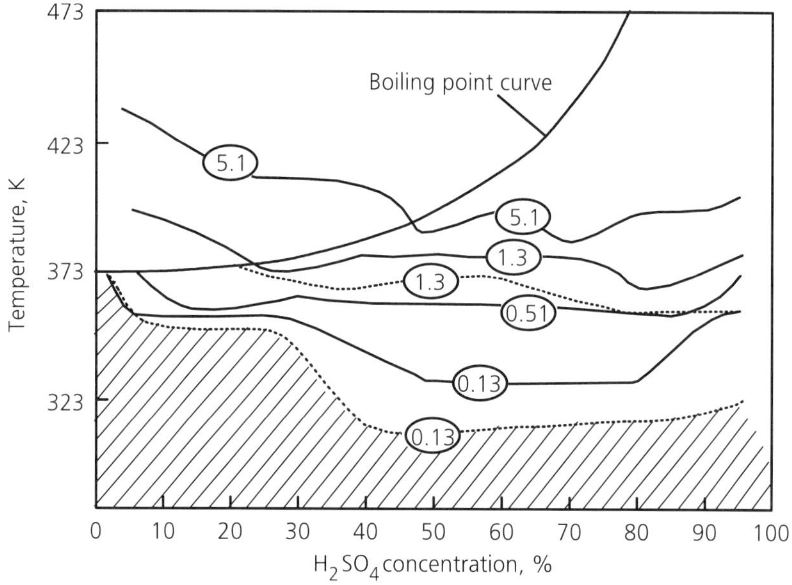

Figure 113: Isocorrosion curves (mm/a) of Hastelloy® C-276 (2.4819) ((%) Ni-16Cr-16Mo-5Fe-4W-max.0.010C) in sulfuric acid [425, 426, 449–452]

——— H_2SO_4

– – – – H_2SO_4 + 200 ppm Cl^- in form of NaCl

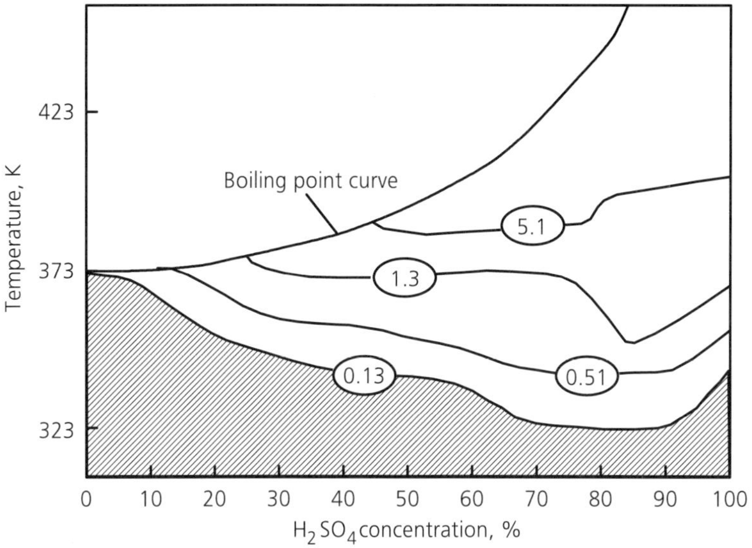

Figure 114: Isocorrosion curves (mm/a) of Hastelloy® C-22 ((%) Ni-22Cr-13Mo-3W-3Fe-max.0.010C) in sulfuric acid [453]

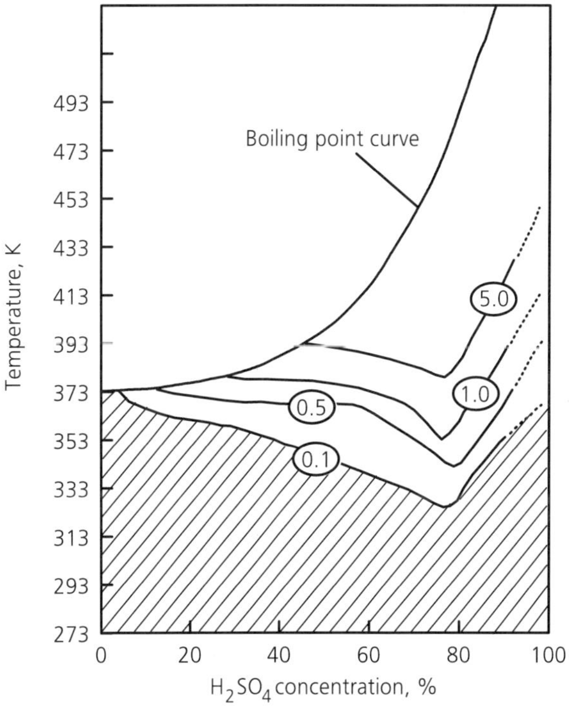

Figure 115: Isocorrosion curves (mm/a) of Chlorimet® 3 ((%) 60Ni-18Cr-18Mo-3Fe-1Si-0.07C) in sulfuric acid [414, 454]

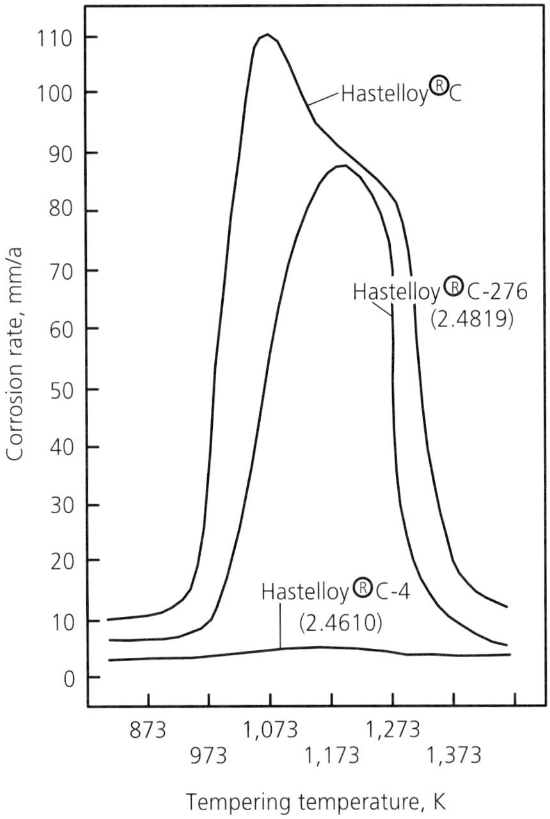

Figure 116: Influence of heat treatment (1 h) on the corrosion behavior (selective corrosion) of NiCrMo-alloys in boiling sulfuric acid (50% H_2SO_4 + 42 g/l $Fe_2(SO_4)_3$) [446, 447]

The alloys Hastelloy® C-4 (2.4610) and Inconel® 625 (2.4856) are described as screw materials in 15 to 70% sulfuric acid at room temperature, with a corrosion rate of less than 0.12 mm/a [418].

In a spent acid concentration of between 10 and 30% at an almost constant temperature of 74°C, specimens of Hastelloy® C-276 introduced into a graphite condenser exhibited a corrosion rate of less than 0.051 mm/a after a test duration of 47 days. The alloy Hastelloy® G proved to be more resistant under these conditions, with a corrosion rate of less than 0.013 mm/a. The corrosion rate of Hastelloy® B-2 of 25 mm/a was only theoretically better than that of Incoloy® 825 with more than 254 mm/a [455].

Enamel has proved to be the better material for use in sulfuric acid, even in comparison with nickel alloys [456].

The use of the Hastelloy® C alloy group becomes critical above 100°C. Figure 113 shows the dependence of the corrosion rate of Hastelloy® C-276 (2.4819) in analytically pure H_2SO_4 on the concentration and temperature. As expected from Figure 113, a sharp increase in the corrosion rate ocurs in the concentration range 50 to

70% sulfuric acid. Only temporary use is possible in the range above 70°C. C-276 can be used for plate heat exchangers in concentrated sulfuric acid up to 92°C [427].

The alloy 2.4819 is used currently in the context of heat recovery (at sulfuric acid temperatures of not more than 90°C in the concentration region of 95% H_2SO_4 under oxidizing conditions using brackish water and sea water as coolants) and proves to be resistant to pitting corrosion in artificial sea water up to 90°C [435].

Figure 101 shows the results of corrosion studies on NiCrMo-alloys in 95% sulfuric acid at 100, 125 and 150°C. As can be seen from the graph, there are more suitable materials for these temperature ranges which, given anodic protection, are practically resistant under these conditions [435].

Figure 117 shows the corrosion rates expected for an NiCrMo-alloy in hot concentrated sulfuric acid after a test duration of 56 days. It exhibits a relatively wide application range and confirms the above-mentioned use as a heat exchanger material. The potential oscillation temperature (the temperature at which the material changes from the passive into the active region of its corrosion behavior) is quoted as 122.5°C [435]. However, high potential oscillation temperatures are no indication that the material is a corrosion-resistant alloy, since under passivity conditions the particular corrosion current density determines the corrosion rate. It increases greatly for Ni-alloys in highly concentrated sulfuric acid as the temperature rises [435]. From this work, the NiCrMo-alloy is more suitable as a heat exchanger material than the steel 1.4361, which, after a test duration of 56 days, gives the corrosion rates shown in Figure 118 (compare with Figure 117).

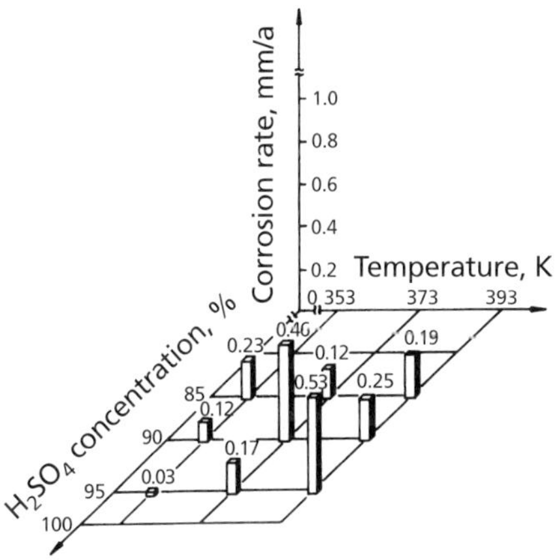

Figure 117: Corrosion behavior of Nicrofer® 5716 hMoW (corresponds to Hastelloy® C-276, 2.4819) in hot sulfuric acid of a high concentration at various temperatures (test duration 56 d) [457]

The chloride content also has an influence on the corrosion resistance of C-276. A corrosion rate of 0.1 mm/a was measured on the alloy Hastelloy® C-276 in 93% sulfuric acid, and a rate of 0.8 mm/a at a chloride content of only 17 ppm [427].

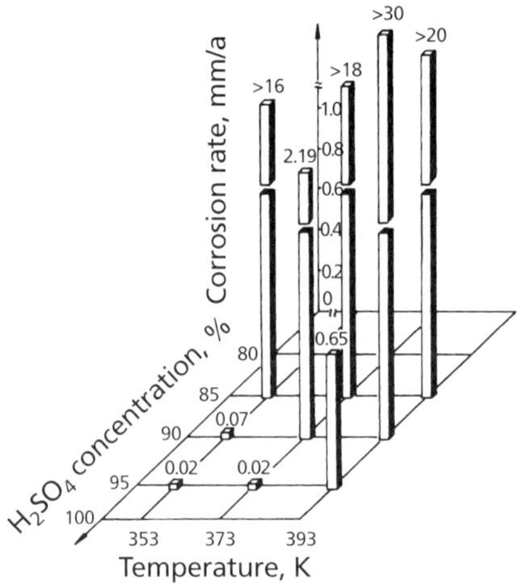

Figure 118: Corrosion behavior of Cronifer® 1815 LCSi (1.4361, X1CrNiSi18-15-4) in hot sulfuric acid of a high concentration at various temperatures (test duration 56 d) [435]

Figure 119 shows the corrosion loss on Hastelloy® C-276 in technically pure 98.5% H_2SO_4 (2.0 ppm Fe-ions, 0.5 ppm F-ions, 0.8 ppm Cl-ions, 0.4 ppm Mg-ions and 0.7 ppm As-ions) as a function of the temperature. As can be seen from this graph, the corrosion rates rise to far in excess of 0.5 mm/a at temperatures above 110°C.

Figure 100 shows the corrosion behavior of various nickel alloys in boiling (probably analytically pure) sulfuric acid up to a concentration of 50% according to four accelerated tests over a period of 24 h [427].

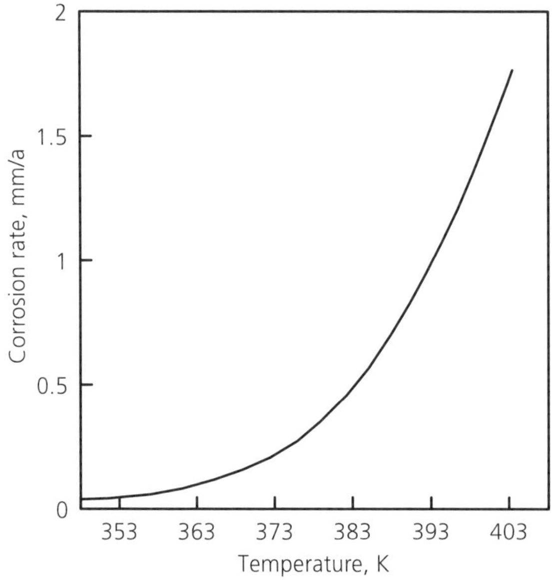

Figure 119: Corrosion rates of Hastelloy® C-276 (2.4819) in technically pure sulfuric acid ((ppm) 2.0 Fe-ions, 0.5 F-ions, 0.8 Cl-ions, 0.4 Mg-ions and 0.7 As-ions) as a function of the temperature (for the test conditions see Figure 120) [427]

In this graph, the Hastelloy® G-30 and G-3 alloys with higher Fe-contents exhibit better corrosion behavior than the materials containing more Mo. The alloy B-2 proved to be the most suitable.

– Influence of chlorides in sulfuric acid on the corrosion behavior –

In the technically pure 98.5% H_2SO_4 described above, the presence of small amounts of chloride has no substantial influence on the corrosion rate of Hastelloy® C-276 at 90°C, but the rate rises sharply above 40 ppm, and reaches its maximum at chloride contents of 50 ppm and above (Figure 120).

Figures 121–124 show accelerated tests over a period of 24 h in pure sulfuric acid (probably analytically pure) with and without the addition of 1% $FeCl_3$. Whilst the alloy Hastelloy® G-3 still exhibits increased susceptibility to pitting corrosion, which decreases as the Mo-content increases, and according to these studies no longer exists in the alloys C-276 and C-22. By comparison, chloride ions as NaCl in sulfuric acid (Figure 99, Figure 113) caused a deterioration in the corrosion resistance, whereas $FeCl_3$ improved the corrosion resistance, as can be seen from Figures 121–124. An exception here is Hastelloy® C-276, whose corrosion curve of 0.13 mm/a with 1% $FeCl_3$ is below that of pure sulfuric acid. It should be pointed out again here that the results are from accelerated tests, in-house studies over a longer period of time give a more reliable indication of suitability for the particular application.

– Corrosion behavior in sulfuric acid-acid mixtures and sulfuric acid solutions –

A comprehensive summary of mixed media in which Hastelloy® C-276 has good to very good resistance appears in [452].

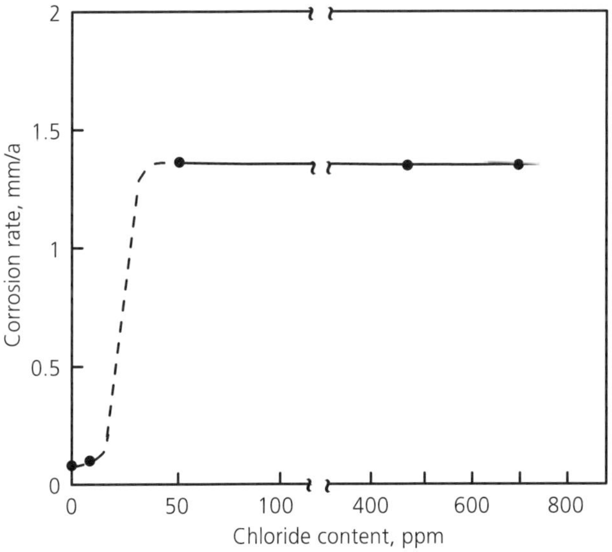

Figure 120: Corrosion rates on Hastelloy® C-276 (2.4819) in 98.5% technically pure H_2SO_4 (Figure 119) at a temperature of 90°C as a function of the chloride content (test duration max. 72 h, rotating disks ø 82 mm, 7 mm thick, 270 rpm, activated with Zn-dust, corrosion rate determined via weight loss determination) [427]

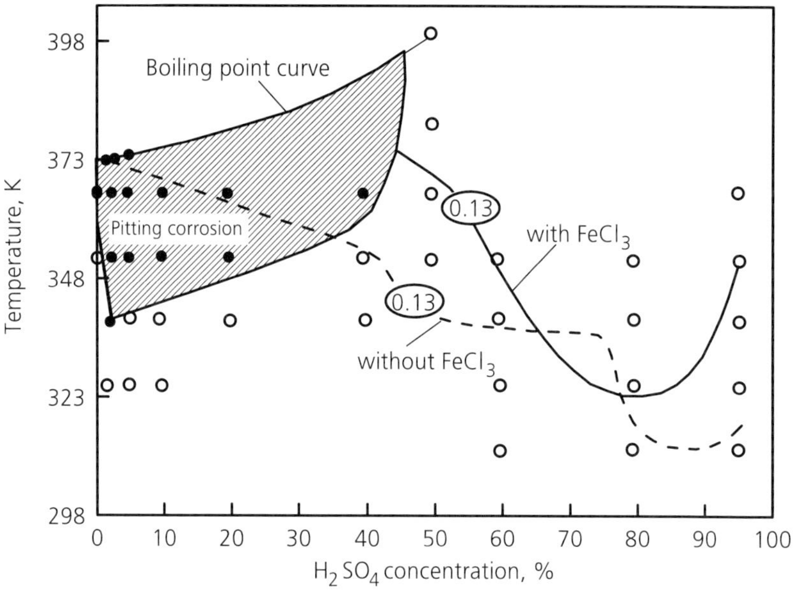

Figure 121: Influence of the addition of 1% $FeCl_3$ to sulfuric acid (analytically pure) on the corrosion behavior of Hastelloy® G-3 (test duration 24 h) [427, p. 64, 458]
● pitting corrosion
○ general corrosion

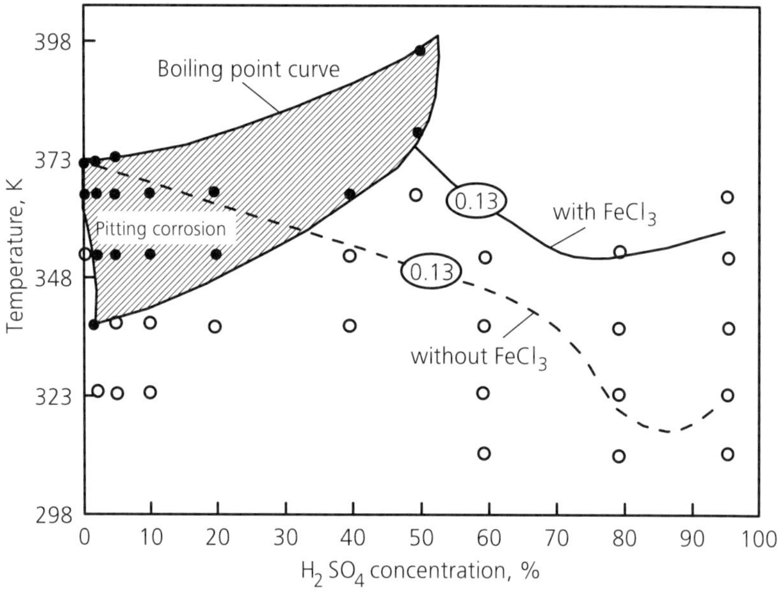

Figure 122: Influence of the addition of 1% $FeCl_3$ to sulfuric acid (analytically pure) on the corrosion behavior of Hastelloy® G-30 (test duration 24 h) [427, p. 65]
● pitting corrosion
○ general corrosion

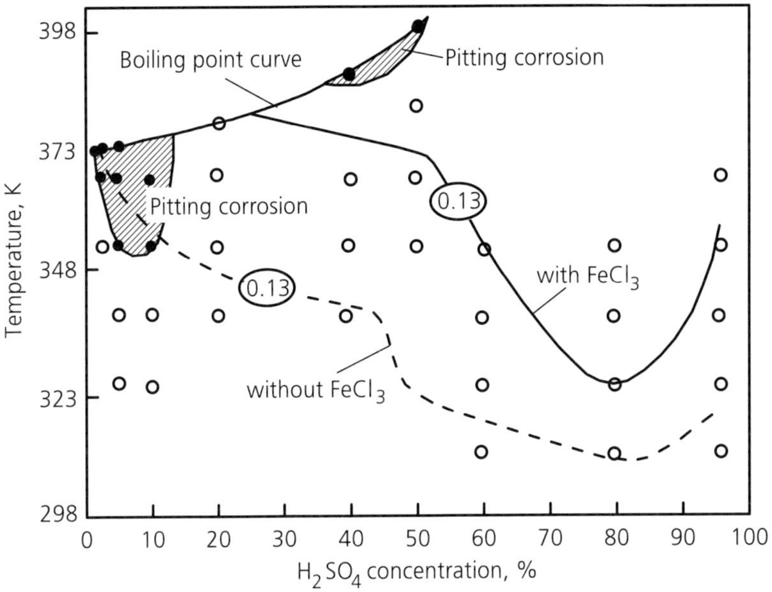

Figure 123: Influence of the addition of 1% $FeCl_3$ to sulfuric acid (analytically pure) on the corrosion behavior of Inconel® 625 (test duration 24 h) [427, p. 65]
● pitting corrosion
○ general corrosion

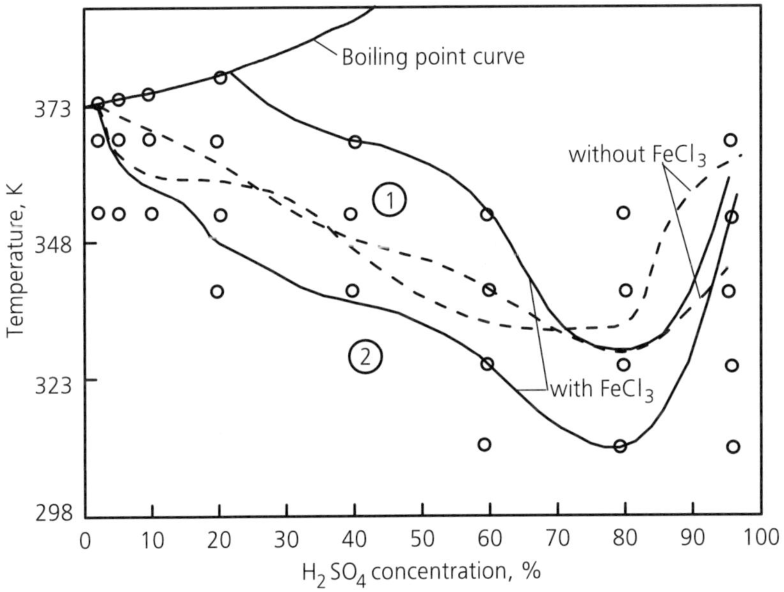

Figure 124: Influence of the addition of 1% $FeCl_3$ to sulfuric acid (analytically pure) on the corrosion behavior of Hastelloy® C-22 and Hastelloy® C-276 (test duration 24 h) [427, p. 65]
○ general corrosion

– Pitting and crevice corrosion –

In comparison with high-alloy steels and nickel alloys, the NiCrMo-alloys have increased resistance to pitting and crevice corrosion [437, 459, 460].

– Intercrystalline corrosion –

The resistance of these alloys to intercrystalline corrosion depends on the degree of precipitation at the grain boundaries. This falls, as described above, as the purity increases. This type of corrosion is largely suppressed in the high-purity alloys (compare with Figure 125).

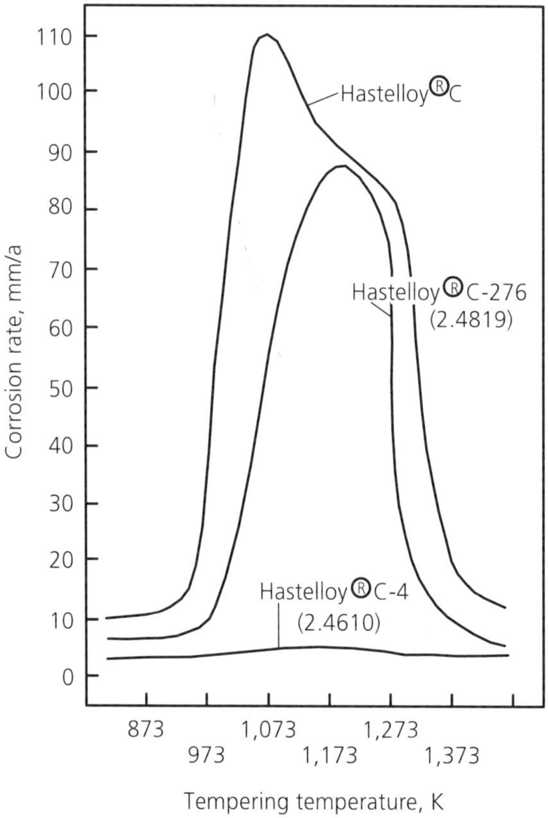

Figure 125: Influence of heat treatment (1 h) on the corrosion behavior (selective corrosion) of NiCrMo-alloys in boiling sulfuric acid (50% H_2SO_4 + 42 g/l $Fe_2(SO_4)_3$) [446, 447]

Nickel-copper alloys

Monel® 400, an alloy containing about 66% Ni and 32% Cu (NiCu 30 Fe, 2.4360), is one of the chief materials used in the chemical industry.

This alloy is even more important when dealing with sulfuric acid than pure nickel; although they are both similar in their chemical resistance to sulfuric acid.

As with nickel, the corrosion of Monel® 400 is increased by the presence of oxygen in sulfuric acid and by iron salts. Its corrosion resistance to all acids, with the exception of highly oxidizing acids, can be described as good [395, 397, 398, 401, 402, 405, 406, 422, 425, 438, 461].

In non-oxidizing acids, the corrosion potentials of Monel® 400 in most cases lie in regions where evolution of hydrogen is avoided. Corrosion can occur, therefore, only if dissolved air or other oxidizing substances are present. Thus the degree of aeration of sulfuric acid is an important factor in the corrosion of NiCu-alloys. For the corrosion behavior of Monel® 400 in sulfuric acid and media containing sulfuric acid see also Figure 90. Figures 126 and 127 show the influence of the oxygen content on the corrosion rates in sulfuric acid at room temperature. Figure 126 shows that hardly any corrosion occurs in air-free sulfuric acid at room temperature, corrosion rates in up to 80% H_2SO_4 with a maximum of 0.3 mm/a prevailing. The corrosion passes through a maximum in air-free H_2SO_4 of approximately 5%. In air-saturated sulfuric acid under the same conditions, the maximum corrosion rate is still 1 mm/a in approximately 5% H_2SO_4. On this basis, Monel® 400 can be used for storage and transportation of sulfuric acid in the concentration range up to 80% H_2SO_4, and is the material used the most for this purpose [395]. Above 80% H_2SO_4 increased attack occurs as a consequence of the oxidizing acid. In the concentration range from 15 to 80%, the corrosion rate falls almost linearly, as a consequence of the decrease in the hydrogen ion concentration and oxygen solubility. Above a concentration of 80%, the sulfuric acid has an oxidizing action and hence causes a higher corrosion rate on Monel® 400 which is then evidently less influenced by the aeration. In very aggressive 5 to 6% sulfuric acid at room temperature, an increase in the temperature has no significant effect on the corrosion rate of Monel® 400 if the acid is air-free. In air-saturated sulfuric acid, in contrast, the corrosion rate increases very sharply and reaches a maximum at 80°C of about 3 mm/a (Figure 127). The weight losses are about 3 times higher in oxygen-saturated acid than in air-saturated acid [461]. The corrosion on NiCu-alloys is intensified as the flow rate increases. As well as fresh acid, the flowing acid also brings dissolved oxygen and oxidizing constituents to the surface of the metal and corrodes this away. Figure 128 shows the conditions for 5% sulfuric acid saturated with air as the flow rate increases.

Figure 129 shows an overview of isocorrosion curves of Monel® 400 in sulfuric acid solutions. Since no further information is available, this graph applies to non-aerated acid. Higher corrosion rates can be expected with aeration, additional agitation of the acid and an increase in temperature, as can be seen from Figure 130 [396, 404, 454, 462].

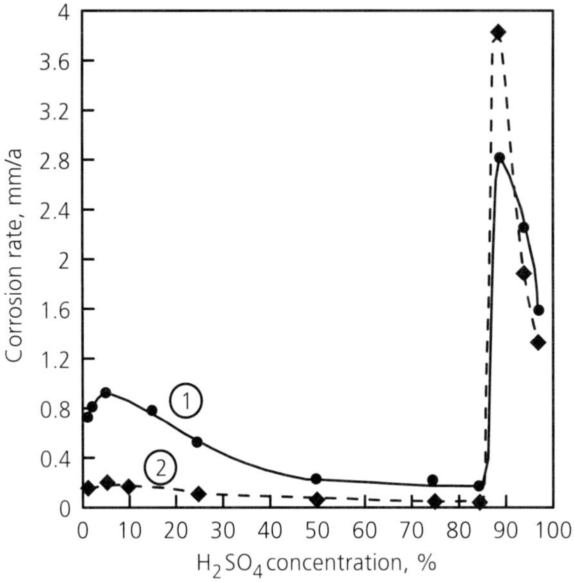

Figure 126: Corrosion behavior of Monel® 400 (2.4360) in sulfuric acid at room temperature as a function of aeration (test duration 48 h) [395, 396, 401, 406, 461, 463]
① saturated with air
② saturated with nitrogen, air-free

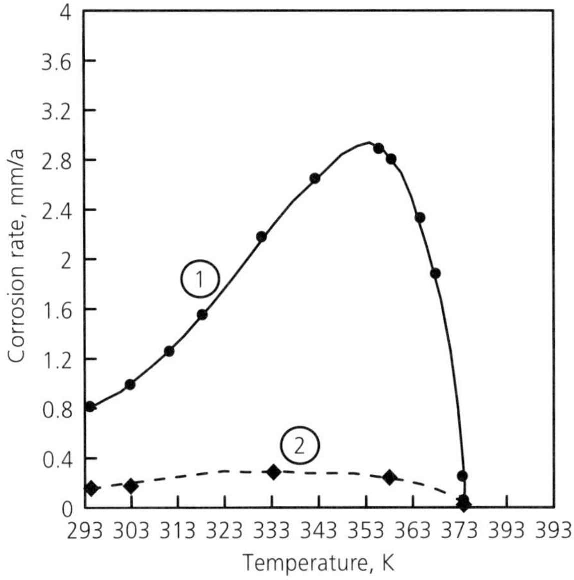

Figure 127: Corrosion behavior of Monel® 400 (2.4360) in 5 to 6% sulfuric acid as a function of aeration and temperature (test duration, 48 h) [395, 396, 401, 406, 461, 462]
① saturated with air
② saturated with nitrogen, air-free

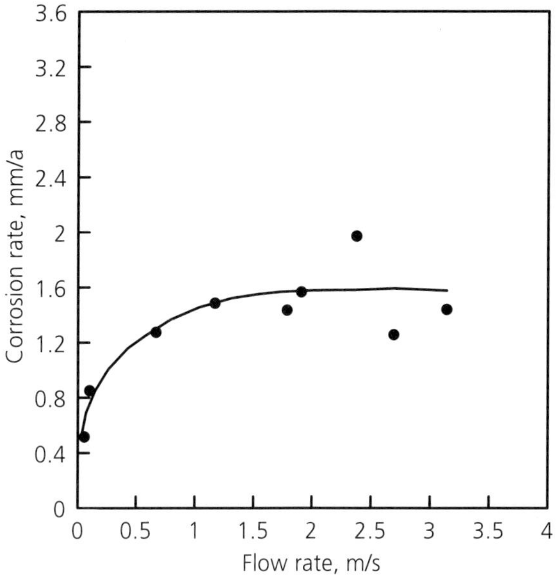

Figure 128: Influence of flow rate on the corrosion rate of Monel® 400 in air-saturated 5% sulfuric acid at room temperature (test duration 28 h) [396]

In comparison with lead materials, for example, Monel® 400 can be used in the concentration range between about 25 and 85% H_2SO_4 [464] with a corrosion rate of not more than 0.1 mm/a up to 40°C, in agreement with [404] (Figure 129 and Figure 131).

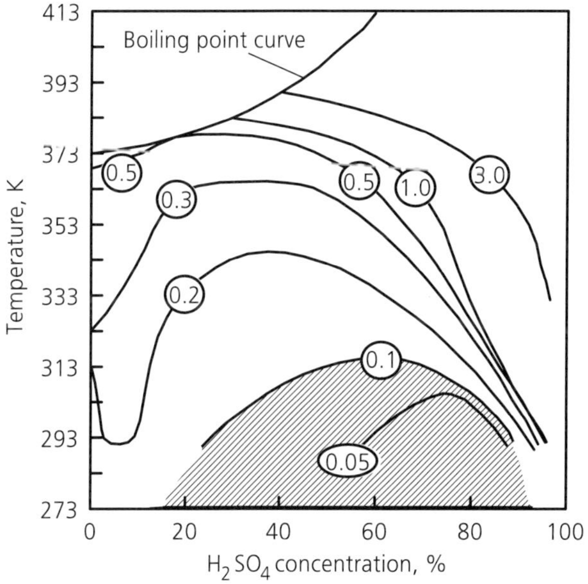

Figure 129: Isocorrosion curves (mm/a) for Monel® 400 ((%) 67Ni-30Cu-1.4Fe-0.15C) in sulfuric acid [404]

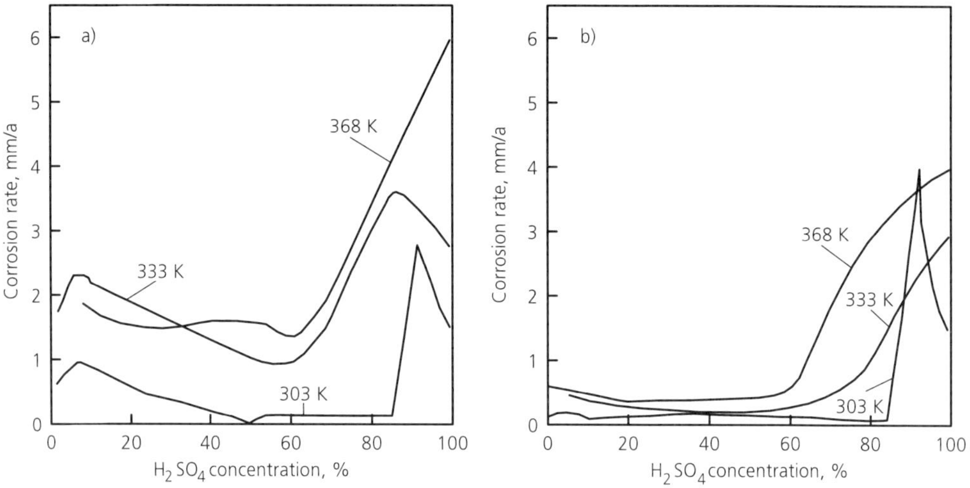

Figure 130: Corrosion rate of Monel® 400 ((%) 67Ni-30Cu-1.4Fe-0.15C) as a function of the air saturation and the temperature of sulfuric acid at a flow rate of 0.08 m/s (test duration 48 h) [396, 454, 463]
a) air-saturated
b) H_2-saturated, air-free

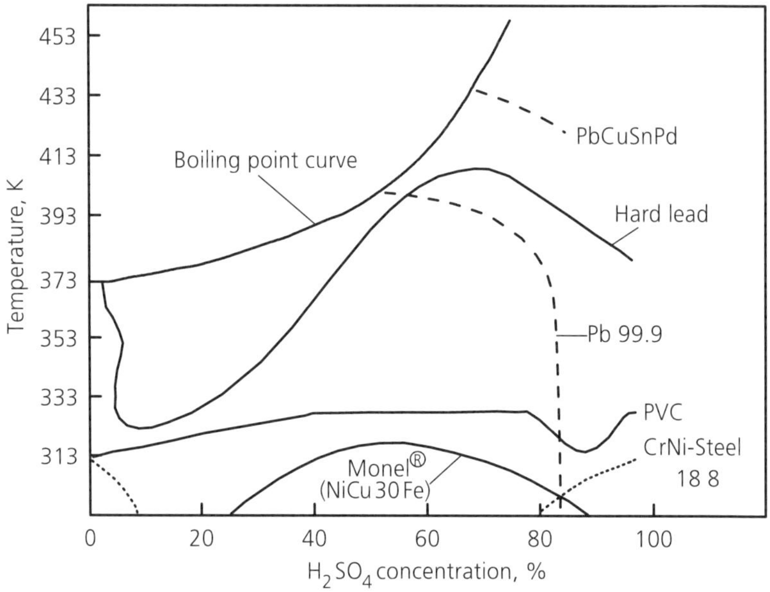

Figure 131: Isocorrosion curves (0.1 mm/a) of various materials in sulfuric acid solutions [464, 465]

From [405], Monel® 400 can be used as a screw material in sulfuric acid at room temperature to only a limited degree.

The corrosion rates in the concentration range between 15 and 70% H_2SO_4 are quoted as being between 0.12 and 1.22 mm/a in this study.

In practice, if the acid is saturated with oxygen or oxidizing constituents are present, consumption occurs, with a subsequent decrease in its aggressiveness towards Monel® 400. Thus corrosion studies under the particular operating conditions are particularly advisable if nickel-copper alloys are to be used; this applies particularly under increased pressure where the acid temperature can be sometimes considerably higher than that found under normal pressure conditions.

The various NiCu-alloys, for example those containing Si, are similar to one another in their corrosion behavior [396].

– Corrosion behavior in sulfuric acid-acid mixtures and sulfuric acid solutions –

It has proved appropriate to use NiCu 30 Fe in the treatment of acid waste in oil refineries and in sulfonation processes in which mixtures of oil and highly concentrated sulfuric acid arise, since oil counteracts the corrosion of this material in highly concentrated sulfuric acid. In a works study of the sulfonation of phenol with 98% H_2SO_4 at 120°C, a corrosion rate of 0.12 mm/a was determined for Monel® 400. The sulfonation apparatus was loaded 180 times [395, 406]. Favorable conditions for Monel® 400 exist in pickling plants. In the pickling of steel, for example, hydrogen is formed constantly and establishes reducing conditions in the pickling bath. The long service lives of pickling baskets, hooks and frames made of Monel® 400 in these solutions – which can extend to 40 years – can be attributed to the reducing conditions of the acid [406].

In sulfuric acid solutions in contact with the material to be pickled, NiCu 30 Fe may be galvanically protected or not at all influenced by dissolution of the latter. In contact with lead, silver solder or charcoal, on the other hand, the corrosion increases (surface conditions) on this alloy. No phenomenon similar to the "dezincification" of brass is found. It is successfully used for sulfuric acid dye liquors, for finishes for the treatment of wool, for leather bating processes, and in all instances where resistance to hot or cold dilute sulfuric acid is required [395, 396, 406].

– Stress corrosion cracking –

According to [396], NiCu-alloys of the type NiCu 30 Fe do not tend to undergo stress corrosion cracking in sulfuric acid. Exceptions occur in the presence of mercury salts or considerable amounts of fluosilicic acid in sulfuric acid solutions [396].

On exposing NiCu 30 Fe to hydrogen, penetration of atomic hydrogen leads to a considerable loss in deformability. Under the influence of stresses the material fractures. As the thickness of the material increases, penetration of hydrogen is made more difficult, and the effect tends to decrease as the thickness of the material increases. Hydrogen can be removed by heat treatment. No optimum temperatures or holding times have been mentioned for NiCu 30 Fe [409, 466].

Nickel-molybdenum alloys

Due to a lack of chromium these alloys cannot be passivated. Their good corrosion resistance in reducing acids is based on a low corrosion rate in the active state caused by molybdenum. The upper limit of the molybdenum content is 30%, since the deformability deteriorates thereafter and the corrosion resistance hardly increases any further. Low C-, Si- and V-contents of around 2% reduce precipitation in these alloys. Ferromolybdenum is the basis for melting alloys for industrial use, which is why these alloys have Fe-contents of between 4 and 7%. In contrast to alloys containing iron, iron-free alloys can be hardened at 700°C without losing their corrosion resistance [467].

Alloys with the commercial name Hastelloy® are used in the chemical industry, in particular Hastelloy® B (20 to 30% Mo, 4 to 7% Fe), comparable with 2.4800 (S-NiMo 30) and 2.4810 (NiMo 30), and Hastelloy® B-2 (26 to 30% Mo, max. 2% Fe), comparable with 2.4615 (SG-NiMo27) and 2.4617 (NiMo28). Other commercial names are Remanit® HB, Superantinit® B, Bergit® B, Euzonit® 70, Chlorimet® 2, Corronel® 210 and 220 and Langalloy® 5R. As already mentioned above, these materials can be used in particular under reducing conditions in sulfuric acid. Figures 90, 100 and 110 give the application ranges and corrosion behavior.

According to [397], higher corrosion losses can be expected in non-aerated H_2SO_4 than in aerated acid. Corrosion rates of less than 0.18 mm/a in sulfuric acid up to 60% at boiling point are found in some cases [395–398, 401, 402, 411, 423, 438, 454, 456, 468–474]. They have better corrosion resistance under these conditions than the NiCrMo-alloys, as can be seen from Figure 100. In 96% H_2SO_4, NiMo 30 reaches a corrosion rate of 1.22 mm/a at 150°C and, therefore, is attacked less than other Ni-alloys [396].

The use of Hastelloy® B and D in the concentration range between about 50 and 78% H_2SO_4 at the boiling point is not to be recommended [396, 401]. According to long-term studies (432 h), an unexpectedly high corrosion rate occurs at the boiling point in this range.

In all cases when Hastelloy® B-2 is deposited by welding onto unalloyed steel, the corrosion rate increases with the iron content in the Ni-alloy [427].

Figure 132 shows the rate of corrosion of Hastelloy® B-3 alloy as a function of the concentration of boiling sulfuric acid [589]. The rates of corrosion of other engineering alloys are shown for comparative purposes. The figure shows that under the strongly reducing conditions of boiling sulfuric acid at concentrations below 60%, B-3 alloy has a low rate of corrosion, espacially compared with other alloys. At concentrations above 60%, sulfuric acid starts to become oxidizing and so the rate of corrosion of B-3 alloy increases more rapidly as the concentration of acid increases.

See also Figure 101 for the use of metallic materials in hot concentrated sulfuric acid.

Salts of copper or iron in H_2SO_4 solutions cause considerable corrosion of these alloys [475].

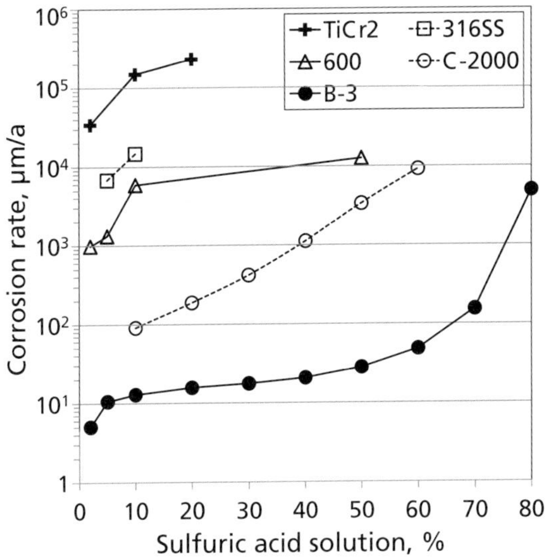

Figure 132: Rate of corrosion of B-3 and other alloys in boiling sulfuric acid solutions (4 × 24 h immersion tests) [589]

After a test duration of 47 d, specimens of Hastelloy® B-2 in flowing 10 to 30% H_2SO_4 at 74°C and an iron concentration of 200 to 500 ppm exhibited a corrosion rate of more than 25.4 mm/a (recycling acid). Hastelloy® C-276 (NiMo16Cr15W) and Hastelloy® G (NiCr22Mo6Cu) are more suitable for use under these conditions with corrosion rates of less than 0.051 mm/a and less than 0.013 mm/a respectively [455].

The isocorrosion curves in Figures 133 and 134 give an overview of the corrosion behavior of the two Hastelloy® alloys B and B-2. According to these graphs, the two alloys differ only slightly in their corrosion behavior in sulfuric acid. The corrosion rates increase as a result of Cl-ions. Table 127 gives a summary of the corrosion rates of various nickel alloys in sulfuric acid.

NiMo-alloys containing 55% Ni and 16% Mo are just as unsuitable as high-alloy steels, Ti and TiPd-alloys and the alloy Ti-5Ta in a reaction mixture consisting of about 25% H_2SO_4, 7% HCl as well as sulfochlorides, dichloroethane and chloro-sulfonic acid at temperatures up to 110°C and pressures up to 0.17 MPa for the production of a cation exchanger, with corrosion rates of 4.9 mm/a. The test durations under operating conditions were 15,000 h in all [476].

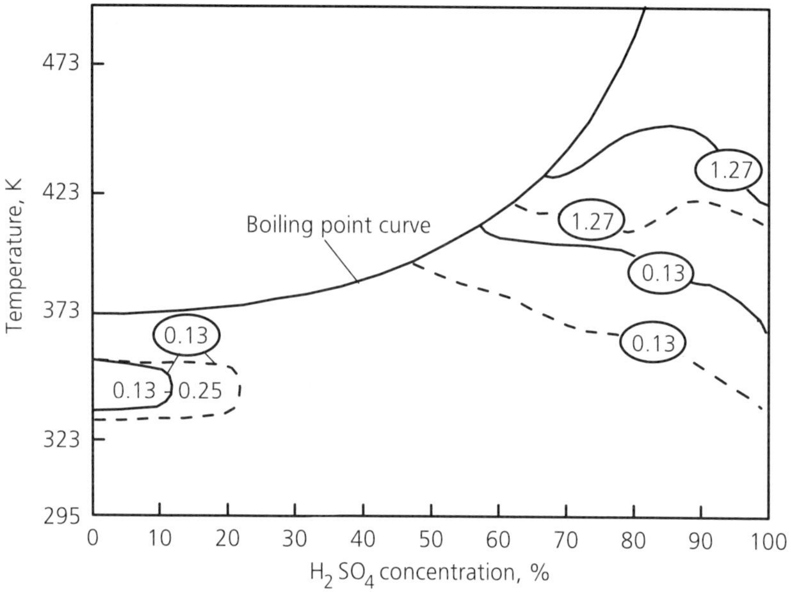

Figure 133: Isocorrosion curves (mm/a) for Hastelloy® B ((%) Ni-26–30Mo 4–7Fe) in sulfuric acid with and without Cl^- [404, 414, 427, 444, 450]

——— H_2SO_4

– – – – H_2SO_4 + 200 ppm Cl^- in the form of NaCl

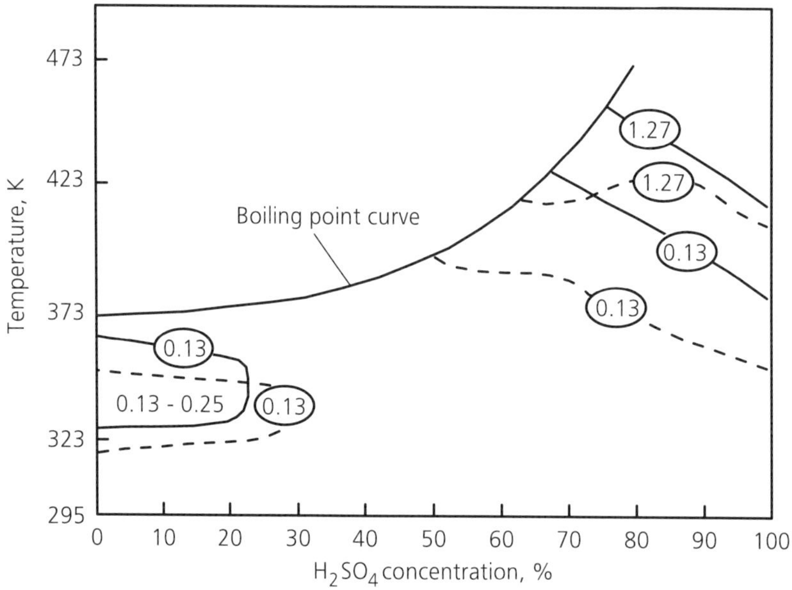

Figure 134: Isocorrosion curves (mm/a) for Hastelloy® B-2 ((%) Ni-26–30Mo-max.2Fe) in sulfuric acid with and without Cl^- [427, 450, 452]

——— H_2SO_4

– – – – H_2SO_4 + 200 ppm Cl^- in the form of NaCl

Other nickel alloys

– NiSi-alloys –

In this section, amongst others, the NiSi-alloys are of interest. Alloys containing about 5% Si have only a slightly improved corrosion resistance compared with nickel. Their structure has not yet changed. Only as the Si-content increases does the corrosion resistance of these alloys increase. Above 5% Si, the brittle gamma phase precipitates out of the alloys so that alloys containing more than 11% Si are no longer deformable. For the reasons mentioned, the Si-contents in these alloys are limited to 10%, and the Cu-content is 3%. Cu-contents in these alloys have a favorable effect on the corrosion behavior. Figure 135 shows the influence of Si and Cu on the corrosion behavior of NiSi alloys [467]. The application ranges of the alloy Hastelloy® D are shown in Figure 110. The corrosion rates in comparison with other nickel alloys are given in Table 127.

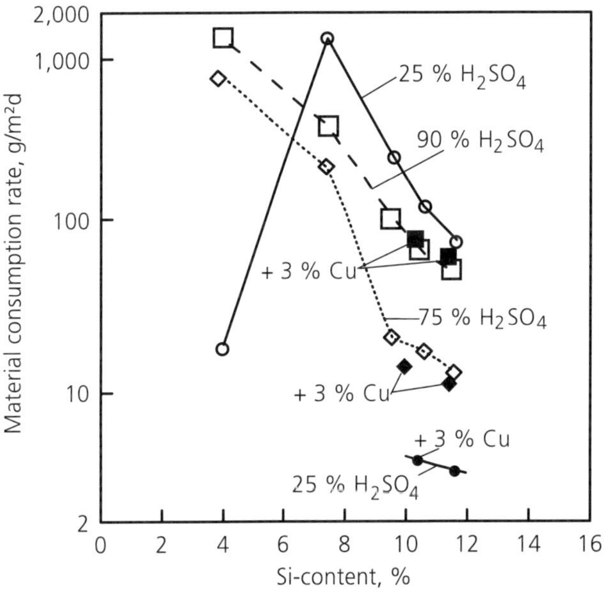

Figure 135: Material consumption rate of NiSi-alloys with and without Cu-additions in boiling sulfuric acid [467]

Figure 136 shows isocorrosion curves for Hastelloy® D in sulfuric acid. From Figure 110, this alloy shows poorer corrosion behavior at 66°C than the alloys Hastelloy® B and C in some cases, but, according to [469], is superior at 70°C in the concentration range between 20 and 90% H_2SO_4 (Figure 137). For this reason, it is important to consult the manufacturer about the corrosion behavior of the alloy regarding the specific application, or to perform in-house corrosion studies under operating conditions, especially if similar alloys are available on the market under different commercial names. The use of Hastelloy® D in the concentration range between 50 and 60% H_2SO_4 at the boiling point is not advised. Corrosion studies in 55% H_2SO_4 at the boiling point showed high corrosion rates after a short test duration, which decreased to 0.33 mm/a after 144 h as a consequence of the formation of Si-containing protective layers, as is the case with all high-silicon alloys. The corrosion rate after a test duration of 384 h was 0.20 mm/a, and this rose to the unacceptable level of 37 mm/a after 432 h. No explanation for this was given in this study. The alloy Hastelloy® D is quoted as containing 8 to 11% Si and Ni as the balance [401], and thus differs in composition from the Hastelloy® D alloy in Figure 136 as well as [477].

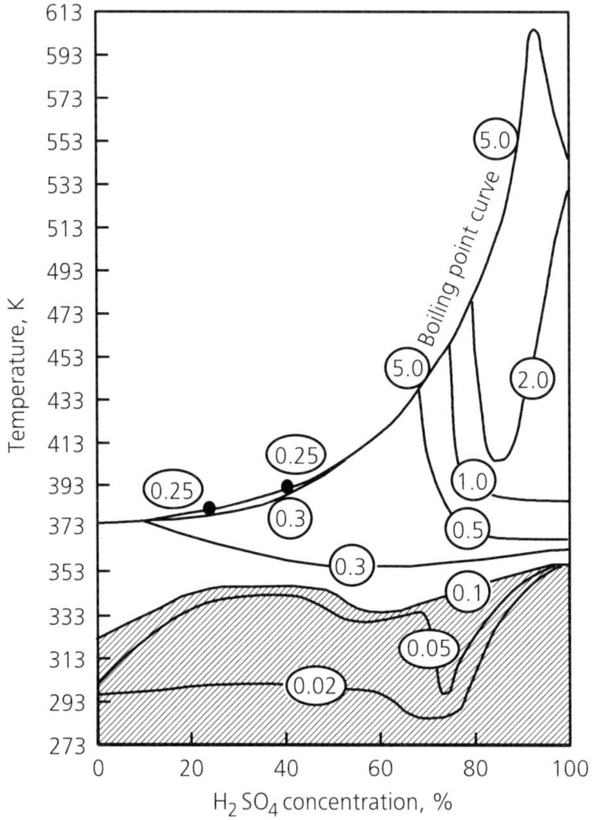

Figure 136: Isocorrosion curves (mm/a) for Hastelloy® D ((%) 82Ni-10Si-3Cu-2Fe-1.5Co-1Cr-0.1C) in sulfuric acid [404]

Temperature °C	*H_2SO_4 concentration %*	*Hastelloy® B ***	*Hastelloy® C*	*Hastelloy® D*	*Hastelloy® F*
		Corrosion rate, mm/a			
RT	2	0.025	0.008	0.025	0.003
100*	2	0.025	0.375	0.100	0.650
RT	10	0.025	0.005	0.025	0.003
102*	10	0.050	1.18	0.325	2.43
RT	25	0.025	0.003	0.025	0.000
106*	25	0.050	2.95	0.225	7.98
RT	60	0.005	0.003	0.025	0.000
140*	60	0.175	unsuitable	0.200	unsuitable
RT to *	65	all alloys unsuitable			
RT	80	0.003	0.003	0.025	0.003
201*	80	unsuitable	unsuitable	0.900	unsuitable
RT	96	0.005	0.000	0.025	0.003
150	96	0.008	0.200	0.025	1.98
290*	96	unsuitable	10.1	2.15	10.6

* boiling point
** see also Table 128
RT: Room temperature

Table 127: Corrosion rates for nickel alloys in sulfuric acid at various temperatures and concentrations (no more precise data) [398] (see also Figure 110)

Overall, NiSi alloys can be expected to have corrosion rates of: less than 0.075 mm/a at room temperature in the concentration range 2 to 96% H_2SO_4, less than 0.15 mm/a at 66°C in 2 to 60% H_2SO_4 and less than 0.33 mm/a at boiling point in the same concentration range as before. In 77 to 96% sulfuric acid at boiling point, the corrosion rates of these alloys rise in the range 0.7 to 4.8 mm/a [444]. According to this study, pump casings for slurry containing sulfuric acid and stirrer blades and shafts are cast chiefly from this alloy.

Hastelloy® Alloy D-205 was developed as a wrought version similar to the original alloy D. Due to the high silicon content, 5%, this alloy is only available as thin sheet and is used in plate heat exchangers for cooling concentrated sulfuric acid [590]. It's chief advantages over high silicon-iron based alloys (e.g. Fe-17Cr-20Ni-5Si) are excellent formability, good resistance to elevated temperature embrittlement, and superior performance in many concentrations of sulfuric acid, particularly high concentations, see Figure 137.

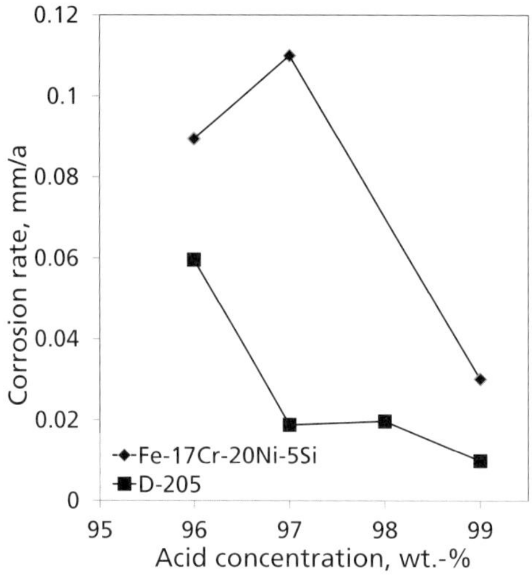

Figure 137: Uniform corrosion resistance of Hastelloy D-205 in concentrated sulfuric acid at 130°C [591]

As with the NiMo-alloys containing no Cr, the corrosive attack on Hastelloy® D is intensified in the presence of oxidizing salts (ferric sulfate, cupric sulfate, chromates, nitrates and peroxides) in sulfuric acid solutions [396]. In air-free sulfuric acid, the alloy Hastelloy® D exhibits no pitting corrosion and uniform general corrosion takes place [407].

The composition of NiSi-alloys is of importance for their corrosion resistance, and an optimal heat treatment is a basic prerequisite [477].

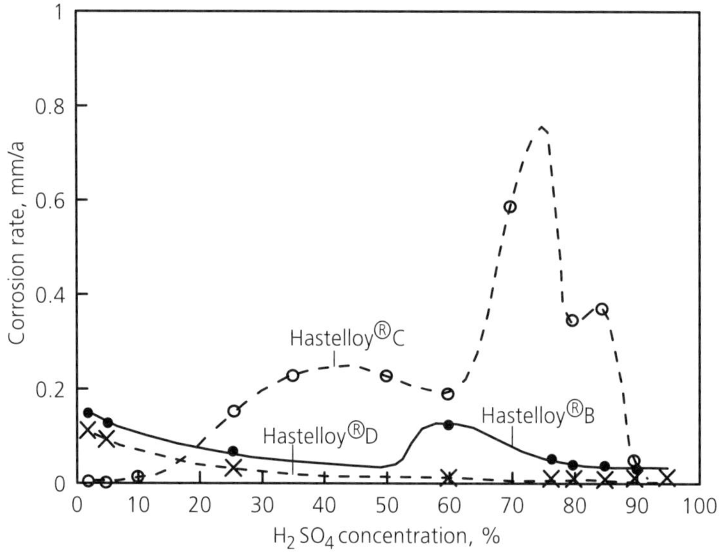

Figure 138: Corrosion rates of nickel alloys in pure sulfuric acid at 70°C [469]

– NiMoW-alloys –

The precipitation-hardening NiMoW-alloys Ni65Mo20W15 ((%) Ni-19.6Mo-16.6W-0.01C-0.04Si-0.004S-0.006P-0.16V-0.83Fe) and Ni55Mo20W25 ((%) Ni–19.4Mo–25.7W–0.02C–0.07Si–0.005S–0.005P–0.18V–0.97Fe) also exhibit very good corrosion resistance in sulfuric acid solutions. Table 128 summarizes the dependence of the corrosion rate on concentration and temperature. These alloys tend not to undergo intercrystalline corrosion, according to studies in [478]. However, their corrosion behavior is dependent upon the tempering state, as is demonstrated in the study involving exposure to boiling 21% HCl (test duration 240 h) as a function of heat treatment. The corrosion minimum of the alloys Ni65Mo20W15 (600 to 1,100°C, 1 h) and Ni55Mo20W25 (800°C, 1 h) is 0.6 mm/a. Higher corrosion rates are measured in all other cases under the conditions mentioned [478].

– NiTi-alloys –

Plasma deposition welds of NiTi4 showed a higher corrosion rate in boiling 20% H_2SO_4, at 4.3 mm/a after a test duration of 20 h, than those of the alloy NiCu 30 MnTi under the same conditions, at 1.8 mm/a. They are therefore less resistant and should not be used under these conditions. The Cu-alloy CuAl 8 was more suitable [479].

Temperature	***H_2SO_4 concentration, %***						
°C	***5***	***10***	***25***	***50***	***60***	***80***	***90***
	Corrosion rate, mm/a						
	Ni65Mo20W15						
20	0.03	0.05	0.02	0.01	0.008	0.005	0.005
60	0.024	0.024	0.024	0.007	0.007	0.0065	0.0065
90	0.04	0.045	0.040	0.026	0.018	0.0016	0.016
boiling point	–	0.058	0.095	0.70	< 0.70	–	–
	Ni55Mo20W25						
20	0.04	0.03	0.028	0.005	0.002	0.002	0.002
60	0.031	0.031	0.031	0.020	0.016	0.009	0.008
90	0.050	0.054	0.050	0.050	0.050	0.058	0.030
boiling point	–	0.65	0.10	0.90	> 0.90	–	–

Table 128: Corrosion rates for the precipitation-hardenable NiMoW-alloys Ni65Mo20W15 and Ni55Mo20W25 in the delivery state in mm/a in sulfuric acid solutions at various temperatures and concentrations (100 h) [478]

– *NiCo-alloys* –

As electrochemical studies at room temperature have shown, the amorphous alloy Ni40Co20Cr12Mo6Fe6B16 is more resistant in 0.005–5.0 mol/l H_2SO_4 (air-free) than the CrNi-steel 18 8 and nickel alloys, according to [480]. On contact with sulfuric acid, spontaneous passivation occurs, which is also retained in the transpassive region. No pitting corrosion has been established in 0.5 mol/l H_2SO_4 + 1 mol/l NaCl at + 1,000 mV, but the alloy is susceptible to crevice corrosion [480].

– *NiBe-alloys* –

Of the alloys investigated in [481], the NiBe-alloy Beryvac® 520 (spring alloy) has the highest bending fatigue strength (± 400 N/mm^2 at 10^8 stress cycles). The corrosion rates quoted in 10 and 96% H_2SO_4 are in the range 0.1 to 1.0 mm/a at 20°C, and 1 to 10 mm/a at 100°C. In 50% acid, values of up to 0.1 mm/a at 20°C and above 10 mm/a at 100°C are observed [481].

– *NiGe-alloys* –

Figure 139 shows the corrosion behavior of nickel alloys with increasing germanium content. As the Ge-content increases, an increase in the corrosion resistance in boiling 5% H_2SO_4 is detectable. This phenomenon is no longer found in more highly concentrated acid (75% H_2SO_4).

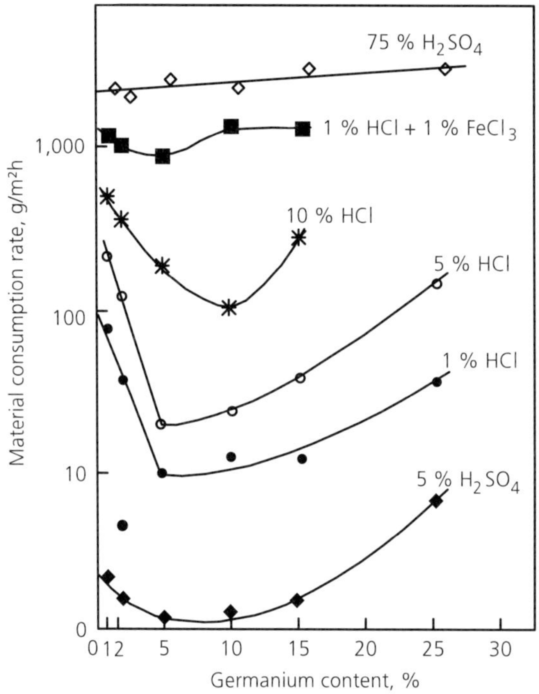

Figure 139: Material consumption rates of NiGe-alloys in various acid solutions at boiling point (test duration 5 × 20 h periods, after each boiling period the specimens were weighed and the acid renewed; only the last 3 boiling periods were taken into account for determination of the weight loss) [482]

The corrosive attack in 5% acid again increases for alloys containing more than 15% Ge. The reason for this is the selective corrosive attack as a result of the two-phase structure of these alloys [482].

Material recommendations for nickel materials

Of the nickel materials, Ni, NiCr- and NiCu-alloys are less suitable on exposure to sulfuric acid than those containing Mo or, Cr and Mo. Since alloys containing Mo cannot be passivated due to the lack of chromium, they are, therefore, more suitable under reducing conditions; whereas materials containing Cr are better for use under oxidizing conditions. The corrosion rates in sulfuric acid are decisive, as can be seen from Figures 110 and 138. Si-containing materials also have a very good resistance. Nickel alloys are more resistant in sulfuric acid than ferrous materials, but are in some cases inferior in hot concentrated acid, especially since the latter can be protected anodically (Figures 90 and 101).

Alkaline Earth Hydroxides

Author: *A. Weser* / Editor: *R. Bender, R. Durham*

Nickel

Nickel and its alloys have very high resistance to alkaline systems.

Nickel is not subject to stress corrosion cracking, except in molten hydroxides and highly concentrated alkali solutions. In 50% boiling NaOH, nickel is attacked and loses 0.06 g/m^2 d (0.0025 mm/a) [483]. No different behavior is expected for alkaline earth hydroxide solutions.

Material designation	*Trade name*	*Chemical composition, %*					
		C max.	*Si*	*Mn*	*Ni*	*Cr*	*Cu*
GGL-NiCuCr 15 6 2	Ni-Resist® 1	3.0	1.0 to 2.8	1.0 to 1.5	13.5 to 17.5	1.0 to 2.5	5.5 to 7.5
GGL-NiCuCr 15 6 3	Ni-Resist® 1 b	3.0	1.0 to 2.8	1.0 to 1.5	13.5 to 17.5	2.5 to 3.5	5.5 to 7.5
GGL-NiCr 20 2	Ni-Resist® 2	3.0	1.0 to 2.8	1.0 to 1.5	18 to 22	1.0 to 2.5	–
GGL-NiCr 20 3	Ni-Resist® 2 b	3.0	1.0 to 2.8	1.0 to 1.5	18 to 22	2.5 to 3.5	–

Table 129: Chemical composition of Ni-Resist® alloys [484]

Nickel-copper alloys

NiCu-alloys do not behave differently from pure nickel and Ni-alloys in alkaline earth hydroxides.

Ammonia and Ammonium Hydroxide

Author: *P. Drodten* / Editor: *R. Bender, R. Durham*

Nickel

The chemical compositions of the pure nickel grades (Table 130) customarily utilized in plant construction are contained in DIN 17740 (09/2002) and differ in carbon content and degree of purity. They combine good mechanical properties with high thermal conductivity and excellent resistance in numerous corrosive media, in particular highly alkaline solutions.

The low carbon (LC) grades with restricted carbon contents of max. 0.02% which prevents the formation of carbide precipitation producing embrittlement at temperatures above 315°C, are utilized preferentially at elevated temperatures. According to the regulations for pressurized containers they are approved for utilization temperatures up to 600°C. A maximum of 1.0 weight% Co is permissible for all grades.

Material No.	*2.4060*	*2.4061*	*2.4066*	*2.4068*
Abbreviation	*Ni99.6*	*LC-Ni99.6*	*Ni 99.2*	*LC-Ni 99*
UNS	*N02200*	*N02205*	*N02200*	*N02201*
C	≤ 0.08	≤ 0.02	≤ 0.10	≤ 0.02
Mn	≤ 0.35	≤ 0.35	≤ 0.35	≤ 0.35
Mg	≤ 0.15	≤ 0.15	≤ 0.15	≤ 0.15
Fe	≤ 0.25	≤ 0.25	≤ 0.40	≤ 0.40
Cu	≤ 0.15	≤ 0.15	≤ 0.25	≤ 0.25
Si	≤ 0.15	≤ 0.15	≤ 0.25	≤ 0.25
Ti	≤ 0.10	≤ 0.10	≤ 0.10	≤ 0.10
V	≤ 0.005	≤ 0.005	≤ 0.005	≤ 0.005

Table 130: Designations and chemical compositions of the commercial pure nickel grades

Nickel is resistant in water-free gaseous ammonia up to temperatures of 300°C and in liquid ammonia up to 50°C [485, 486]. However, nickel is not resistant to moist ammonia with water contents over 1%.

In aqueous ammonia solutions the stability of nickel depends strongly on the concentration and the temperature. In a 1% ammonia solution a corrosion rate of 0.025 mm/a is reported but in 13% solution the rate is > 12 mm/a [487].

Nickel-chromium alloys

Nickel-chromium-iron alloys (without Mo)

Nickel-chromium-molybdenum alloys

The compositions of the nickel-chromium alloys, that can also contain larger amounts of iron and molybdenum, are specified in DIN 17742 (09/2002).

The NiCrMo alloys are listed partly in DIN 17742 (09/2002) and also in DIN 17744 (09/2002). Some frequently utilized NiCrMo alloys are listed in Table 131.

Material No.	*2.4668*	*2.4858*	*2.4619*	*2.4856*	*2.4605*
Abbreviation	*NiCr19NbMo*	*NiCr21Mo*	*NiCr22Mo7Cu*	*NiCr22Mo9Nb*	*NiCr23Mo16Al*
UNS	*N07718*	*N08825*	*N06985*	*N06625*	*N06059*
Designations	Alloy 718	Alloy 825 Incoloy® 825 Nicrofer® 4221 Uranus® 825	Alloy G-3 Inconel® G-3 Nicrofer® 4823 hMo	Alloy 625 Inconel® 625 Nicrofer® 6020 hMo Uranus® 625	Alloy 59 Nicrofer® 5923 hMo
Ni	50–55	38–46	balance	balance	balance
Cr	17–21	19.5–23.5	21.5–23.5	21–23	22–24
Mo	2.8–3.3	2.5–3.5	6–8	8–10	15–16.5
C	0.02–0.08	≤ 0.025	≤ 0.015	≤ 0.025	≤ 0.010
Fe	balance	balance	18–21	≤ 3	≤ 1.5
Cu		1.5–3	1.5–2.5		
Ti		0.6–1.2			
Nb				3.2–3.8	

Table 131: Nickel-chromium-molybdenum alloys

Ni-Cr-Mo alloys are resistant to dry gaseous ammonia up to high temperatures [488]. Depending on the alloying contents, temperatures of 400–600°C are specified as utilization limits in plants for ammonia production [486]. For the nickel-chromium-molybdenum alloys a corrosion rate of less than 0.05 mm/a is reported in the temperature range up to 316°C [488].

Nickel-copper alloys

The NiCu alloys listed in DIN 17743 (09/2002) with a copper content of about 30% (Table 132) are known under the trade names of “Monel”, “Nicorros” or “Silverin” and are characterized by good resistance to reducing media. The NiCu alloy 2.4375 which can be hardened by the addition of aluminum and titanium has considerably higher mechanical values.

Material No.	2.4360	2.4361	2.4363	2.4375
Abbreviation	NiCu 30 Fe	LC-NiCu 30 Fe	NiCu30FeS	NiCu 30 Al
UNS	N04400	N04402		N05500
Designations	MONEL® alloy 400			MONEL® alloy K-500
Cu	28–34	28–34	28–34	27–33
C	≤ 0.15	≤ 0.04	≤ 0.3	≤ 0.2
Mn	≤ 2.0	≤ 2.0	≤ 2.0	≤ 1.5
Fe	1–2.5	1–2.5	≤ 2.5	0.5–2
Al	≤ 0.5	≤ 0.5	≤ 0.5	2.2–3.5
Ti	≤ 0.3	≤ 0.3	≤ 0.3	0.3–1.0
Si	≤ 0.5	≤ 0.3	≤ 0.5	≤ 0.5
V	≤ 0.02	≤ 0.02	0.025–0.060	≤ 0.015
Ni	balance	balance	balance	balance

Table 132: Nickel-copper alloys

In liquid and gaseous ammonia with very low water contents the nickel-copper alloys have good stability [488], but they are strongly attacked when higher water contents are present and in aqueous ammonia solutions [489].

The corrosion rate of 2.4360 as a function of ammomnium concentration in aqueous solution is reported in Table 133 [490].

Concentration of the solution %	Temperature °C	Material consumption g/m^2 d
2.7	20	≤ 0.025
3.6	20	25–75
15	30	25–75
28.5	20	2.5–25

Table 133: Stability of NiCu 30 Fe (Material No. 2.4360) in aqueous ammonia solution [490]

In the technical regulations for steam boilers (TRD), in the TRD 451 that is applicable to plants for the storage of ammonia liquefied under pressure, the utilization of the nickel alloy NiCu 30 Fe (Material No. 2.4360) is not allowed for parts in contact with the medium [491].

Nickel-molybdenum alloys

The nickel alloys with high molybdenum content are listed in DIN 17744 and can also contain higher amounts of chromium. The more important NiMo(Cr) alloys are listed in Table 134.

Material No.	*2.4819*	*2.4610*	*2.4617*
Abbreviation	*NiMo16Cr15W*	*NiMo16Cr16Ti*	*NiMo28*
Standard	*DIN 17744*	*DIN 17744*	*DIN 17744*
UNS	*N10276*	*N06455*	*N10665*
Designations	*Alloy C-276* *Hastelloy® C-276* *Nicrofer® 5716 hMoW*	*Alloy C-4* *Hastelloy® C-4* *Nicrofer® 6616 hMo*	*Alloy B-2* *Hastelloy® B-2* *Nimofer® 6928*
Mo	15–17	14–17	26–30
Cr	15–16.5	14.5–17.5	0.4–1
C	≤ 0.010	≤ 0.009	≤ 0.01
Fe	4–7	≤ 3	1.5–2
Co	≤ 2.5	≤ 2	≤ 1
others	W: 3–4.5	Ti: ≤0.7	
Ni	balance	balance	balance

Table 134: Nickel-molybdenum alloys

The corrosion rates of 2.4819 and 2.4617 in ammonium environments are reported in Table 135 [492]. The low corrosion resistance of the material 2.4617 in aqueous ammonia solution at higher temperatures is also confirmed in other investigations [493].

Medium	*Concentration %*	*Temperature °C*	*Corrosion rate mm/a 2.4819*	*Corrosion rate mm/a 2.4617*
Ammonia gas	27 6% CO_2, 65% H_2O	85	< 0.05	< 0.05
Ammonia gas		(200–250)	0.05–0.25	
Liquid ammonia	100	up to 310	> 0.25–< 0.51	> 0.25–< 0.51
Liquid ammonia	100	90	< 0.05	0.05–0.25
Ammonia solution	concentrated	300	> 0.25–< 0.51	> 1.27
Ammonia solution		120	0.05–0.25	
Ammonia solution		66	< 0.05	
Ammonia solution	27	50–90	< 0.05	

Table 135: Stability of the materials 2.4819 and 2.4617 to ammonia and aqueous ammonia solution [492]

Lithium Hydroxide

Author: *K. John* / Editor: *R. Bender, R. Durham*

Nickel

Unalloyed nickel is a material which is particularly resistant to attack by lithium hydroxide solutions and molten lithium hydroxide. A passive layer 0.3 to 0.6 nm thick is formed in 2 mol/l lithium hydroxide solution at 25°C [494].

In addition to platinum and gold, nickel is also one of the most resistant materials to corrosion by molten lithium hydroxide. Its material consumption rate in the melt (80 mol percent LiOH, 20 mol percent LiF, 800°C) is 0.16 g/m^2 h, corresponding to about 0.28 mm/a. Moreover, under these conditions nickel exhibited no signs whatsoever of intergranular corrosion. The corrosion rate of nickel 200 is calculated to be about 0.24 mm/a from 577 hour tests at 650°C. On partial immersion, corrosion in the vapor phase is slightly less than 0.19 mm/a [495].

The dependence of the material consumption rate of nickel in molten lithium hydroxide on temperature is comparable with that of the titanium-stabilized steel Ch18N9T (c.f. 1.4541) [496].

The good resistance of nickel in molten lithium hydroxide may be attributable to the incorporation of lithium ions into the NiO lattice. For electrical neutrality reasons this reduces the concentration of the nickel ion vacant sites which determine transport of the nickel ions through the dense NiO layer and therefore the corrosion rate. Oxidation tests with nickel at 1,000°C in oxygen containing Li_2O at 0.1 MPa confirm the validity of this assumption. The lithium oxide reduces the oxidation rate of nickel from 0.9×10^{-6} to 1.2×10^{-7} g^2/cm^4 h [497].

Nickel-chromium alloys

Nickel-chromium-iron alloys

Nickel-chromium-molybdenum alloys

As already mentioned, the water in the primary circuit of boiling water reactors is adjusted to a pH of 10, for which purpose lithium hydroxide and additionally 25 to 50 ppm of hydrogen are used so as to reduce radiation problems by keeping the quantity of radiolytically produced oxygen as small as possible. Inconel® 600 is used in addition to SAE 304 steel for this water circuit. Figure 140 shows the chronological pattern of corrosion [498, 499]. The influence of surface preparation and heat treatment was noticeable amongst the samples with ground samples exhibiting higher rates of attack. Comparison of the curves with borate-free water (1 and 2) seemed to indicate that the addition of boric acid increased the corrosion rate regardless of the surface preparation.

After damage due to stress corrosion cracking had occurred in the primary circuit of various pressurized water reactors using Inconel® 600 as a material, the behavior of this material in lithium hydroxide was investigated more closely. In this context, in 1 mol/l lithium hydroxide solution at 220°C no stress corrosion cracking had occurred in Inconel® 600 even after 266 days, whereas at 335°C up to 10 cracks per cm^2 were counted after 88 days.

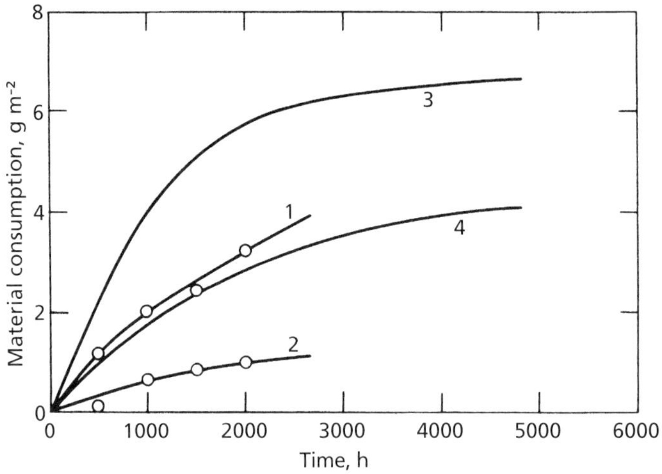

Figure 140: Chronological pattern of material consumption per unit area from Inconel® 600 [499]
1, 2 – in water containing lithium hydroxide at 288°C (1: ground, 2: bright annealed)
3, 4 – in water containing boric acid at 316°C (3: ground, 4: bright annealed)

Cracking in the pressurized water was completely prevented by the addition of lithium hydroxide and hydrogen as previously reported [500].

According to this, stress corrosion cracking of Inconel® 600 in the primary circuit of a nuclear reactor can be largely prevented provided that a pH of 10 is set by the addition of lithium hydroxide in combination with boric acid and hydrogen.

The water in the primary circuit of the ship "Otto Hahn" with nuclear propulsion was adjusted with lithium hydroxide to a pH of 9, which is weaker than usual, and without boric acid. Corrosion of Inconel® 600 in the steam generator and of Zircaloy® 4, the cladding material for the fuel elements, was slight. The principal corrosion products of chromium, nickel and iron compounds were mainly present in undissolved form and were therefore not deposited on the tubes that had been clad [501].

Table 136 lists nickel alloys resistant to the melt already mentioned (80 mol percent LiOH, 20 mol percent LiF, 800°C). According to this, Inconel® 600 and Nimonic® are the most suitable nickel alloys [495]. Although nickel, with a corrosion rate of only 0.25 mm/a, proved to be the most resistant in a 550 hour test of this melt, at 650°C and exposed to high-pressure steam, its high-temperature strength leaves something to be desired; although it loses 2.5 mm/a under the same conditions, Inconel® 600 is to be preferred because of its better high-temperature strength. All other nickel-chromium alloys, with the exception of Nimonic®, are not sufficiently resistant [495].

Tubes of Inconel® 600 partially immersed in the abovementioned melt (80 mol percent LiOH, 20 mol percent LiF) are attacked 10 to 20% more severely on the outside and in the vapor phase. At 500°C after 100 hours, however, no corrosion can be detected (Figure 141) [495].

Nickel alloy	*Composition, %*					*Corrosion rate mm/a*
	Ni	*Cr*	*Fe*	*Mo*	*Other elements*	
Hastelloy® C	57	16	5	16	2Co, 4W	595
Hastelloy® F	46	22	19	6	3Co, 2Mn, 1Si	2,678
Hastelloy® G	46	22	20	7	2Cu	2,529
Hastelloy® N	71	7	4	16	–	37.2–64
Hastelloy® X	63	5	4	26	–	> 2,380
Inconel® 600	75	16	8		1Mn	1.5–37.2
Inconel® 702	80	16	–	–	3Al	> 74.4
Inconel® 718	55	18	18	3	5Nb	> 1,636–2,529
Inconel® W	75	15	6	–	2Ti	44.6
Haynes® R-41	50	19	5	10	11Co, 3Ti, 2Al	2,529
Chlorimet® 3	60	18	2	18	–	> 148
Chromel®	80	20	–	–	–	> 13.6
Nichrome® V	80	20	–	–	–	13.6
Nichrome®	60	16	14	–	–	22.3
Nimonic®	78	21	1	–	–	2.2
Incoloy® 901	43	13	34	7	3Ti	20.8
Tophet® 30	70	30	–	–	–	> 59

Table 136: Corrosion rates of some nickel-chromium alloys in molten LiOH/LiF (80/20) at 800°C, calculated from 3 hour tests [495]

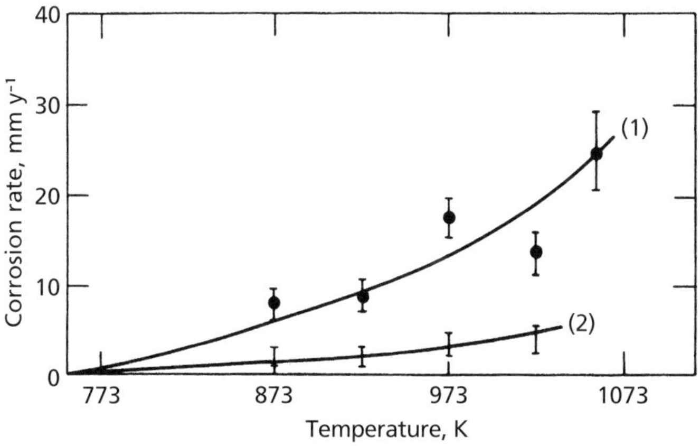

Figure 141: Corrosion rate of a tube of Inconel® 600 as a function of temperature in a 100 hour test [495]
1 – on the outside
2 – on the inside

At higher temperatures corrosion of the outer surface becomes increasingly greater in comparison with that on the inner surface.

Nickel-copper alloys

Nickel-molybdenum alloys

Nothing has been published in the relevant literature on the corrosion behavior of nickel-copper and nickel-molybdenum alloys in solutions containing lithium hydroxide.

The corrosion behavior of these alloys in contact with lithium hydroxide solutions can nevertheless be estimated from their behavior in contact with sodium and potassium hydroxides.

Table 137 shows the corrosion rates of these alloys in the previously mentioned melt (80 mol percent LiOH, 20 mol percent LiF, 800°C). As may be seen, the corrosion behavior of Monel® K-500 is similar to that of Inconel® 600. It can be further discerned that the otherwise highly-resistant NiMoFe-alloy Hastelloy® B is attacked the most severely. It needs to be assessed whether the Ni-15Mo-15Fe alloy, which is more resistant than other nickel-molybdenum and nickel-molybdenum-iron alloys in molten sodium hydroxide at 825°C is also superior in lithium hydroxide [502].

Alloy	*Composition, %*					*Corrosion rate mm/a*
	Ni	*Cr*	*Mo*	*Fe*	*Other metals*	
Monel®	66	32	–	1	1Mn	10.4–34
Inconel® 600	75	16	8	–	1Mn	1.5–37.2
Monel® K-500	65	30	–	1	1Mn, 3Al	0.3–59
Hastelloy® B	64	–	28	5	1Cr	104

Table 137: Corrosion rate of a number of nickel alloys in molten LiOH-LiF (80/20) at 800°C, calculated from 3 hour tests [495]

According to Figure 142, the material consumption rate of nickel between 497 and 797°C is virtually independent of temperature, whereas that of Monel® very much increases with temperature. Whereas Monel loses 0.19 g/m^2 h (corresponding to about 0.18 mm/a) at 500°C; the material consumption rate at 800°C is approximately 0.4 g/m^2 h (corresponding to a corrosion rate of about 0.38 mm/a). Moreover, Monel® alloys, in contrast to nickel, exhibit slowly progressive intergranular attack, this being particularly marked in the case of Monel® K-500 [495].

As comparative tests with nickel containing 2.8% TiO_2 and carbon have shown, the material consumption rate in molten lithium hydroxide at 775°C of 0.3 g/m^2 h corresponding to a corrosion rate of about 0.29 mm/a, is significantly greater than that of pure nickel [495].

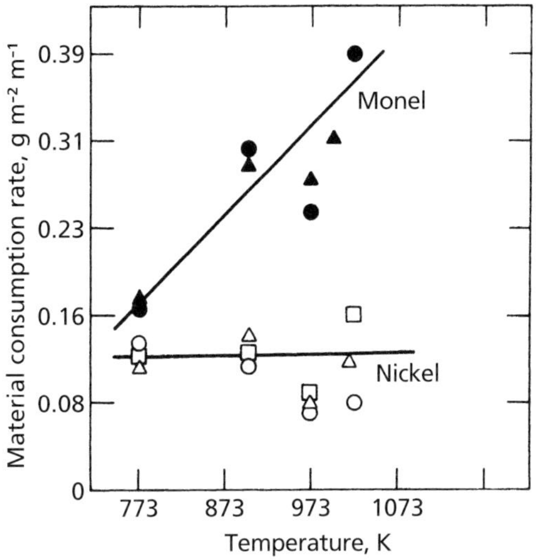

Figure 142: Material consumption rates of Monel® and nickel in molten LiOH-LiF (80/20) as a function of temperature (test duration 100 hours) [495]

● Monel®
▲ Monel® 400
□ Nickel 200
△ Nickel 201
○ pure nickel

Potassium Hydroxide

Author: *P. Drodten* / Editor: *R. Bender, R. Durham*

Nickel

The chemical compositions of pure nickel grades (Table 138) customarily utilized in plant construction are contained in DIN 17740 [503]. They differ in carbon content and purity. These materials combine good mechanical properties with high thermal conductivity and excellent resistance in numerous corrosive media, in particular alkaline solutions. After silver, nickel shows the highest resistance to alkaline solutions among the metals, also at high concentrations and temperatures.

Material No.	*2.4060*	*2.4061*	*2.4066*	*2.4068*
Abbreviation	Ni99.6	LC-Ni99.6	Ni 99.2	LC-Ni 99
UNS	N02200	N02205	N02200	N02201
C	≤ 0.08	≤ 0.02	≤ 0.10	≤ 0.02
Mn	≤ 0.35	≤ 0.35	≤ 0.35	≤ 0.35
Mg	≤ 0.15	≤ 0.15	≤ 0.15	≤ 0.15
Fe	≤ 0.25	≤ 0.25	≤ 0.40	≤ 0.40
Cu	≤ 0.15	≤ 0.15	≤ 0.25	≤ 0.25
Si	≤ 0.15	≤ 0.15	≤ 0.25	≤ 0.25
Ti	≤ 0.10	≤ 0.10	≤ 0.10	≤ 0.10
V	≤ 0.005	≤ 0.005	≤ 0.005	≤ 0.005

Table 138: Designations and chemical compositions of the commercial pure nickel grades

The LC grades (low carbon types) have a limited carbon content of max. 0.02%, so that carbide precipitation leading to embrittlement at temperatures above 315°C is prevented. These grades are preferentially utilized at elevated temperatures. According to the regulations for pressurized containers they are approved for utilization temperatures up to 600°C. A maximum of 1.0 weight-% Co is permissible for all grades [504].

In a 6 M KOH solution in the temperature range from 25 to 30°C nickel is passive and the corrosion rate for uniform surface corrosion is negligibly small.

In Table 139 investigation results are summarized which were obtained in the laboratory for pure nickel (nickel 200, Material No. 2.4066) compared with the nickel alloy Inconel® 600 (Alloy 600, Material No. 2.4816), in various KOH solutions under various test conditions [505].

Material	***Corrosion rate mm/a***							
	Boiling point				***150°C***			
	30% KOH		***47% KOH***		***50% KOH***	***70% KOH***	***50% KOH***	***70% KOH***
	saturated with KCl 0.05% potassium chlorate		***saturated with KCl 0.078% potassium chlorate***		***6.6 m/min 1)***		***106 m/min 1)***	
	26 days				***7 days***		***35 days***	
	liquid phase	***vapor phase***	***liquid phase***	***vapor phase***	***liquid phase***			
Nickel 200	0.005	0.0025	0.0025	0.0075	2)	0.0100	2)	0.040
Alloy 600	0.0025	0.0025	0.0100	0.0025	0.0125	0.0175	0.0125	0.143

1) Rotation speed of the specimens (for 8 hours per day and at the weekends the specimens were at rest). This corresponds to ten 24-hour days for the specimen at 106 m/min
2) Slight weight increase

Table 139: Corrosion rates of specimens of pure nickel and alloy 600 in KOH solutions under various test conditions [505]

The addition of sulfur compounds or chlorides can strongly accelerate the corrosion of nickel [506].

Advanced water electrolysis systems that operate with potassium hydroxide solution as the electrolyte solution and are designed for operating temperatures of 160°C or even 200°C are subject to high material stress. Against this background exposure tests and electrochemical investigations of the anodic dissolution behavior of nickel and the three nickel based alloys Alloy 600 (NiCr15Fe, Material No. 2.4816), Alloy 690 (NiCr29Fe, Material No. 2.4642) and Monel® 400 (NiCu30Fe, Material No. 2.4360) were carried out [507]. The electrolyte solution was simulated with a 40% KOH solution with oxygen partial pressures from 2 to 18 bar. 160 and 200°C were chosen as the test temperatures. The following materials and operating parameters were taken into consideration:

- Chemical composition (C and S contents)
- Heat treatment
- Welding
- Presence of crevices and water lines.

From the test results it is concluded that in electrolysis systems with operating temperatures higher than 160°C Nickel 200 (Ni 99.6, Material No. 2.4060) and Nickel 201 (LC-Ni 99.6, Material No. 2.4061) are the only utilizable materials, at least for relatively thin walled components such as electrodes, gas separators etc.

The corrosion rate of these materials is negligible also in the region of weld seams under normal conditions, but it can increase strongly in crevices or at water boundary lines. This should be taken into consideration for optimized design. For the three nickel alloys corrosion rates of 0.05 to 0.3 mm/a were found, whereby thick but poorly adherent oxide layers appeared. Contamination of the solution with chloride ions led to a strong increase in the corrosion rate of the material Alloy 600, which already had the poorest resistance of the three investigated materials.

As Figure 141 shows, the current density of a nickel coating deposited on platinum in 0.1 to 0.5 M potassium hydroxide solutions was less than 5 $\mu A/cm^2$ in the passive range between $U_{sce} = -1$ to 0.3 V, wherefrom a corrosion rate of less than 0.008 mm/a is determined by calculation [508].

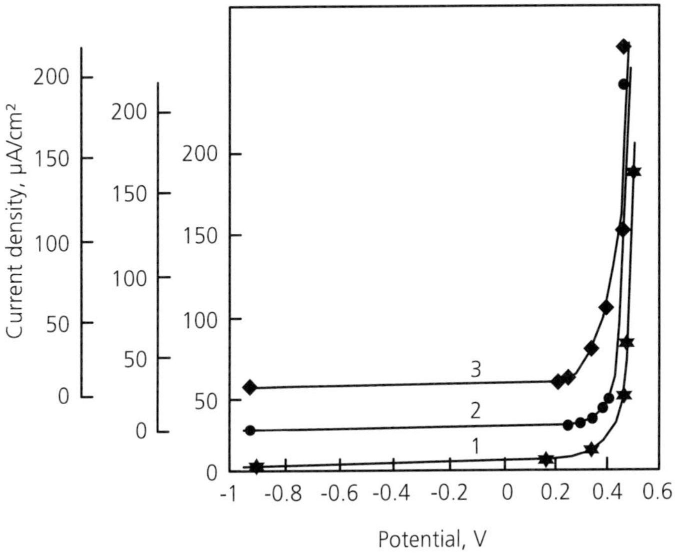

Figure 143: Current density/voltage curves of nickel coatings deposited on platinum in KOH solutions of various concentrations at 30°C (1 = 0.1 M KOH, 2 = 0.3 M KOH, 3 = 0.5 M KOH) [508]

Electroless deposited nickel layers have proven successful for protecting unalloyed steel against attack by alkaline solutions [509]. Such nickel layers have in some investigations shown even better behavior than Nickel 200 (Ni 99.6, Material No. 2.4060) [510] (see also [511]).

An example of this is described in [512]: An evaporator boiler for NaOH and KOH developed leaks on the circumferential and longitudinal butt-welded joints. The design of the boiler required that the material combination was resistant to scaling on the exterior and corrosion resistant by the above media on the interior. An explosive cladding of nickel plate was deposited on the chosen base material, a high-alloy steel Material No. 1.4541. During evaporation of the caustic solutions, the necessary critical temperatures and concentrations for stress corrosion cracking of nickel were reached. The stress necessary for the generation of SCC damage was reached at the high temperatures because of the differing heat expansion coefficients of the two materials 1.4541 and nickel with a ratio of 16 to 13.3. At the operating temperature of more than 300°C this gave rise to a tensile stress of approximately 150 N/mm^2 in the nickel. To this stress the residual stress of the butt-welds from welding was added to give an overall critical level of stress.

Information concerning the influence of chloride ions on the pitting corrosion of pure nickel in KOH solutions is provided by Figure 144 [513].

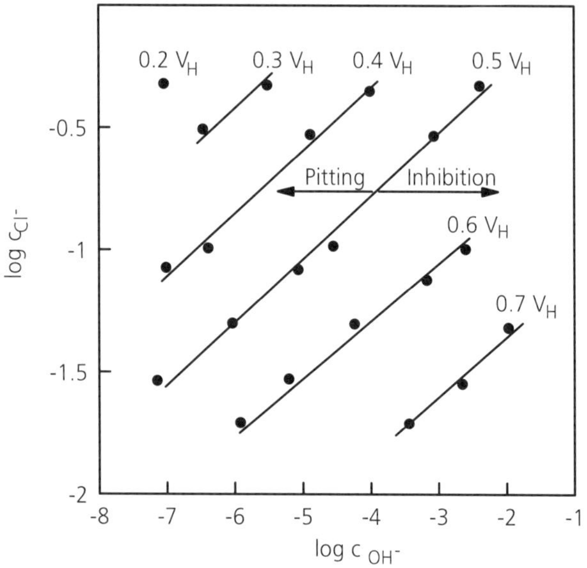

Figure 144: Conditions for the appearance of pitting corrosion on nickel depending on the potential and on the concentration of chloride and hydroxyl ions in KOH solutions [513]

Stress corrosion cracking of nickel can take place at high alkali concentrations and high temperatures (Figure 145). The stresses necessary to trigger SCC at critical concentrations and temperatures can also be caused by construction or manufacture.

According to existing experience, nickel can be considered as resistant to potassium hydroxide solutions also at high temperatures and concentrations as well as in KOH melts up to about 500°C. By virtue of its favorable electrochemical properties and excellent corrosion resistance, nickel is also utilized as an electrode material in alkaline batteries with KOH as an electrolyte solution.

Nickel-chromium alloys

Nickel-chromium-iron alloys (without Mo)

Since in the presence of chromium other alloying elements, such as iron, have a positive influence on the corrosion resistance of nickel, pure nickel-chromium alloys without the addition of other alloying elements are practically not utilized in alkaline solutions or alkali metal hydroxide melts.

Nickel-chromium-iron alloys and in particular the most important representative of this range of alloys, NiCr15Fe (Material No. 2.4816, Inconel® 600, Alloy 600) are distinguished by their high resistance to uniform surface corrosion and to stress corrosion cracking in alkaline solutions [514]. However, at high temperatures significantly stronger attack can take place in KOH solutions of higher concentration in critical places such as crevices [507].

The damage observed, for example, in pressurized water reactors by stress corrosion cracking on steam generating pipes made of Inconel® 600 in contact with water containing low concentrations of potassium hydroxide, is due to local concentration of the potassium hydroxide, mostly in crevices [514–517].

The temperature and concentration fields shown in Figure 145 for the occurrence of stress corrosion cracking on nickel and NiCr15Fe are based on a systematic assessment of damage on components made of these materials in NaOH solutions, but they are certainly also helpful for assessing the behavior in potassium hydroxide solutions [518].

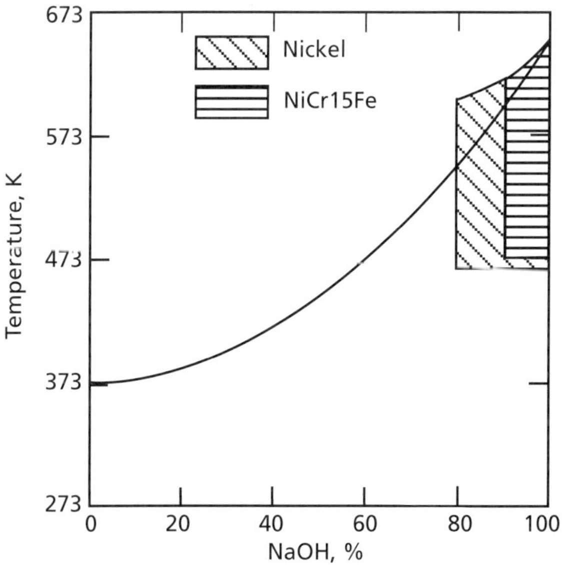

Figure 145: Regions of sensitivity for stress corrosion cracking of nickel and NiCr15Fe in NaOH solution depending on the temperature and concentration of the alkaline solution (the continuous line corresponds to the boiling point curve) [518]

Since the stress corrosion cracking behavior of the alloy Incoloy® 600 in alkaline solutions is strongly influenced by chromium carbide precipitations at the grain boundaries, extensive investigation results of the influence of the carbon content, of the heat treatment and of the carbide precipitations are available [519–526]. However, these investigations have been carried out exclusively in NaOH solutions [511]. It can be assumed that the results described there can be applied analogously to the behavior of the material Incoloy® 600.

In comparative investigations, Alloy 625 exhibited the greatest crack depths and also noticeably more cracks than the alloys Alloy 600 and Alloy 690. This is explained by the high molybdenum content of these alloys. After a corresponding heat treatment that led to chromium-depleted regions and in spite of the high chromium content, alloy 690 exhibited good behavior that corresponds to alloy 600 which contains less chromium.

Investigations of the influence of the elements C, S and P on the stress corrosion cracking behavior of the alloy Alloy 600 and the various Fe-Ni-Cr alloys produced in the laboratory in alkaline solutions showed that the contents of the elements C, S and P as well as their distribution depending on the heat treatment play a decisive role for the initiation and the course of cracks. Sulfur in amounts above 20 ppm always has a strongly negative influence [527].

The materials Inconel® 600 and Incoloy® 800 (X2NiCrAlTi32-20, Material No. 1.4558) are also utilized for the pipe systems in steam generators of pressurized water reactors. For the corrosion damage observed on U-pipes of such steam generators, conditioning substances of the secondary water , among others, could be responsible. With this conditioning, strong concentration of the alkalizing agents can take place in crevices and other places with critical heat transfer.

Against this background numerous investigations were made of the uniform surface corrosion, of the inter-crystalline corrosion and of the inter-crystalline stress corrosion cracking on the materials Inconel® 600 (Material No. 2.4816) and Incoloy® 800 (1.4558, with the compositions shown in Table 140), in solutions containing alkalizing agents, whose concentrations were increased by the factor of 10^4 relative to the directives for water conditioning [528, 529].

Material No.	*% C*	*% Si*	*% Mn*	*% Cr*	*% Ni*	*% Cu*	*% Al*	*% N*	*% Ti*	*% Fe*
1.4558	0.018	0.54	0.57	21.2	35.6	0.10	0.16	0.016	0.58	bal.
2.4816	0.016	0.36	0.80	16.28	72.80	0.004	0.18	0.017	0.39	9.38

Table 140: Chemical composition of the investigated materials [528]

Further details of the investigations mostly carried out in NaOH solutions as well as further extensive investigations of the stress corrosion cracking behavior of the NiCrFe alloys in NaOH solutions described in [530–537], that can certainly also provide information on the behavior in potassium hydroxide solution, are reported in [511].

Nickel-chromium-molybdenum alloys

As Figure 146 shows for the example of a NaOH solution, with increasing molybdenum content of the nickel-chromium-molybdenum alloys the corrosion resistance in boiling alkaline solutions increases.

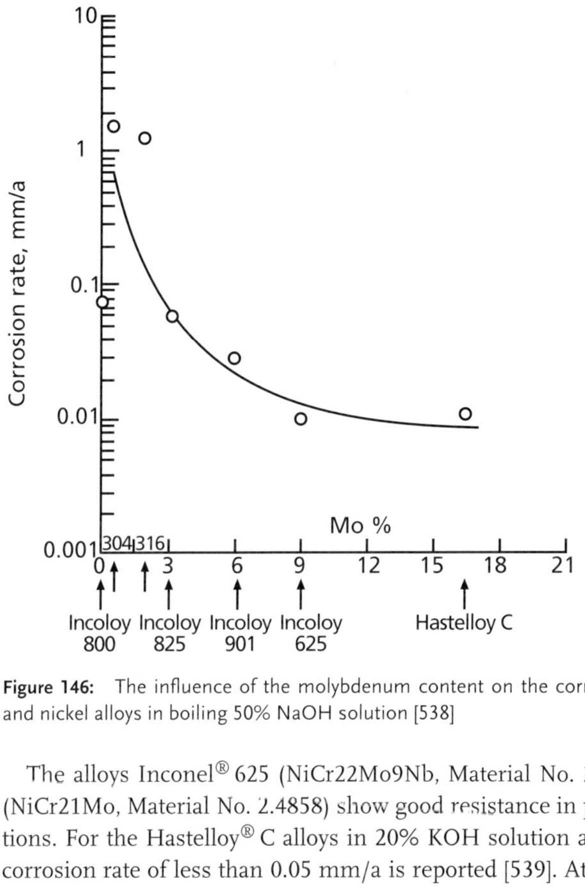

Figure 146: The influence of the molybdenum content on the corrosion of chrome nickel steels and nickel alloys in boiling 50% NaOH solution [538]

The alloys Inconel® 625 (NiCr22Mo9Nb, Material No. 2.4856) and Incoloy® 825 (NiCr21Mo, Material No. 2.4858) show good resistance in potassium hydroxide solutions. For the Hastelloy® C alloys in 20% KOH solution at a temperature of 95°C a corrosion rate of less than 0.05 mm/a is reported [539]. At higher caustic alkali concentrations and at boiling temperatures the corrosion rates increase to values between 0.25 and 0.51 mm/a. Nothing is known regarding the incidence of stress corrosion cracking of these alloys in alkaline solutions.

The application ranges in sodium hydroxide solutions dependent upon the temperature and concentration of the caustic solution are given in Figure 147 for some nickel-based materials, including the alloys Ni20Cr15Mo and Ni21CrW. Thus, pure nickel can be used in all ranges and shows superior corrosion resistance compared to its alloys [540]. Since for most materials NaOH solution is a more critical medium than KOH solution, the limit lines shown in this diagram constitute a safe conservative estimate for the behavior in potassium hydroxide solution.

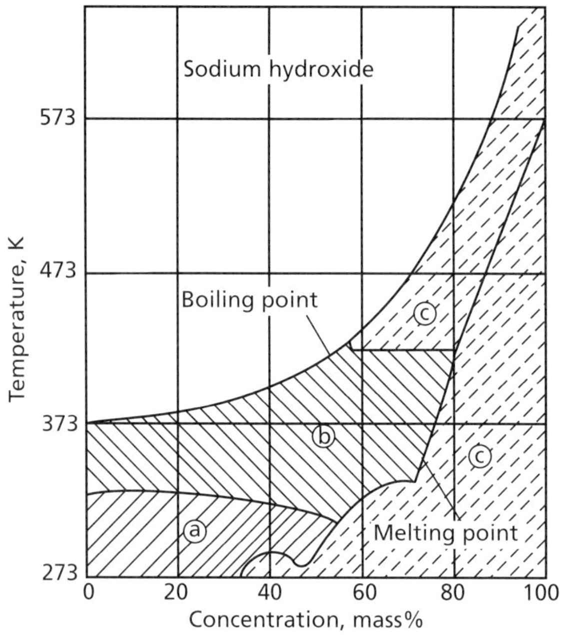

Figure 147: Application fields of pure nickel and nickel alloys in sodium hydroxide solutions [540].
(a): LC-Ni 99, NiCu30Fe,
(b): LC-Ni 99, NiCu30Fe, NiCr20Mo15, NiCr15Fe, NiCr21Mo and
(c): LC-Ni 99, NiMo30

Nickel-copper alloys

The nickel-copper alloy Monel® 400 (Ni-32.7Cu-1.3Fe-0.15Co-1.1 Mn-0.15C, Material No. 2.4360, NiCu30Fe) is practically resistant to uniform surface corrosion in alkaline solutions [504]. Monel® 400 also shows better resistance to stress corrosion cracking in alkaline solutions than Inconel® 600 [541]. Heat treatment for 5.5 h at 600°C has a beneficial effect on the resistance to stress corrosion cracking [541].

Monel® 400 along with nickel LC-Ni 99 are two of the most corrosion-resistant materials in hot concentrated alkaline solution. Monel® 400 is also quite resistant in molten NaOH, although LC-Ni 99 is even better [540].

Nickel-molybdenum alloys

The nickel-molybdenum alloys (e.g. the Hastelloy® B types) have been developed preferentially for utilization in reducing acids. No experience is available regarding their behavior in aqueous alkaline solutions.

Other nickel alloys

Nickel-niobium alloys with nickel contents from 47% are resistant to potassium hydroxide solution at room temperature and are possible materials for the electrolytic production of hydrogen in plants with higher performance [542]. With lower nickel contents not only strong corrosion, but also embrittlement of the alloy takes place due to hydride formation. Nickel alloys based on Ni50-Fe50 and Ni50-W50 are suitable for electrodes for the electrolysis of 30 to 40% potassium hydroxide solution. Also, for electrodes made from Ni-50W and Ni3Ti a thin extremely low conducting nickel oxide film is produced in a 30% KOH solution at 80°C, that gives good protection against corrosion, whereas corresponding Ni-Ru and Ni-Ir alloys have been found to be non-resistant in KOH solutions [543, 544].

Sodium Hydroxide

Author: *P. Drodten* / Editor: *R. Bender, R. Durham*

Nickel

After silver, nickel shows the highest resistance of metals to highly concentrated NaOH solutions. The corrosion data given in Table 141 provide preliminary information [545].

NaOH *%*	*Temperature* *°C*	*Corrosion rate* *mm/a*
10–15	80–100	0.0005–0.005
30–50	90–140	0.005–0.03
60–75	140–200	0.02–0.04
75–100	260– 450	0.05–0.06

Table 141: Corrosion rates of nickel in NaOH solutions at various temperature ranges [545].

The isocorrosion diagram for commercially pure nickel, Nickel 200 and the low carbon grade, Nickel 201, is shown in Figure 148. Very low corrosion rates are shown over most concentration and temperature ranges below the boiling point curve. The maximum rate at higher concentrations and temperatures is 0.13 mm/a [546].

As shown in Table 142, sulfur compounds in the NaOH solution can strongly accelerate the corrosion of nickel [545].

Additive	*Corrosion rate*	
	g/m² d	*mm/a*
no additive	0.35	0.02
0.75% Na_2S	14.1	0.57
0.75% $Na_2S_2O_3$	4.90	0.20
0.75% Na_2SO_3	3.20	0.13
0.75% Na_2SO_4	0.40	0.02

Table 142: Corrosion rates of nickel in 75% NaOH solutions with various additives at 130°C [545]

As a result of the good corrosion resistance of nickel, even in concentrated NaOH solutions at elevated temperatures (70 to 170°C), nickel is used at 70°C as an anode in cleaning cells with 50% NaOH solution and a cathode made of carbon or graphite [547].

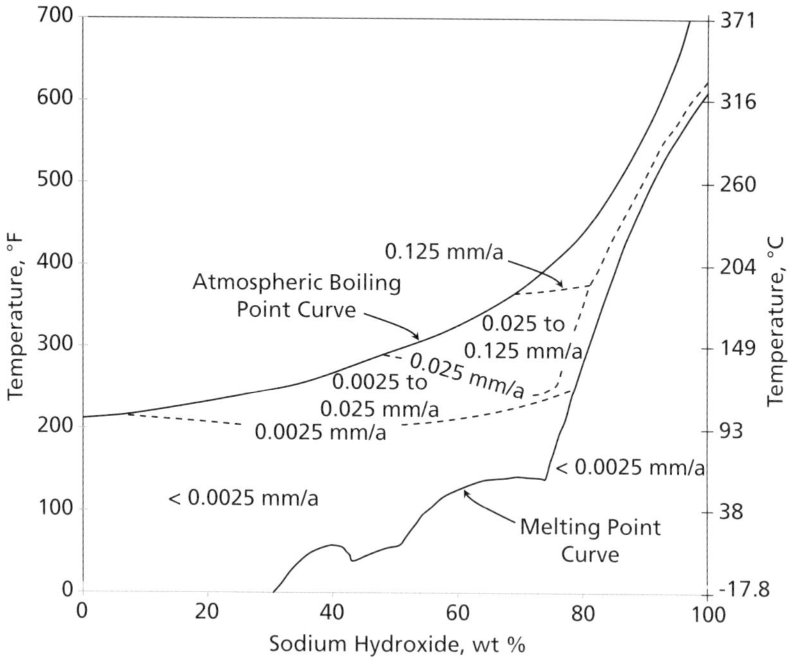

Figure 148: Isocorrosion chart for pure Nickel, Nickel 200 and Nickel 201 in sodium hydroxide [546]

Nickel-200 (Ni 99.6, DIN-Mat. No. 2.4060) corrodes in 50% NaOH solution at 55 to 75°C with an average corrosion rate of approximately 0.0005 mm/a. Nickel-201 (LC-Ni 99, DIN-Mat. No. 2.4068), which contains only max. 0.02% carbon compared to Nickel-200 with max. 0.15% C, should be preferred for use at temperatures above 310°C to avoid embrittlement which can occur for Nickel-200 as a result of carbide precipitation. Nickel-201 is the standard material for downdraft evaporators, that are used to produce up to 75 t/d NaOH [548].

A corrosion rate of approximately 0.0024 mm/a was measured on nickel evaporator plants in which NaOH solutions are concentrated from 30 to 50% at 240°C. On evaporating from 60 to 75% NaOH (at 150 to 180°C) or from 60 to 100% NaOH (at 150 to 260°C), the corrosion rates were 0.026 to 0.079 mm/a [549]. This is in agreement with the data reported in [545] in which a corrosion rate of approximately 0.075 mm/a is given for nickel in boiling NaOH solution of more than 75%.

For the storage of NaOH solutions in nickel tanks, a corrosion rate of only 0.00053 mm/a is reported for a 50% NaOH solution between 55 and 65°C while a rate of 0.040 mm/a is reported for a 75% NaOH solution at 135°C [545].

Electroless deposition of nickel coatings can be used for the protection of unalloyed steel for valves. Tempering experiments in 10% NaOH solutions at room temperature were carried out on steel specimens with such nickel coatingscoating and no corrosion attack was observed while the corrosion rates given in Table 143 were measured for the control specimens of unalloyed steel and of stainless 18-8-CrNi steel [550].

Material	Corrosion rate mg/m² h
Ni cladding	0
18-8-CrNi steel	0.6
unalloyed steel	18.6

Table 143: Corrosion rates of electrolessly deposited nickel coatings in 10% NaOH solution at 20 to 25°C compared to unalloyed steel and stainless 18-8-CrNi steel [509]

The favorable behavior of electrolessly deposited nickel coatings in NaOH solutions is also reported in [551]. Considerable damage was observed on the steel valves of railway tank wagons used to transport NaOH solutions. This damage was only brought under control by the use of nickel-plated valves. As the results given in Table 144 demonstrate, the studies carried out concerning this problem on electrolessly deposited nickel coatings showed better behavior than nickel alloy 200 (Ni 99.2, DIN-Mat. No. 2.4066).

Temperature °C	Flow rate m/s	Corrosion rate µm/a	
		Ni coating	Ni200
25	static	0.7	1.3
	1.83	0.8	1.8
60	static	0.7	1.3
	1.83	0.8	1.5
93	static	1.5	2.0
	1.83	3.6	4.0

Table 144: Corrosion rates of electrolessly deposited nickel coatings in static and flowing 50% NaOH solutions at 25 to 93°C compared to nickel alloy 200 [510]

Stress corrosion cracking of nickel can take place at high NaOH concentrations and high temperatures (Figure 145). The stresses necessary to trigger SCC at critical concentrations and temperatures can also be caused by construction or manufacture. An example of this is described in [552]: an evaporator boiler for NaOH and KOH developed leakages on the circumferential and longitudinal butt-welded joints. The design of the vessel required that the material combination was resistant to scaling on the exterior and resistant to corrosion by the above media on the interior. An explosive cladding of nickel plate was deposited on the chosen base material, a high-alloy steel DIN-Mat. No. 1.4541. During evaporation of the caustic solutions, the necessary critical temperatures and concentrations for stress corrosion cracking of nickel was reached. The stress necessary for the generation of SCC damage was reached at the high temperatures because of the differing heat expansion coefficients of the two materials 1.4541 and nickel with a ratio of 16 to 13.3. At the operating temperature of more than 300°C this gave rise to a tensile stress of approximately 150 N/mm^2 in the nickel. The residual stress of the butt-welds was added to this stress to give an overall critical stress level.

Nickel-chromium alloys

Nickel-chromium-iron alloys (without Mo)

Pure nickel-chromium alloys without addition of other alloying elements are not used in NaOH solutions or NaOH melts.

Corrosion resistance data of various nickel alloys has been reported in [554] in an effort to determine the useful temperature-concentration ranges with respect to general corrosion rate and stress corrosion cracking susceptibility. Details of the investigated alloy compositions are given in Table 145.

Alloy name	*UNS No.*	*Fe*	*Cu*	*Ni*	*Cr*	*Ti*	*Mo*	*Nb*	*W*
400	N04400	1.7	32.5	6	–	–	–	–	–
600	N06600	9.5	–	7	15.8	–	–	–	–
690	N06690	10.0	–	6	2	–	–	–	–
686	N06686	0.5	–	5	20.5	0.06	16	–	4
625	N06625	3.0	–	6	2	–	–	3.5	–
C-276	N10276	5.5	–	5	16	–	–	–	–
C-4	N06455	2	–	65	16	–	–	–	–
22	N062622	2.5	–	57	22	–	–	–	–
2000	N06200	3*	1.6	60	23	–	–	–	–

* maximum

Table 145: Nominal chemical compositions of selected nickel alloys [554]

General corrosion test results from this are shown in Table 146.

In 10% NaOH, boiling at 103°C all alloys examined displayed very low corrosion rates regardless of nickel content. In 50% NaOH boiling at 143°C and 70% NaOH boiling at 182°C, the investigated alloys still showed very low corrosion rates. Alloys 600 and 400 are common materials for handling hot concentrated caustic. These alloys continue to exhibit good corrosion resistance at all concentrations, as would be expected.

In boiling 50% NaOH there is an expected beneficial effect of increasing nickel content as shown in Figure 149. These results indicate that the critical nickel content for resistance to corrosion in this solution is about 50%. However, 20–30% nickel levels also provide a dramatic improvement compared to stainless steel 304 [554].

Alloy	*10% NaOH boiling at 103°C*	*50% NaOH boiling at 143°C*	*70% NaOH boiling at 182°C*
600	< 0.025	< 0.025	< 0.025
686	< 0.025	< 0.025	0.18
625	< 0.025	< 0.025	< 0.025
400	–	< 0.025	–
600*	–	< 0.025	–
C-4*	–	< 0.025	–
2000*	–	0.05	–
C-22*	–	0.076	–
C-276	–	< 0.025	–

* Results from reference [553]

Table 146: General corrosion rates in boiling NaOH solutions. Average of duplicate specimens, 168 hr test (mm/a) [554]

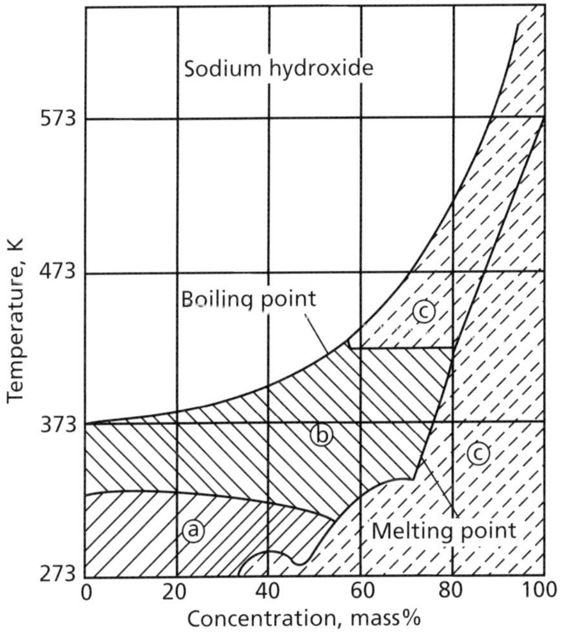

Figure 147: Effect of nickel content on corrosion rate in boiling 50% NaOH [554]

Nickel-chromium-iron alloys, particularly the most important representative of this type of alloys, NiCr15Fe, DIN-Mat. No. 2.4816 (Inconel® 600, alloy 600), are distinguished by their high resistance to uniform surface corrosion and to stress corrosion cracking in NaOH solutions. During investigations of Inconel® 600 in 10% NaOH solutions at a temperature of 325°C and a pressure of 275 MPa, there was no indication of crack formation after 100 h. Isolated cracks had an average depth between 0.05 and 0.08 mm after 200 days [555].

U-bend specimens of the same alloy exhibited no crack formation after 2 weeks in NaOH melts at 370°C. Additions of 4% Na_2SO_4, of 5% $Na_2Cr_2O_7$ and of 5% NaCl did not have any effect [555].

Based on the systematic evaluation of the damage to components made of nickel and NiCr15Fe, the susceptibility to stress corrosion cracking of the two materials is shown in Figure 150 for the given temperature and concentration range [556]. The area corresponding to the application of nickel is larger than that for NiCr15Fe.

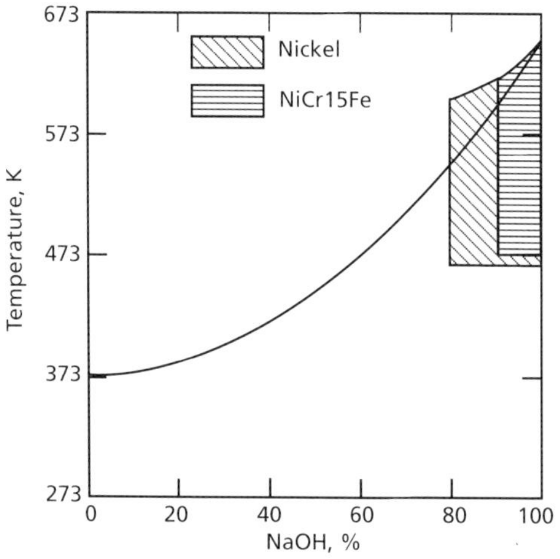

Figure 150: Range of susceptibility to stress corrosion cracking in NaOH solutions of nickel and NiCr15Fe in dependence on the temperature and concentration of the solution [518]
(The boiling point curve of the NaOH solution at 1 bar is also shown)

The carbon content and the heat treatment can have a marked effect on the stress corrosion cracking behavior of Inconel® 600. Figure 151 shows the results of stress corrosion cracking tests using slow strain rates in 10% NaOH solutions at 288°C [557]. The investigated Inconel® 600 pipes had C contents of 0.02, 0.03, 0.04 and 0.05%. They were tested in the following three heat-treated states:

- As-delivered-state, i.e. cold-drawn and tempered
- As-delivered-state state + tempering at 593°C for 4 h
- As-delivered-state state + tempering at 704°C for 15 h

The percent proportion of intergranular fracture on the fracture surface of the failed specimens was chosen as a measure of the susceptibility to stress corrosion cracking. For the specimens in the as-delivered state, the intergranular fracture proportion i.e. the SCC susceptibility, decreased with increasing C content to reach a minimum at 0.04 to 0.05% C. The SCC resistance was markedly increased by subsequent heat treatment and the carbon content had practically no effect.

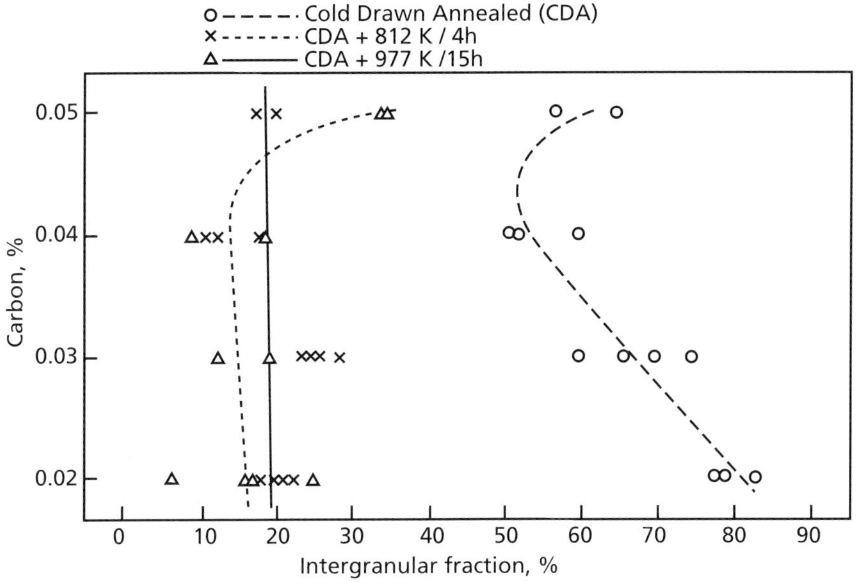

Figure 151: Influence of the carbon content and heat treatment on the stress corrosion cracking behavior of Inconel® 600 in 10% NaOH solutions at 288°C [557]

Further investigations of Inconel® 600 have shown that this alloy can be sensitized to intergranular crack formation in highly pure water and in NaOH solutions by precipitates on the grain boundaries [558, 559]. The alloys listed in Table 147 were subjected to special heat treatments in order to produce different microstructures on the grain boundaries.

Designation	*% Cr*	*% Fe*	*% C*	*% Ni*
AFU 188	16.0	9.0	0.0290	bal.
AFU 41	15.5	9.0	0.0350	bal.
AFR 12	16.8	9.0	0.0008	bal.
AFX 67	16.4	9.3	0.002	bal.
AFZ 207	16.6	9.1	0.032	bal.
AFV 277	16.4	9.1	0.063	bal.
AFR 45	16.6	9.3	0.004	bal.
AFR 46	16.4	9.1	0.025	bal.
AFT 210	5.22	9.11	0.0001	bal.
AGA 104	29.7	9.10	0.0019	bal.

Table 147: Chemical composition of the investigated NiCrFe alloys [558, 559]

The tests were carried out as slow strain-rate tests with a nominal strain rate of 25.4 x 10^{-6} mm/s in 50% NaOH solutions at 140°C and a potential of $U_H = -0.65$ V. The susceptibility to stress corrosion cracking was evaluated at the end of the test by means of the maximum crack depth outside the fractured area. The susceptibility to intergranular stress corrosion cracking increased with decreasing chromium content and increasing carbon content. For the various heat-treated states, the highest susceptibility was found for the presence of carbide precipitates on the grain boundaries in combination with chromium depletion of the neighbouring regions. Although longer tempering at 700°C increased the carbide precipitates, it also decreased the chromium depletion and thus the SCC susceptibility. The lowest susceptibility to stress corrosion cracking was observed for the almost carbon-free alloys AFR 12 and AFT 210.

The influence of a heat treatment on the SCC behavior of pipes made of Inconel® 600 in 1% NaOH solutions at 315 to 360°C was investigated by C-ring specimens and by slow tensile tests (strain rate 10^{-6} 1/s) [561]. The results of the two methods were in good agreement: the stress corrosion cracking behavior of the pipes was very strongly dependent on the carbide precipitates on the grain boundaries. A subsequent heat treatment for 50 h at 650°C is thus an effective method to improve the resistance of the material to stress corrosion cracking in NaOH solutions [560, 561].

The influence of a heat treatment and of the test potential on the stress corrosion cracking of Inconel® 600 is also reported in [562]. Specimens of two alloys (compositions given in Table 148) were subjected to the heat treatments detailed in Table 149 and then tested by means of slow strain-rate tests with constant extension (CERT tests) in 25 M NaOH solutions at 140°C. Tables 150–152 give the testing potentials, the measured mechanical properties as well as the results of the CERT tests.

Alloy	Grain size	%Ni	%Cr	%Fe	%C	%S	%P
ML	90 µm	73.8	16.23	8.32	0.077	0.002	0.006
MS	15 µm	73.7	16.25	8.31	0.068	< 0.001	0.008

Table 148: Major alloying contents and grain size of the investigated Inconel® 600 specimens [562]

Alloy/Specimen	Heat treatment	Alloy/Specimen	Heat treatment
ML	Mill-Annealed	SA	ML + solution annealed 1120°C/15 min
ML / ML2	ML + 700°C/1h	SA / SA1	SA + 700°C/20 min
ML / ML4	ML + 700°C/10h	SA / SA2	SA + 700°C/1h
ML / ML6	ML + 700°C/100h	SA / SA3	SA + 700°C/3h
MS	Mill-Annealed	SA / SA4	SA + 700°C/10h
MS / MS2	MS + 700°C/1h	SA / SA5	SA + 700°C/24h
MS / MS4	MS + 700°C/10h	SA / SA6	SA + 700°C/100h
MS / MS6	MS + 700°C/100h		

Table 149: Heat treatment conditions of the investigated Inconel® 600 specimens [562]

Heat treatment	Applied potential mV_{SCE}	R_m MPa	A %	Z %	True failure stress Fracture stress MPa	Fracture area ratio (FAR)	Failure morphology
ML	argon	610	44.8	75.4	1,676	0.0	ductile
	–980	542	45.7	45.4	677	0.42	TG
	–1250	649	33.7	54.7	1,128	0.10	TG
	–900	417	21.9	33.2	434	0.69	IG
	–800	449	27.6	20.6	429	0.51	IG
	–750	587	43.4	36.4	454	0.22	IG
	–600	352	23.5	25.0	168	0.76	IG
	–450	442	23.6	21.6	459	0.70	IG
	–300	592	47.6	57.2	1,011	0.08	ductile
	–150	582	52.3	52.0	841	0.25	IG
	0	593	40.8	34.0	639	0.54	TG
ML2	glycerine	632	51.9	57.3	1,166	0.0	ductile
	–900	579	36.4	35.7	727	0.40	IG
	–600	607	51.6	60.6	1,158	0.15	IG
ML4	glycerine	678	40.4	61.5	1,423	0.0	ductile
	–980	689	33.4	–	–	0.05	ductile
	(OCP)	655	34.4	52.0	1,097	0.11	IG
	–900	649	38.9	59.4	1,255	0.08	ductile
	–600						
ML6	glycerine	671	40.3	58.5	1,416	0.0	ductile
	–900	648	32.8	44.4	936	0.14	IG
	–600	658	35.8	40.2	948	0.02	ductile
	0	628	34.1	30.6	750	0.23	TG

R_m: Ultimate tensile strength; A: Elongation; Z: Reduction of area; IG: intergranular; TG: transgranular; FAR: stress corrosion cracked area divided by the final cross-sectional area; OCP: Open circuit potential

Table 150: Test parameters, mechanical values and results of the CERT tests of specimens of the material ML in the as-delivered state (coarse grain) [562]

Heat treatment	Applied potential mV_{SCE}	R_m MPa	A %	Z %	True failure stress Fracture stress MPa	Fracture area ratio (FAR)	Failure morphology
SA	air	640	55.9	63.0	1,329	0.0	ductile
	–980	601	54.5	57.0	1,088	0.16	TG
	–1250	584	51.2	49.4	1,052	0.15	TG
	–900	527	52.6	49.4	752	0.40	IG
	–800	610	58.9	67.5	1,251	0.06	IG
	–600	578	51.5	58.0	1,098	0.13	IG
	–400	599	62.3	56.3	974	0.08	IG
	–200	579	64.5	68.5	1,343	0.18	IG
	0	554	40.9	38.6	672	0.46	TG
	+150	632	61.6	61.0	1,168	0.0	ductile

* boiling point
** see also Table 128
RT: Room temperature

Table 127: Continued

Heat treatment	*Applied potential mV_{SCE}*	*R_m MPa*	*A %*	*Z %*	*True failure stress Fracture stress MPa*	*Fracture area ratio (FAR)*	*Failure morphology*
SA1	air	621	50.0	67.0	1,371	0.0	ductile
	–980 (OCP)	544	42.5	51.3	844	0.34	TG
	–900	510	45.3	34.5	574	0.43	IG
	–800	580	53.7	59.4	1,082	0.03	ductile
	–600	588	54.9	62.3	1,098	0.10	IG
	0	487	33.5	32.1	567	0.37	TG + IG
SA2	air	630	53.5	62.1	1,287	0.0	ductile
	–980 (OCP)	540	47.4	54.3	900	0.29	TG
	–1250	627	54.6	53.1	1,185	0.0	ductile
	–900	573	47.1	45.0	816	0.39	IG
	–600	582	61.7	60.5	1,165	0.05	ductile
	0	571	40.6	47.3	726	0.34	TG + IG
	+150	635	55.6	59.1	1,232	0.0	ductile
SA3	air	621	54.1	59.0	1,234	0.0	ductile
	–980 (OCP)	610	54.9	54.1	1,101	0.03	ductile
	–900	515	44.7	29.1	606	0.33	IG
	–800	614	54.5	58.4	1,182	0.0	ductile
	–300	583	55.1	60.2	1,217	0.0	ductile
	0	440	28.5	21.3	385	0.46	IG
SA4	air	642	38.9	57.5	1,273	0.0	ductile
	–980 (OCP)	615	35.7	56.6	1,172	0.02	ductile
	–900	546	29.2	36.7	755	0.33	IG
	0	513	20.2	24.8	550	0.49	IG
SA5	air	638	40.3	57.0	1,139	0.0	ductile
	–980 (OCP)	619	39.1	49.4	993	0.10	ductile
	–1250	644	41.6	41.1	985	0.0	ductile
	–900	595	37.5	34.8	774	0.29	IG
	–600	644	41.7	53.0	1,118	0.08	ductile
	0	535	20.8	21.4	459	0.45	IG
	+150	653	43.8	57.6	1,134	0.0	ductile
SA6	air	644	44.2	56.5	982	0.0	ductile
	–980 (OCP)	625	47.4	62.8	1,338	0.02	ductile
	–900	630	47.1	63.7	1,069	0.18	ductile
	–400	622	42.3	60.1	1,209	0.09	ductile
	0	533	24.9	36.7	576	0.50	IG

R_m: Ultimate tensile strength; A: Elongation; Z: Reduction of area; IG: intergranular; TG: transgranular; FAR: stress corrosion cracked area divided by the final cross-sectional area; OCP: Open circuit potential

Table 151: Test parameters, mechanical properties and results of the CERT tests of specimens of the solution-annealed material SA [562]

Heat treatment	*Applied potential mV_{SCE}*	*R_m MPa*	*A %*	*Z %*	*True failure stress Fracture stress MPa*	*Fracture area ratio (FAR)*	*Failure morphology*
MS	glycerine	711	45.1	62.8	1,247	0.0	ductile
	–1050	727	46.0	62.1	1,430	0.0	ductile
	–950	640	50.8	54.3	1,018	0.03	ductile
	–900	517	27.4	22.2	525	0.51	TG
	–850	630	34.0	40.9	834	0.29	IG
	–700	683	41.0	57.8	1,193	0.07	IG
	–600	596	37.4	39.4	817	0.43	ductile
	–400	525	25.0	28.9	555	0.39	IG
	–300	530	22.7	27.4	577	0.49	IG
	–200	565	28.8	29.3	681	0.22	IG
	0	577	22.2	15.5	553	0.33	TG
	+150	709	45.5	66.2	1,519	0.0	ductile
MS2	glycerine	690	43.4	63.3	1,358	0.0	ductile
	–900	598	34.8	43.9	867	0.25	IG
	–800	681	41.2	61.4	1,307	0.02	ductile
	–600	680	42.2	60.7	1,282	0.04	ductile
	0	621	26.2	22.4	662	0.40	IG
MS4	glycerine	667	41.3	65.2	1,412	0.0	ductile
	–980 (OCP)	679	57.0	66.9	1,487	0.08	ductile
	–900	677	39.9	62.9	1,353	0.05	ductile
	–800	684	42.2	62.8	1,368	0.0	ductile
	0	612	24.9	23.8	576	0.31	IG
MS6	glycerine	681	39.7	61.1	1,309	0.0	ductile
	–900	672	42.3	64.9	1,393	0.08	ductile
	–400	669	40.8	60.1	1,168	0.08	ductile
	0	661	34.5	37.6	813	0.21	IG

R_m: Ultimate tensile strength; A: Elongation; Z: Reduction of area; IG: intergranular; TG: transgranular; FAR: stress corrosion cracked area divided by the final cross-sectional area; OCP: Open circuit potential

Table 152: Test parameters, mechanical properties and results of the CERT tests of specimens of the material MS in the as-delivered state (fine grain) [562]

According to the results in Tables 150–152, Inconel® 600 is susceptible to stress corrosion cracking in 25 M NaOH solutions in three different potential ranges: at U_{cal} = –0.98 V, at U_{cal} = –0.90 and U_{cal} = –0.40 V. These correspond to the free corrosion potential, the first active peak and the second anodic peak. At the free corrosion potential, there is transgranular cracking while there is intergranular cracking in the other cases. A heat treatment at 700°C decreases the susceptibility to stress corrosion cracking in the first passive range and for free corrosion; however, it does not influence the behavior in the second passive range. The most favourable influence of the heat treatment cannot be explained by chromium depletion and nickel enrichment alone, but it is attributed to an alteration of the sliding processes on the grain boundaries as a result of chromium carbide precipitates [562].

The test results reported in [563] also confirm that a heat treatment at 700°C for 1 h or longer improves the resistance of alloy 600 to intergranular attack and stress corrosion cracking in a solution of 30% NaOH + 10% Na_2SO_4 at 350°C. According to these results, the susceptibility to this attack is also dependent on the final heat treatment during rolling. If this treatment does not bring enough of the carbon into solution, small particles or colonies of $M_{23}C_6$ are formed on the grain boundaries during subsequent heat treatment. These particles do not decrease the attack along the grain boundaries. The temperature of the final heat treatment should be high enough to dissolve sufficient carbon so that in the subsequent heat treatment, semi-continuous precipitates of M_7C_3 are formed on the grain boundaries. These provide greatly improved corrosion resistance, probably because they prevent the formation of a continuous film of precipitated impurities along the grain boundaries.

The influence of chromium carbide precipitates on the grain boundaries on the SCC behavior of alloy 600 in hot NaOH solutions (40% NaOH at 325°C) was investigated in detail in [564]. The favourable influence of a heat treatment in the temperature range of 600 to 700°C is explained by the fact that chromium carbide precipitates and other impurities at the grain boundary are prevented by the elements C, B and P and that chromium-depleted regions are formed near the grain boundaries in which there are correspondingly higher contents of nickel and iron that exhibit improved resistance in NaOH solutions. In contrast to the statements made in [559], reference [564] attributes a positive influence to the chromium-depleted regions.

Similar results are reported in [565]: U-bend specimens of nickel alloys alloy 600, alloy 625 and alloy 690 (compositions given in Table 153) were exposed for 500 h in 10% NaOH solutions at 307°C.

Alloy *DIN-Designation*	*DIN-Mat. No.*	*%C*	*%Mn*	*%Si*	*%P*	*%S*	*%Cr*	*%Mo*	*%Nb*	*%Fe*	*%Ti*	*%Al*	*%Co*
alloy 625* NiCr22Mo9Nb	2.4856	0.03	0.10	0.20	0.012	0.003	21.86	9.00	3.50	4.02		0.20	
alloy 625**	2.4856	0.01	0.21	0.21	0.012	0.001	22.47	8.51	3.54	3.85	0.22	0.16	
alloy 690* NiCr29Fe	2.4692	0.03	0.14	0.20	0.001	0.001	29.20			7.90	0.19	0.25	0.08
alloy 600* NiCr15Fe	2.4816	0.04	0.19	0.22	0.005	0.002	15.29			8.36	0.17	0.14	0.05

* = pipe ** = sheet

Table 153: Chemical composition of the investigated nickel alloys [565]

All three alloys 600, 625 and 690 exhibited cracks after the test period [565]. The heat treatments of specimens and the measured crack depths are summarized in Table 154.

Alloy	*Heat treatment*	*Stress relief*	*Crack depths µm*
alloy 625*		566°C / 7 h	439
alloy 625**	1,093°C / 24 h + 871°C / 8 h + 649°C / 24 h	566°C / 7 h	1,232
alloy 690*	1,052°C / 4 h + 718°C / 10 h	607°C / 7 h + 566°C / 7 h	58
alloy 600*		607°C / 7 h	137

* = pipe ** = sheet

Table 154: Heat treatment of the stress corrosion cracking specimens and maximum crack depths after a 500 h test period [565]

Alloy 625 exhibited the greatest crack depths and also noticeably more cracks than the alloys 600 and 690. This is explained by the high molybdenum content. In the region of the crack tip and on the fracture surface, microstructural and micro-analytical investigations were carried out. Chromium-rich oxides were found at and close to the region of the crack tip [565]. Outside the region of the crack tip there were areas with increased nickel contents. A corresponding heat treatment led to chromium-depleted areas and in spite of the high chromium content, alloy 690 exhibited good behavior compared to alloy 600 which contains less chromium.

Investigations on the influence of the elements C, S and P on the stress corrosion cracking behavior of FeNiCr alloys are also reported in [566]. Slow strain-rate tests were carried out in NaOH solutions with 750 g/l at 350°C on samples of the commercial alloy 600 and laboratory melts with the compositions given in Table 155. The results indicate that the contents of the elements C, S and P as well as their distribution (dependent on the heat treatment) play a decisive role in the generation of cracks and their propagation. Sulfur in amounts above 20 ppm always has a strongly negative influence, however the combination of C + P is particularly interesting for 600 alloys in the quenched state.

					Group 1					
No.	***1***	***2***	***3***	***4***	***5***	***6***	***7***	***8***	***9***	***10***
%Fe	70	63	53	33	46	31	12	28	10	0
%Ni	13	33	31	32	55	52	52	72	76	100
%Cr	17	0	18	35	0	18	36	0	14	0
ppm C	50	26	50	80	20	33	70	32	60	180
ppm S	50	40	50	80	80	70	70	30	<20	<20
ppm P	<15	<30	<30	<30	<30	<30	<30	<30	<15	<15

Table 127: Corrosion rates for nickel alloys in sulfuric acid at various temperatures and concentrations (no more precise data) [398] (see also Figure 110)

Table 151: Continued

					Group 2					
No.	***11***	***12***	***13***	***14***	***15***	***16***	***17***	***18***	***19***	***20***
%Fe	8	8	8	12	8	8	8	8	8	8
%Ni	74	74	74	74	74	74	74	74	74	74
%Cr	18	18	18	14	18	18	18	18	18	18
ppm C	20	20	20	60	120	120	400	140	20	20
ppm S	<10	22	230	25	20	70	20	<10	<10	230
ppm P	<15	<15	<15	<15	<15	<15	<15	72	74	70
alloy 600	%Fe 8	%Ni 77	%Cr 15	ppm C 600	ppm S 30	ppm P 120				

Table 155: Contents of the major elements in the investigated FeNiCr alloys [566]

Investigations are also described in [567] with regards to tests of alloy 600 and alloy 690 on the formation of carbides and precipitation of impurities on the grain boundaries in dependence on the heat treatment. The resulting microstructures were characterized. The compositions of the investigated alloys are given in Table 156.

Type	*Heat treat-ment*	*Cr*	*Fe*	*C*	*Mn*	*S*	*P*	*Si*	*Al*	*Cu*	*Ti*	*N*	*B*
600	A1	15.7	8.45	0.011	0.23	0.003	0.003	0.10	0.20	0.31	0.20	nb	0.0002
600	B1	15.7	8.88	0.018	0.23	0.025	0.025	0.11	0.21	0.31	0.21	nb	0.0002
600	C1	15.7	8.7	0.021	0.23	0.014	0.003	0.12	0.20	0.31	0.20	nb	0.0001
690	A2	30	9.39	0.025	0.23	0.002	0.0025	0.17	0.24	0.17	0.23	0.013	0.0005
690	B2	29.4	9.5	0.026	0.25	0.001	0.019	0.21	0.16	0.22	0.31	0.004	0.0002
690	C2	30.1	8.94	0.028	0.25	0.008	0.001	0.21	0.18	0.26	0.29	0.012	0.0005
600	Spec.	15–17	7–11	0.015–0.025	≤ 0.5	≤ 0.01	≤ 0.025	≤ 0.5	–	≤ 0.5	–	≤ 0.05	≤ 0.01
690	Spec.	28–31	7–11	0.015–0.025	≤ 0.5	≤ 0.01	≤ 0.025	≤ 0.5	–	≤ 0.5	—	≤ 0.05	≤ 0.01

Table 156: Chemical composition of the investigated alloys (mass-%) [567]

The specimens for the metallographic, analytical and corrosion-chemical investigations were subjected to different heat treatments (Table 157) to produce either a wide spectrum of microstructures and precipitates or to simulate the usual heat treatments for steam turbine rotor blades.

Designation	Treatment
A1, B1, C1/SA	solution-annealed 925°C/1h, WQ solution-annealed 1,000°C/1h, WQ
A1, B1, C1/SA + TT	solution-annealed 925°C/1h, + 704°C/16h, WQ solution-annealed 1,000°C/1h, + 704°C/16h, WQ
A2, B2, C2/SA	4 x solution-annealing of each melt 1,070°C, 1,150°C, 1,200°C, 1,250°C each 1h, WQ
A2, B2, C2/SA + TT	4 x solution-annealing of each melt 1,070°C, 1,150°C, 1,200°C, 1,250°C each 1h, + 704°C/16h, WQ

WQ = Water quenched; TT = Thermally treated

Table 157: Heat treatments of the investigated samples of alloy 600 and alloy 690 [567]

Table 158 gives the compositions of the grain boundaries of some samples determined by Auger electron spectroscopy. These are the average values of the concentrations from at least twenty measurements of each sample [567].

As an example, Figure 152 shows the obtained profiles of the contents of some alloying elements in the grain boundary region.

Type	Heat treatment	Ni	Cr	Fe	Ti	N	C	B	S	P
A1	SA 1,000°C	72.58	15.59	8.50	–	–	0.58	1.40	–	0.20
A1	SA + TT 1,000°C	68.73	19.51	5.59	–	–	1.79	0.68	–	0.70
B1	SA 1,000°C	74.45	15.50	8.13	–	–	0.068	–	–	1.23
B1	SA + TT 1,000°C	68.32	19.14	7.92	–	–	2.23	–	–	2.11
A2	SA 1,150°C	60.21	29.01	9.38	0.27	–	0.49	0.34	0.25	–
A2	SA + TT 1,150°C	51.62	33.46	9.64	1.77	0.60	1.89	0.19	0.59	0.40
B2	SA 1150°C	58.80	28.00	9.70	1.00	–	0.44	0.47	0.32	1.37
B2	SA + TT 1,150°C	49.00	37.00	9.74	0.43	–	1.54	–	–	2.50

SA = Solution annealed; TT = Thermally treated

Table 158: Alloy compositions on the grain boundaries as determined by AES (mass-%) [567]

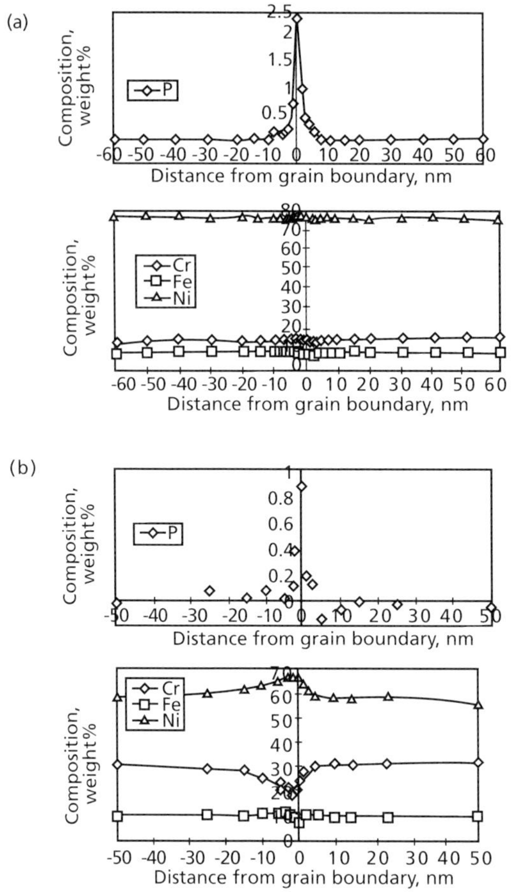

Figure 152: Concentrations of the alloying contents on the grain boundaries of (a) alloy 600 1,000°C/1h + 704°C/16h and (b) alloy 690 1,150°C/1h + 704°C/16h [567]

The susceptibility to intergranular corrosion was studied on unstressed specimens in 30% NaOH solutions with added 10% Na_2SO_4 at 350°C. This solution was chosen to simulate the conditions that may occur locally under unfavourable circumstances in steam generators of pressurised-water reactors. Here, impurities from the secondary water cycle can be concentrated up to 50% NaOH and up to 20% Na_2SO_4 in overheated gaps between the tube and the tube plate. This test temperature was selected to accelerate attack rather than the operating temperature of 325°C. The results of the corrosion tests are presented in Figure 153 and in Figure 154.

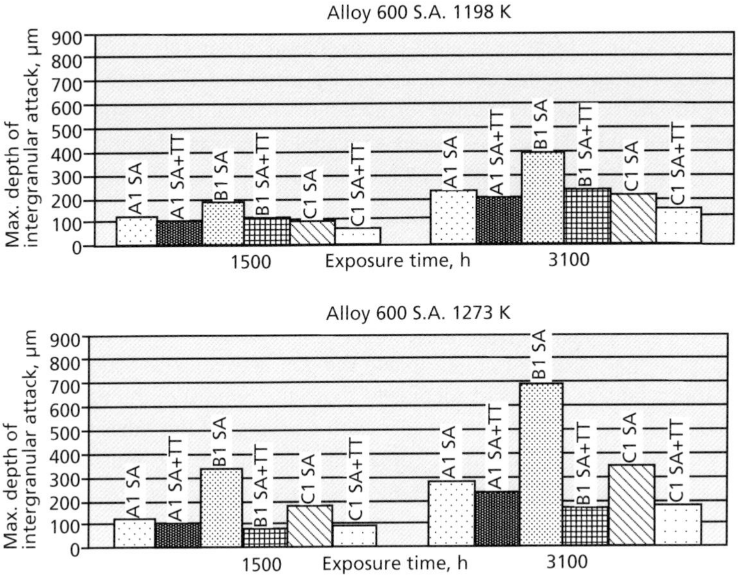

Figure 153: Maximum depth of the intergranular attack on the specimens of alloy 600 after a test period of 3,140 h at 350°C in 30% NaOH solutions with an addition of 10% Na_2SO_4 [567]

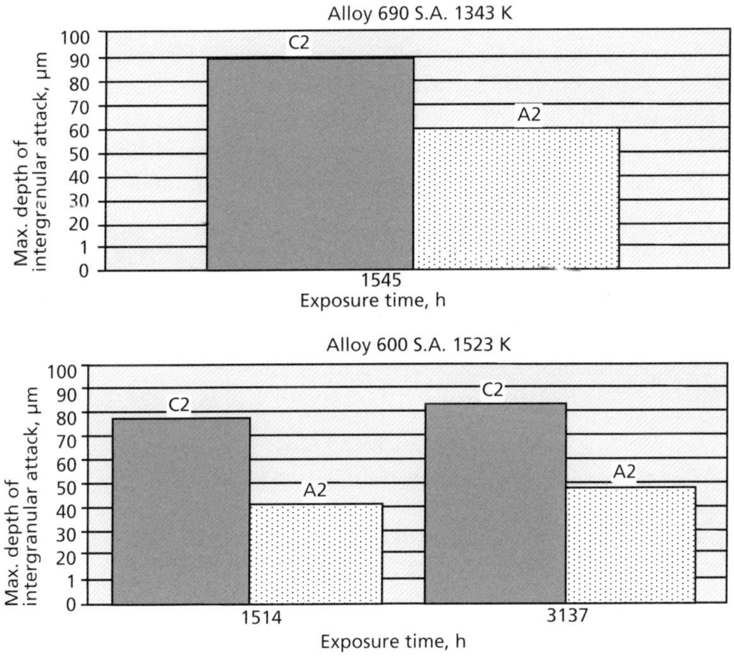

Figure 154: Maximum depth of the intergranular attack on the specimens of alloy 690 after a test period of 3,140 h at 350°C in 30% NaOH solutions with an addition of 10% Na_2SO_4 [567]

The precipitation of impurities on the grain boundaries, particularly phosphorus, markedly decreases the resistance of alloy 600 to intergranular corrosion. The formation of semi-continuous chromium-rich carbides improved the corrosion resistance. In alloy 690 there was preferential formation of chromium-rich sulfide particles whose presence on the grain boundaries appeared to increase the intergranular corrosion rate. This should be prevented by a reduction of the sulfur content in the alloy [567].

The tube system in steam generators is one of the most critical systems in pressurized-water reactors. The types of corrosion damage to U-tubes of such steam generators may be intergranular corrosion, intergranular stress corrosion cracking and surface corrosion of the exterior of the tubes and on the secondary side of the steam generator. They are found, in particular, inside the gap between the bottom of the tube and the tube plate, as well as immediately above the bottom of the tube in the region of deposits and in U-bends (stress corrosion cracking) of the tubes. Intergranular corrosion and intergranular stress corrosion cracking are the most frequent types of damage in steam generator tubes of pressurised-water reactors. Materials used for the tubing in steam generators are Inconel® 600 (NiCr15Fe, DIN-Mat. No. 2.4816) and also Incoloy® 800 (X10NiCrAlTi32-20, DIN-Mat. No. 1.4876). An important aspect of the problem is the conditioning of the secondary water, i.e. the controlled addition of chemicals to minimize corrosion in the water/steam cycle. If volatile or solid materials are used to make the solution alkaline, these materials can be concentrated by a factor of 10^4 in gaps where there is a heat transfer or delayed boiling on heat-transfer surfaces.

Investigations were carried out on Inconel® 600 and Incoloy® 800 with regards to surface corrosion, to intergranular corrosion and to intergranular stress corrosion cracking in solutions of the alkalifying agents NaOH, Na_2HPO_4 and Na_3PO_4 whose concentration was increased by factor of 10^4 with respect to the guidelines on water conditioning [568, 569]. The composition of the two materials is given in Table 159.

DIN-Mat. No.	%C	%Si	%Mn	%P	%S	%Cr	%Ni	%Cu	%Al	%N	%Ti	%Fe
1.4558 Alloy 800	0.018	0.54	0.57			21.2	35.6	0.10	0.16	0.016	0.58	bal.
2.4816 Alloy 600	0.016	0.36	0.80	0.006	0.003	16.28	72.80	0.004	0.18	0.017	0.39	9.38

Table 159: Chemical composition of the investigated materials [568]

The specimen tubes were tested in the as-delivered state and after a long-term annealing for 1,000 and 10,000 h at 400°C. Table 160 gives the individual strengths of the materials. Stress corrosion cracking tests were carried out using capsule specimens filled with the test solution. The wall thickness of the tubes was machined down so that at the testing temperature an average circumferential stress of 40% of the yield strength at room temperature was attained as a result of the internal pressure. The test duration was 3,000 h. The results obtained in NaOH solutions are summarized for the material no. 1.4558 (Inconel® 800) in Table 161 and for material no. 2.4816 (Inconel® 600) in Table 162.

Material	Designation	Heat treatment at 400°C h	$R_{p0.2}$ N/mm²	R_m N/mm²	A %	Z %
1.4558	a	0	395	655	29	48
Alloy 800	b	1,000	410	683	31	49
	c	10,000	545	855	23	44
2.4816	a	0	329	706	43	47
Alloy 600	b	1,000	368	780	39	46
	c	10,000	345	755	39	45

Table 160: Yield strength ($R_{p0.2}$), tensile strength (R_m), percentage elongation after fracture (A) and reduction of area (Z) of the tested materials in the as-delivered state and after long-term heat treatment [568]

Test solution	Designation	Test temperature					
mol/l NaOH		270°C	310°C		350°C		
		Δ_{mA} mg/cm²	Δ_{mA} mg/cm²	IS µm	Δ_{mA} mg/cm²	IS µm	IC µm
0.42	a	6.2	6.8	< 20	7.8	< 20	< 20
	b	6.3	6.4	< 20	12.3	230	57
	c		6.8	< 20	9.2	62	< 20
0.84	a	6.4	6.9	< 20	26.0	> 1,000	
	b	6.5	6.5	< 20	27.6	495	
	c		6.9	< 20	25.3	> 1,000	
1.26	a	6.6	9.7	< 20	19.1	25	
	b	7.1	7.1	< 20	20.2	*	
	c		9.5	< 20	29.9	*	

* = Capsule failed

Table 161: Mass loss (Δ_{mA}) and depth of intergranular corrosion (IC) in coupon specimens as well as the crack depth (IS) of the intergranular stress corrosion cracking in capsules in dependence on the test conditions and state of the material for 1.4558 (Incoloy® 800) in the three heat-treated states (a, b, c, see Table 160)

In NaOH solutions, Incoloy® 800 only showed stress corrosion cracking at the highest test temperature of 350°C. In the solution with 1.26 M NaOH and after long-term annealing (heat-treated states b and c), none of the capsules held long enough, even in repeated tests, for the evaluation of the crack depths. Obviously, the material in this state is particularly susceptible to stress corrosion cracking.

Test solution	*Designation*	*Test temperature*						
mol/l NaOH		*270°C*	*310°C*			*350°C*		
		Δ_{mA} *mg/cm²*	Δ_{mA} *mg/cm²*	*IS* *µm*	*IC* *µm*	Δ_{mA} *mg/cm²*	*IS* *µm*	*IC* *µm*
0.42	a	4.9	5.8	23	< 20	18.0	119	59
	b	5.0	6.4	35	< 20	17.7	183	81
	c		5.9	< 20		11.1	< 20	< 20
0.84	a	64.9	6.9	53	24	12.7	198	66
	b	5.4	7.9	25	< 20	17.9	190	75
	c		6.8	< 20		12.5	85	38
1.26	a	5.2	8.3	62	21	25.7	264	112
	b	6.4	8.2	56	22	20.5	461	27
	c		8.3	< 20		13.0	93	< 20

Table 162: Mass loss (Δ_{mA}) and depth of intergranular corrosion (IC) in coupon specimens as well as the crack depth (IS) of the intergranular stress corrosion cracking in capsules in dependence on the test conditions and state of the material for 2.4816 (Inconel® 600) in the various heat-treated states (a, b, c, see Table 160)

In the case of Inconel® 600, cracks appeared at the test temperatures of 310 and 350°C and the crack depth increased with increasing concentration of the NaOH solutions.

Silicon compounds are frequently found and in large amounts in deposits in steam generator pipes. Investigations on the influence of silicon compounds on stress corrosion cracking behavior of alloy 600 in NaOH solutions are reported in [570]. The composition of the investigated material is given in Table 163 and was heat-treated for 4 min at 927°C.

%C	*%Mn*	*%S*	*%Si*	*%Cr*	*%Ni*	*%Fe*	*%Cu*
0.038	0.27	0.001	0.30	15.88	74.83	8.2	0.01

Table 163: Composition of the investigated material alloy 600 [570]

The stress corrosion cracking tests of C-ring specimens with 2% deformation taken from pipes of 19 mm diameter were carried out in the following solutions:

- 10% NaOH
- 10% NaOH + 7.5% SiO_2 added as amorphous silicon oxide
- 10% NaOH + 7.5% SiO_2 + 15% Na_2SiO_3

The tests were performed under potentiostatic control at 315°C. The test conditions are given in Table 164.

Test solution		*Testing potential mV (Ni electrode)*		
	free corrosion	*active range*	*passivation*	*passive range*
10% NaOH	2	100	150	200
10% NaOH + 7.5% SiO_2	2	80	130	180
15% Na_2SiO_3	2	80	130	180
10% NaOH + impurities	3	130	170	220
10% NaOH + 7.5% SiO_2 + impurities	3	100	150	200
15% Na_2SiO_3 + impurities	2	130	170	220

Table 164: Test solutions and testing potentials employed in the stress corrosion cracking tests [570]

The results of the stress corrosion cracking tests are given in Figure 155 and Figure 156.

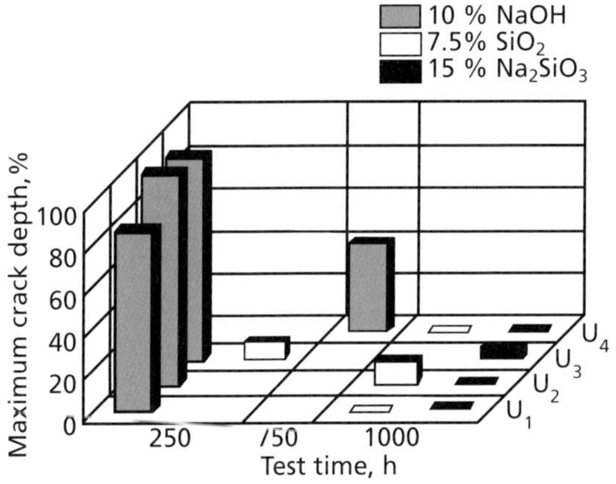

Figure 155: Results of the stress corrosion cracking tests in the given test solutions at 315°C
U_1: potential in the passive range
U_2: potential at the active-passive transition
U_3: potential in the active range
U_4: potential of free corrosion

According to these results, the test potential has no decisive influence on the corrosion behavior in NaOH solutions. Both silicon compounds studied exhibited an inhibiting effect on both the intergranular corrosion as well as the stress corrosion cracking. These inhibiting effects of the silicon compounds appear to be caused by a different structure of the layers on the surface. The Auger depth profiles show the presence of silicon and sodium in the outer layers. Impurities that were added to the NaOH solution did not exhibit a negative effect on the susceptibility to intergranular corrosion or stress corrosion cracking of the alloy.

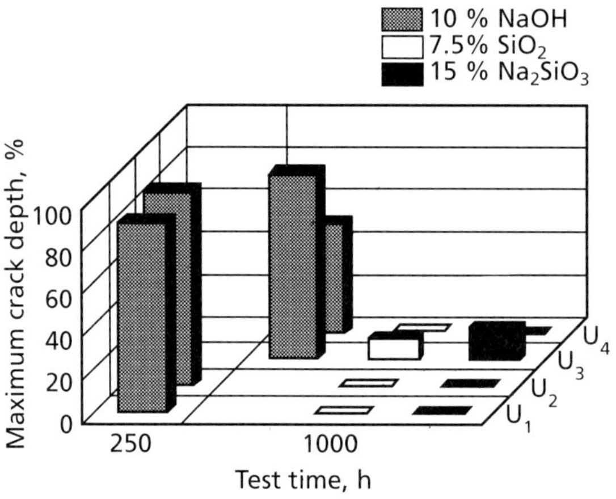

Figure 156: Results of the stress corrosion cracking tests in the given test solutions, plus impurities at 315°C
U_1: potential in the passive range
U_2: potential at the active-passive transition
U_3: potential in the active range
U_4: potential of free corrosion

Because of the susceptibility of the alloy Inconel® 600 (used for tubing in steam generators in pressurised-water reactors) to stress corrosion cracking in NaOH solutions in the temperature range 260 to 430°C, investigations on stress corrosion cracking are often carried out at NaOH concentrations that are higher than those which may occur under operating conditions. SCC experiments are described in [571] in which the influence of the NaOH concentration on the crack behavior of Inconel® 600 was investigated using fracture mechanics specimens. The composition of the material is given in Table 165.

The material was tested in the as-delivered state as well as after additional heat treatment. The processing steps and the individual heat treatments are summarized in Table 166.

%C	%Si	%Mn	%P	%S	%Cr	%Fe	%Ti	%Al	%Ni
0.025	0.30	0.32	0.007	0.001	15.32	8.08	0.18	0.13	74.85

Table 165: Chemical composition of the investigated Inconel® 600 [571]

The CT specimens (Compact Tension specimens) were tested in NaOH solutions of differing concentrations (0.04%, 0.4%, 4%, 10%, 40%) under the conditions given in Table 167.

Figure 157 shows the influence of the NaOH concentration between 0.04 and 40% on the dependence of the crack propagation rate (da/dt) and on the stress intensity factor K_{max} for the specimens in heat-treated state I (as-delivered state). As the NaOH concentration increased, the crack propagation rate increased and the shape of the da/dt – K_{max} curve is convex above 4% NaOH.

Processing	*Heat treatment I*	*Heat treatment II*
Forging	950–1,180°C	
1st heat treatment	1,025°C, 1 h water quenching	1,045°C, 45 min water quenching
Cold rolling	30%	30%
2nd heat treatment	1,100°C, 1 h air cooling	925°C, 1 h water quenching
3rd heat treatment		700°C, 30 h air cooling

Table 166: Heat treatment states of the investigated specimens [571]

Temperature °C	*Pressure MPa*	*Potential mV*	*Stress ratio*	*Frequency Hz*
325	12.3	0.1	0.85	0.5

Table 167: Test conditions in NaOH solution [571]

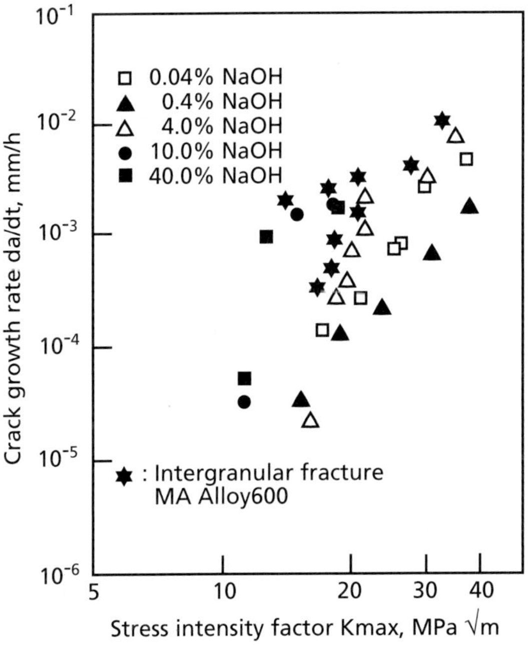

Figure 157: Influence of the NaOH concentration on the crack propagation rate of the specimens in the mill-annealed (MA)state [571]

Figure 158 shows the corresponding results for the NaOH concentrations between 0.04 and 10% for the specimens in heat-treated state II. The values of the crack propagation rate are all lower than those of the specimens in heat-treated state I and at higher NaOH concentrations, and the curve for the thermally treated Alloy 600 is straight, even in a 4% NaOH solution.

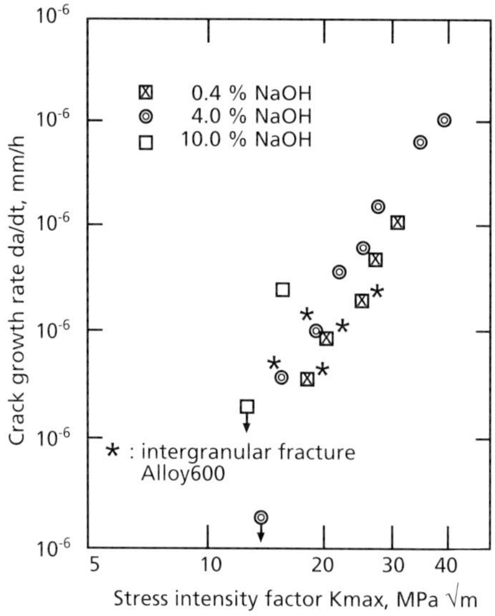

Figure 158: Influence of the NaOH concentration on the crack propagation rate of the specimens subjected to thermal treatment (TT) [571]

In the absence of tensile stress, precipitates on the grain boundaries in alloy 600 at 140°C in 50% NaOH solutions lead to intergranular attack of the grain boundaries, which, in contrast to intergranular stress corrosion cracking, readily increases in heat-treated specimens [572]. The appearance of the damage observed in steam generators caused by intergranular corrosion of Inconel® 600 can be reproduced well in the laboratory by the attack of hot NaOH solutions (10 to 50% NaOH at approx. 340°C) [573].

A modified Inconel® 600 alloy with a chromium content increased to 19% did lead to an improvement of the SCC resistance in high-temperature water; however, in tests in a NaOH solution with a NaOH concentration of 10 g/l at 350°C it did not provide significantly improved resistance [574].

Fracture mechanics investigations of intergranular corrosion and stress corrosion cracking of alloy 600 in NaOH at high temperatures are reported in [575]. Since components of steam-generating plants mostly exhibit damage in the gaps between tubes and tube plates, both the plates and the pipes (1.27 mm diameter; 22.23 mm wall thickness) of alloy 600 were investigated. This material had the composition given in Table 168. The plate material was heat-treated at 925°C for 30 min and the tube material for 2 min.

	%C	%Si	%Mn	%P	%S	%Cr	%Fe	%Cu	%Ni
Plate	0.021	0.33	0.34	0.004	0.002	15.40	7.24	0.002	76.66
Tube	0.026	0.29	0.29	0.007	0.001	15.90	8.20	0.01	74.90

Table 168: Composition of the investigated samples of alloy 600 [575]

For the investigations on the crack propagation of the sheet material, DCB (Double Cantilever Beam) specimens were used with the dimensions given in Figure 159.

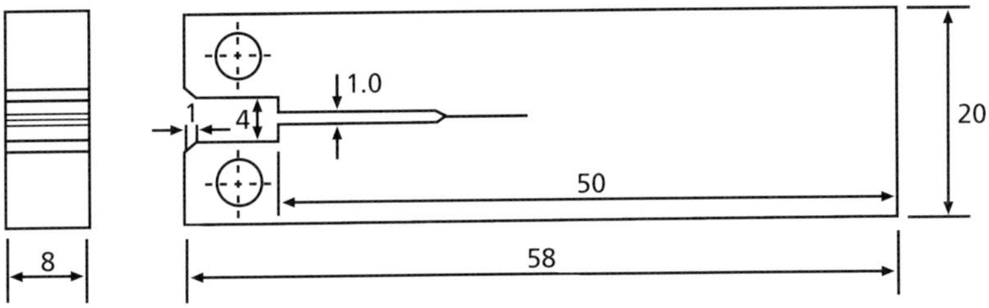

Figure 159: Shape and dimensions (mm) of the DCB specimens used for measurements of the crack propagation [575]

The tests of the plate samples were carried out in solutions with 20% NaOH + 4% Na_2CO_3 at 280, 320 and 350°C. The content of dissolved oxygen in the solution was < 5 ppb. The pipes were tested in a model cycle of a steam generator in order to investigate the crack propagation behavior under heat-transfer conditions and simulated crevice corrosion conditions. The shape and arrangement of the cracks in the pipe samples as well as their position is shown in Figure 160.

Figure 161 gives the flow diagram of the model cycle of the steam generator for the tests of the pipe specimens.

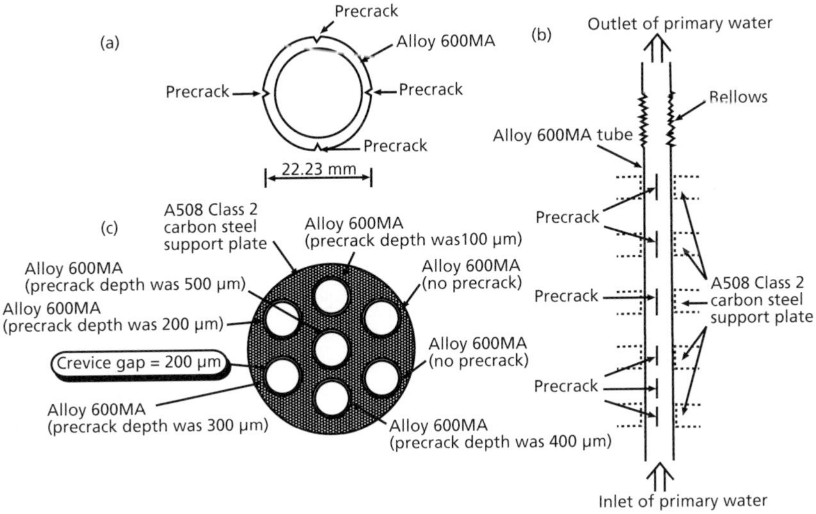

Figure 160: Sketch of the pipe specimens with cracks for the model cycle of a steam generator [575]

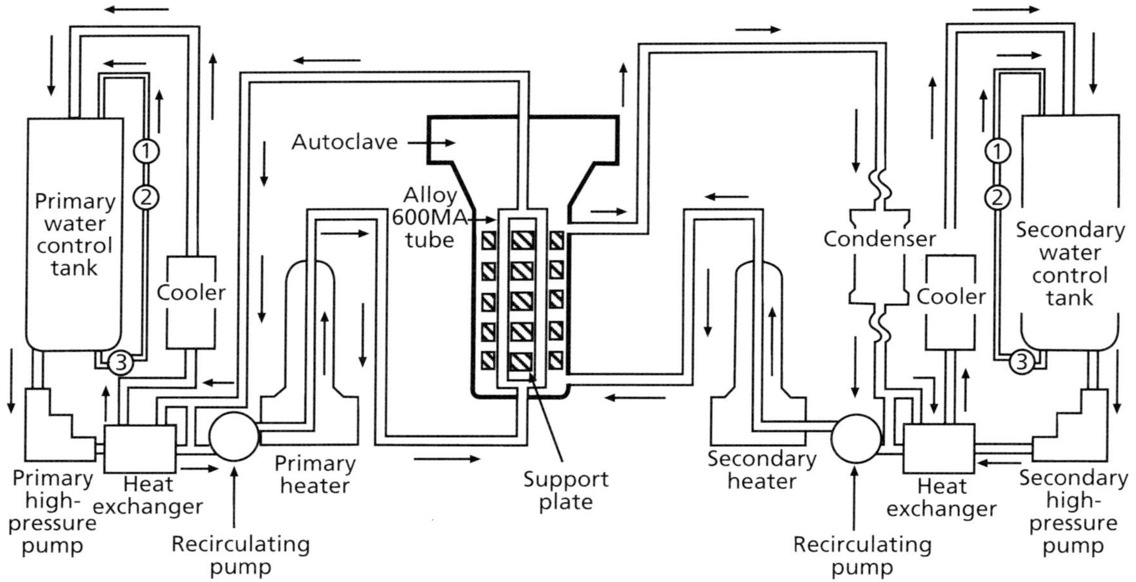

measuring point for: ① dissolved O_2; ② conductivity; ③ flow rate

Figure 161: Flow diagram of the model cycle of a steam generator [575]

The test conditions are given in Table 169.

	Primary side	*Secondary side*
Medium	distilled water	distilled water + 10 ppm NaOH
Temperature (°C)	330	290.6
Pressure	16.7 MPa	5.59 MPa
Dissolved oxygen	< 5 ppb	< 5 ppb
Flow rate	50 l/min	25 l/min

Table 169: Test conditions for the model cycle of a steam generator [575]

The specimens were loaded by both the pressure difference as well as the temperature difference between the interior and exterior of the pipes.

The results of the crack propagation measurements of the DCB specimens are given in Figure 162 for the three test temperatures. At all three test temperatures, the crack propagation rate was very small when the stress intensity factor was K < 10 MPa √m. Rapid crack growth was not observed, even up to a K value of 60 MPa √m. Only above higher K values did the crack propagation rate increase at the highest test temperature of 350°C.

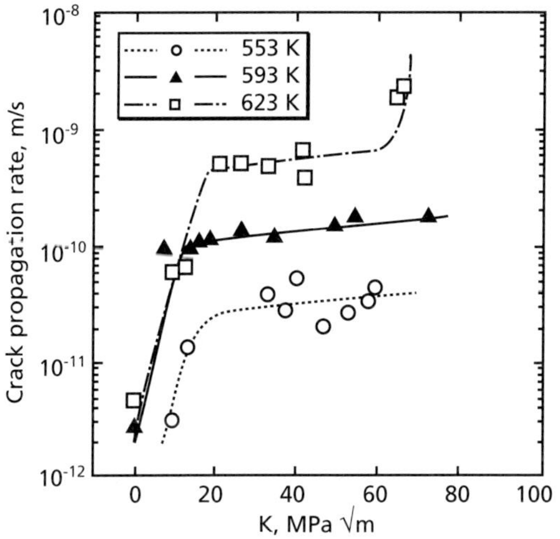

Figure 162: Measured crack propagation rate (m/s) of the DCB specimens in dependence on the stress intensity factor K (MPa √m) [575]

The results obtained for the tube specimens in the model cycle of a steam generator are summarized in Figure 163. Under these test conditions, the crack propagation rates were essentially constant at 10^{-12} to 10^{-11} m/s for the range of the stress intensity factor K of 0 to 6 MPa √m.

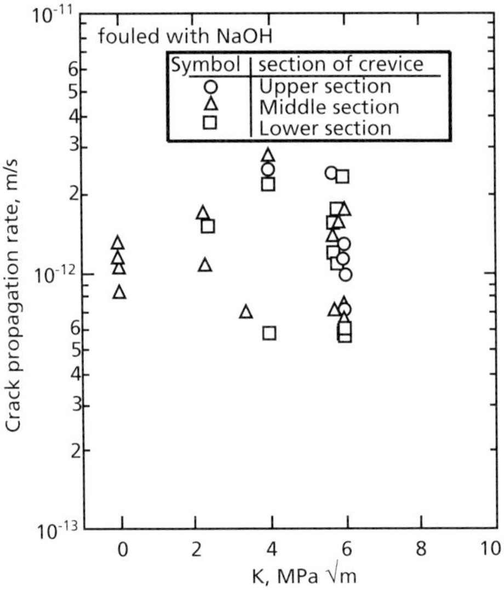

Figure 163: Influence of the stress intensity factor K (MPa √m) at the crack tip on the crack propagation in the tube specimens in the model cycle of a steam generator under conditions of heat transfer [575]

It has already been reported that the alloy 690 (NiCr29Fe, DIN-Mat. No. 2.4642) which has a much higher Cr content compared to alloy 600 (NiCr15Fe8, DIN-Mat. No. 2.4816) is less susceptible to SCC in NaOH solutions. Figure 164 shows the influence of chromium and nickel on the stress corrosion cracking behavior of nickel alloys tested in 10% NaOH solutions at 325°C [576].

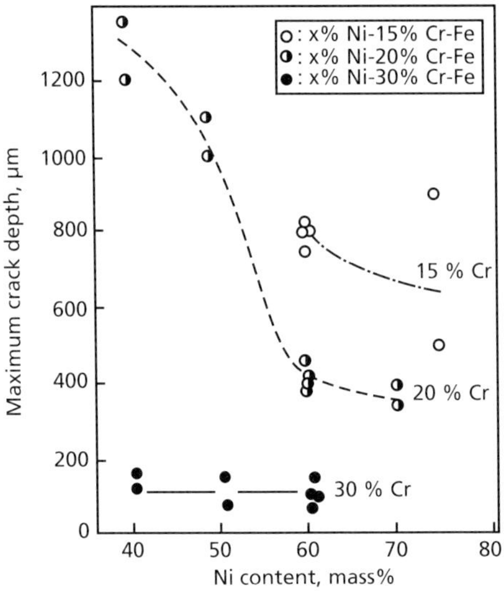

Figure 164: Influence of chromium and nickel on the SCC resistance of Ni alloys in 10% NaOH solutions at 325°C (test duration 500 h) [576]

For investigations of the stress corrosion cracking behavior of NiCr30Fe9 (Inconel® 690, DIN-Mat. No. 2.4642), 4 melts of this material with different compositions (Table 170) were investigated and compared to a melt of NiCr15Fe9 (Inconel® 600 DIN-Mat. No. 2.4816) [577]. The materials were tested in the as-delivered state and after a subsequent tempering at 704°C for 15 h.

Material	*%C*	*%Si*	*%Mn*	*%S*	*%Fe*	*%Cr*	*%Al*	*%Ti*	*%Cu*	*%Ni*
Inconel® 690	0.01	0.10	0.15	0.003	10.22	27.27	0.17	0.25	0.15	bal.
Inconel® 690	0.01	0.43	0.33	0.001	8.85	29.20	0.26	0.46	0.12	bal.
Inconel® 690	0.02	0.10	0.17	0.003	8.66	29.82	0.15	0.26	0.01	bal.
Inconel® 690	0.03	0.15	0.17		9.38	29.82	0.22	0.22	0.01	bal.
Inconel® 600	0.03	0.21	0.35	0.007	7.60	15.11	0.50	0.26	0.29	bal.

Table 170: Chemical composition of the investigated alloys Inconel® 690 and Inconel® 600 [577]

Slow strain-rate tests with different extension rates between 1×10^{-6} and 9×10^{-7} 1/s were carried out in an air-free 10% NaOH solution at 288 and 316°C. In contrast to the previously described test results of Inconel® 600 [527, 557–559, 564, 565], both the carbon content as well as the heat treatment and the test temperature did not have a noticeable effect on the stress corrosion cracking behavior of Inconel® 690. The maximum crack depths in the specimens measured after the end of the tests were always less than 0.13 mm. Figure 165 shows as an example, the stress – time curves for slow strain-rate tests on specimens from Inconel® 690 melt 1 for both heat-treated states at a strain rate of 4.1×10^{-6} 1/s in air and in NaOH solution.

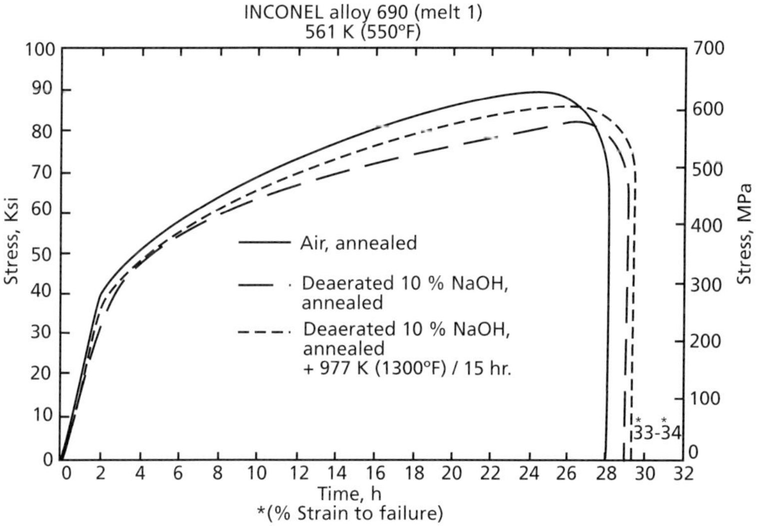

Figure 165: Slow strain-rate tests of Inconel® 690 in air and in 10% NaOH solutions at 288°C [577]

The maximum crack depths of the Inconel® 600 melts at a test temperature of 288°C were also below 0.13 mm; however, they increased to 0.89 mm at a test temperature of 316°C. The improved resistance of the alloy Inconel® 690 is attributed to a markedly higher chromium content.

The higher chromium content in alloy 690 led to modified behavior compared to alloy 600 with respect to the carbon solubility and carbide precipitation. Thus, for alloy 690 a different optimum heat treatment is expected for the improvement of the SCC resistance in NaOH solutions. Corresponding test results are reported in [578] on U-bend specimens of both materials after different heat treatments in 10% NaOH solutions at 350°C. Table 171 lists the materials, their carbon contents, their heat treatments and the test duration in NaOH solutions until cracks or failure occurred. According to these results, after shorter annealing times at higher temperatures, alloy 690 showed better resistance to stress corrosion cracking in NaOH solution than alloy 600.

Alloy	C content %C	Heat treatment		Time h	
		Temperature °C	Duration h	Crack initiation	Fracture
alloy 690	0.01	none	–	1,440	3,120
	0.01	704	1	nc	nf
	0.01	704	5	nc	nf
	0.01	704	15	nc	nf
	0.01	760	1	nc	nf
	0.01	816	0.17	nc	nf
	0.01	871	0.17	nc	nf
	0.01	607	8	3,120	nf
	0.03	none	–	1,440	4,125
	0.03	704	1	nc	nf
	0.03	760	1	nc	nf
alloy 600	0.03	none	–	720	720
	0.03	704	15	3,120	nf

nc = no cracking nf = no fracture

Table 171: Influence of the heat treatment of alloy 600 and alloy 690 on the stress corrosion cracking behavior of U-bend specimens in 10% NaOH at 350°C [578]

The improved behavior of alloy 690 compared to alloy 600 could not be confirmed by other investigations [579]. In these studies, the behavior of plates and tubes of the two above alloys in oxygen-free 10% NaOH solutions at 316 and 332°C as well as in 50% NaOH solutions at 316°C was investigated and compared to that of a martensitic Cr steel, a ferritic CrMo steel and two austenitic steels. The chemical compositions of the investigated materials are given in Table 172. The tests were carried out in different heat-treated states of the materials with both U-bend and C-ring specimens. The C-ring specimens were pre-stressed to 90% of their yield strength. In the case of the U-bend specimens, the exact stress is not known; however, there was plastic elongation of at least 10% on the outer surface.

Material	*Type*	*%C*	*%Si*	*%Mn*	*%S*	*%P*	*%Fe*	*%Cr*	*%Mo*	*%Ti*	*%Ni*
410 (DIN-Mat. No. 1.4024)	plate	0.13	0.37	0.39	0.009	0.018	bal.	12.5	–	–	–
	tube	0.09	0.37	0.46	0.005	0.014	bal.	12.3	0.14	–	0.48
18Cr2MoTi	plate	0.04	0.28	0.49	–	–	bal.	18.34	1.90	0.36	–
	tube	0.04	0.63	0.42	0.018	0.016	bal.	18.33	2.30	0.37	–
304 (DIN-Mat. No. 1.4301)	plate	0.06	0.49	1.89	0.006	0.022	bal.	18.21	–	–	8.95
	tube	0.05	0.47	1.67	0.020	0.027	bal.	18.92	0.24	–	9.14
alloy 800 (DIN-Mat. No. 1.4876)	plate	0.03	0.36	0.84	0.007	–	bal.	20.18	–	0.46	31.5
	tube	0.01	0.32	0.88	0.007	–	44.3	21.70	–	0.36	31.7
alloy 690 (DIN-Mat. No. 2.4642)	plate	0.03	0.15	0.17	0.007	–	9.38	29.82	–	–	60.4
alloy 600 (DIN-Mat. No. 2.4816)	plate	0.03	0.28	0.22	0.007	–		15.92	–	–	74.7
	plate	0.03	0.22	0.17	0.007	–	7.92	15.40	0.29	–	76.1
	tube	0.03	0.27	0.17	0.007	–	8.02	15.23	0.19	0.20	76.1

Table 172: Chemical composition of the investigated materials [579]

For most of the investigated alloys, the U-bend specimens were the more critical test condition compared to the C-ring specimens. The ferritic and martensitic stainless steels showed remarkable resistance to stress corrosion cracking; however, the strong general corrosion and the risk of embrittlement restricts their use under the given conditions. For the austenitic alloys, the more highly stressed U-bend specimens showed that the resistance to stress corrosion cracking increased with increasing nickel content. The alloys with low nickel contents failed very quickly in concentrated NaOH solutions. From the test data it is concluded that for high stresses, alloy 600 is expected to exhibit the best behavior in NaOH solutions.

Nickel-chromium-molybdenum alloys

The corrosion rates of two nickel-based alloys and two stainless steels in a concentrated NaOH solution (45 mass-%) containing NaCl between 180 and 240°C were studied (experimental details are given in Table 173). Results showed that the corrosion rate of Alloy C-276 in high-alkali solution with chloride at high temperature is the lowest, about 0.022 to 0.073 mm/a (see Table 174) [580].

Condition	*Temperature °C*	*NaOH wt%*	*$Al(OH)_3$, g/l*	*Bauxite, g/l*	*NaCl, g/l*	*Magnetite, g/l*
1	180	45	0	0	16.5	100
2	180	45	0	360	16.5	100
3	180	45	350	0	16.5	100
4	180	45	33	0	99.7	100
5	180	45	0	0	16.5	100

Table 173: Compositions and temperatures of the five kinds of test solutions [580]

Corrosion condition	*Alloy 400*	*Alloy C-276*	*Type 304*	*Type 316L*
1	0.097	0.022	4.19	4.10
2	0.091	0.073	1.06	1.90
3	0.603	0	4.17	4.84
4	0.452	0.045	2.48	3.19
5	0.302	0.025	14.88	16.40

Table 174: Corrosion rates in various test solutions (mm/a) (for corrosion conditions see Table 173)

According to Figure 166, as the molybdenum content increases in the nickel-chromium-molybdenum alloys, the corrosion resistance increases in boiling 50% NaOH solutions. Under these test conditions, both Inconel® 625 (Ni-22.0Cr-8.8Mo-2.0Fe-4.3(Nb+Ta)-0.03C, DIN-Mat. No. 2.4856) and Hastelloy® C (Ni-16.0Cr-16.5Mo-5.0Fe-4.0W-0.12C) showed a corrosion rate of approximately 0.013 mm/a [581]. Data is unavailable about the occurrence of stress corrosion cracking in these alloys.

The application ranges in sodium hydroxide solutions in dependence on the temperature and concentration is given in Figure 167 for some nickel-based materials, including the alloys Ni20Cr15Mo and Ni21CrMo. Thus, pure nickel can be used in all ranges and shows a superior corrosion resistance compared to its alloys [582].

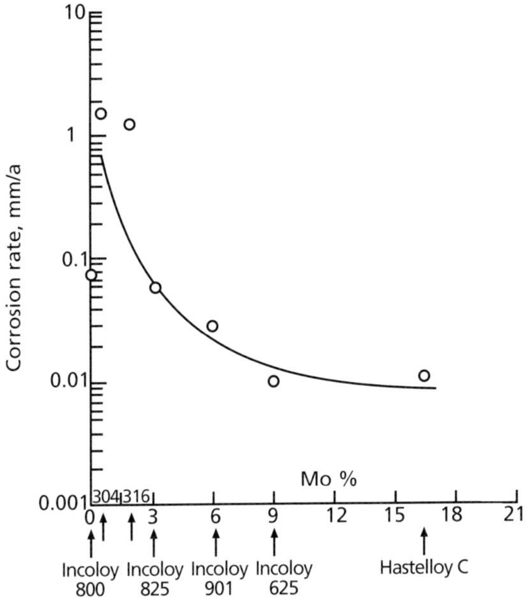

Figure 166: Influence of the molybdenum content on the corrosion of chromium-nickel steels and nickel alloys in boiling 50% NaOH [581]

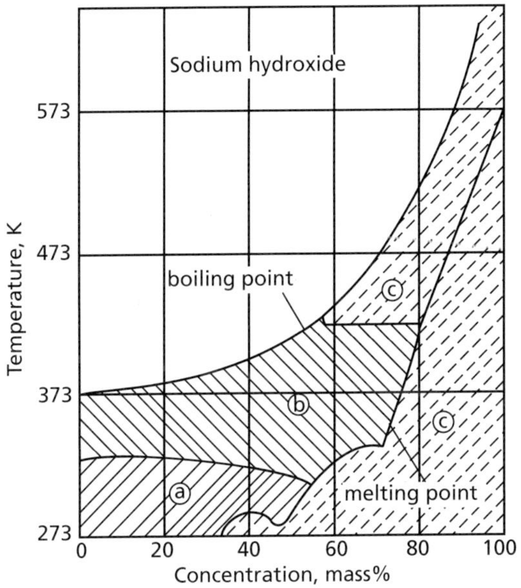

Figure 167: Application ranges of pure nickel and nickel alloys in sodium hydroxide [582]
(a): LC-Ni 99, NiCu30Fe,
(b): LC-Ni 99, NiCu30Fe, NiCr20Mo15, NiCr15Fe, NiCr21Mo and
(c): LC-Ni 99, NiMo30

Nickel-copper alloys

The nickel-copper alloy Monel® 400 (Ni-32.7Cu-1.3Fe-0.15Co-1.1 Mn-0.15C. DIN-Mat. No. 2.4360) is practically resistant to uniform surface attack by NaOH solutions. For a test period of 30 days, an average corrosion rate of 0.00075 mm/a was measured in 50% NaOH solutions at 5 to 65°C [548].

Monel® 400 also shows a better resistance to stress corrosion cracking in NaOH solutions than Inconel® 600. Capsules of Monel® 400 that had been subjected to heat treatment for 5.5 h at 600°C were filled with NaOH solution with NaOH contents between 4% and 20% and then tested at 300°C. They showed no stress corrosion cracking (Table 175) [583].

Stress MPa	*NaOH %*	*Test duration days*	*As delivered state*)*	*5.5 h/600°C*)*
170	4	250	1/5	0/2
	10	250	2/5	0/4
	20	250	0/5	0/2
110	4	250	0/3	0/2
	10	250	0/5	0/2
	20	250	0/5	–

*) Number of fractured samples / number of tested samples

Table 175: Results of stress corrosion cracking tests of Monel® 400 in NaOH solutions of various concentrations at 300°C [583]

The critical limiting stress for the occurrence of stress corrosion cracking in Monel® 400 in NaOH solutions of between 4 and 20% at 300°C is at approximately 140 MPa. Heat treatment for 5.5 h at 600°C has a beneficial effect on the resistance to stress corrosion cracking [583].

Due to its good corrosion resistance, even to concentrated NaOH solutions at elevated temperatures, Monel® 400 is used as a material for autoclaves which are filled with 40% NaOH solutions at 310°C [584].

As shown in Figure 167, Monel® 400 along with nickel LC-Ni99 are two of the most corrosion-resistant materials in hot concentrated NaOH solutions. Even in molten sodium hydroxide NaOH melts, Monel® 400 is quite resistant, although in this case LC-Ni99 is superior [582].

The corrosion behavior of the different nickel-based materials (Table 176) in 30 to 60% NaOH solutions at temperatures up to 166°C is reported in [585]. The dependence of the free corrosion potential measured at the boiling point on the concentration of the NaOH solution is shown in Figure 168 for pure nickel, austenitic 25-20CrNi steel 310, NiCu alloy Monel® 400 and CrFeNi alloy Inconel® 600.

Material	*DIN-Mat. No.*	*%C*	*%Si*	*%Mn*	*%Ni*	*%Cr*	*%Fe*	*%Mo*	*%Cu*	*other*
304	1.4301	0.06	0.54	0.81	8.48	18.91	bal.			
310	1.4841	0.02	0.37	1.132	19.26	24.21	bal.			
alloy 825	2.4858	0.03	0.23	0.4	40.9	21.2	30.5	3.0	2.2	0.1Al;0.8Ti
Monel® 400	2.4360	0.1	0.1	1.0	65.1		1.0		32.6	
alloy 600	2.4816	0.03	0.3	0.2	75.8	15.3	7.8		< 0.05	
Ni		< 0.15	< 0.35	< 0.35	> 99.0		< 0.40		< 0.25	

Table 176: Major alloying contents of the investigated nickel-based materials [585]

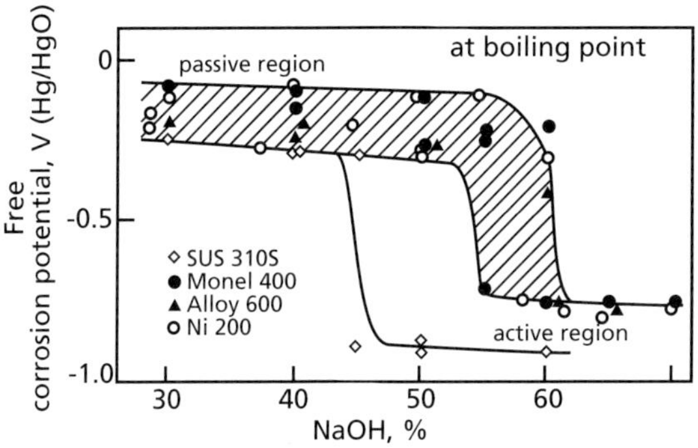

Figure 168: Free corrosion potentials of various nickel-based materials in boiling NaOH solutions of different concentrations [585]

Up to a concentration of approximately 55% NaOH, the free corrosion potentials of the nickel-based materials lie in the passive range, while this is valid for steel only up to approximately 40% NaOH [585]. Figure 169 shows the corrosion rates of the materials determined at corrosion potentials in the passive range. Figure 170 shows the corresponding values for the corrosion in the active potential range.

The corrosion rates are strongly affected by the temperature and decrease with increasing nickel content. When the solution was flushed with oxygen, the attack in the passive range increased noticeably, while flushing with hydrogen reduced the attack in the active range.

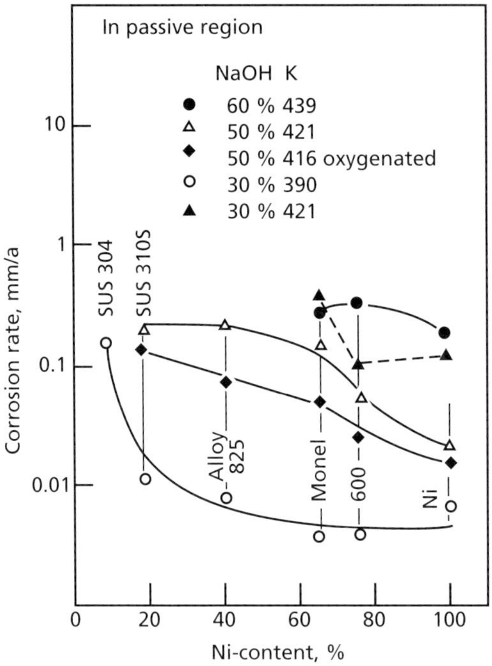

Figure 169: Corrosion rates in NaOH solutions as a function of the nickel content of the materials in the passive range of the individual corrosion potential [585]

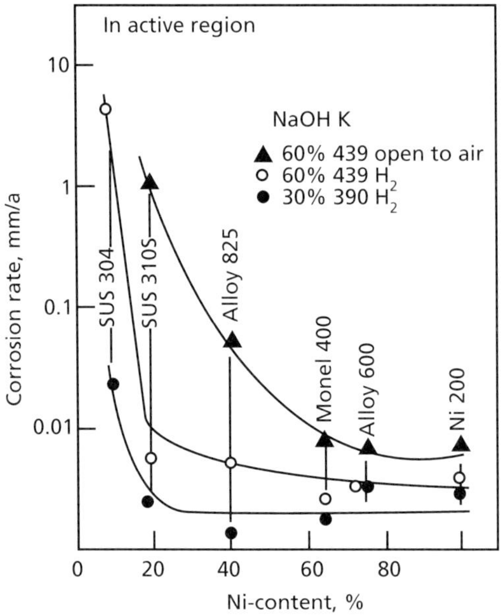

Figure 170: Corrosion rates in NaOH solutions as a function of the nickel content of the materials in the active range of the individual corrosion potential [585]

Nickel-molybdenum alloys

In two older reports [586, 587] the corrosion behavior of the following nickel-molybdenum alloys was investigated in a NaOH melt at 815°C:

Ni-10Mo; Ni-15Mo; Ni-20Mo; Ni-25Mo

Ni-10Mo-10Fe; Ni-15Mo-15Fe; Ni-20Mo-20Fe.

After a test period of 100 h and independent of the molybdenum content, a corrosion rate of approximately 0.0011 mm was found for these alloys. This corresponds to an average corrosion rate of approximately 0.096 mm/a.

Bibliography

[1] Dechema-Werkstoff-Tabelle "Essigsäure" (Dechema corrosion data sheets "Acetic acid") (in German)
DECHEMA, D-6000 Frankfurt am Main, 1957

[2] Rabald, E.
Corrosion Guide, 2nd. ed., p. 3
Elsevier Publishing Company, Amsterdam-London-New York, 1968

[3] Abdel Ael, M. S.; Osman, A. H.
The electrochemical behavior of tin, nickel and 65/35 tin-nickel alloy in acidic media
J. Electrochem. Soc. India *28* (1979) 3, p. 129

[4] Ritter, F.
Korrosionstabellen metallischer Werkstoffe (Corrosion charts of metallic materials) (in German)
4th ed., p. 88
Springer-Verlag, Wien, 1957

[5] Bayaes, M.; Jackson, B.; Shawhan, G.
The corrosion performance of electroless nickel: the role of accelerated test methods compared to actual process condition evaluations in oil and gas environments
Trans. Inst. Met. Finsih. *62* (1985) 4, p. 145

[6] Teeple, H. O.
Corrosion by some organic acids and related compounds
Corrosion *8* (1952) 1, p. 14

[7] Nelson, G. A.
Corrosion Data Survey 1967 – Corrosion charts A-1/A-2 "Acetic acid"
NACE, Houston (Texas/USA), 1968

[8] Hamner, N. E.
Corrosion Data Survey
5th ed., p. 2, p. 198
NACE, Houston (Texas/USA), March 1974

[9] Dubrisay, R.
Chimie et Industrie *44* (1940) Oct. p. 277 (in French)

[10] Le Monnier, E.
in: Kirk-Othmer Encyclopedia of Chemical Technology
2nd ed., vol. 8, p. 386
John Wiley & Sons, Inc., New York-London-Sydney, 1965

[11] Volokhova, V. I.; Vakhidov, R. S.; Lukjanica, A. I.
Corrosion stability of nickel-phosphorous galvanic coatings (in Russian)
Zashch. Met. *11* (1975) 3, p. 370

[12] Anonymous
The corrosion resistance of nickel-platings in the absence of current
Mater. Design. Eng. *54* (1961) 6, p. 141

[13] Anonymous
Chemically deposited nickel alloy coatings
Corrosion *13* (1957) 8, p. 122

[14] Dorozhkin, N. N.; Zavistovskii, S. E.
Corrosion-mechanical wear of a powdered coating of PG-Sr3 nickel in acetic acid (in Russian)
Trenie Iznos. *4* (1983) 5, p. 822

[15] Dillenburg, H.
Elektrolytische Entfernung von Nickel-Deckschichten von Stählen
(Electrolytic removal of nickel coatings from steel) (in German)
German DOS 1 963 415 (Bergische Metallwarenfabrik Villenberg & Co. KG) (24.6.1971)

[16] Product Information
Bulletin 10 M 4-63, 3839, A 280
The International Nickel Company, Inc., New York (USA)

[17] Product Information
Wiggin nickel alloys in the chemical process engineering

[18] Dechema-Werkstoff-Tabelle "Essigsäureanhydrid"
(Dechema corrosion data sheets "Acetic anhydride") (in German)
DECHEMA, D-6000 Frankfurt am Main, 1957

[19] Product Information
Die Eigenschaften von Nickel
(The characteristics of nickel) (in German)
1st ed., Nov. 1971
International Nickel Deutschland GmbH, D-4000 Düsseldorf

[20] Anonymous
Mater. Eng. *88* (1978) 6; Mater. Selector 1979, p. C63

[21] Ward, E.; Maylor, J. B. – Product Information
NI-O-NEL – A new corrosion-resistant alloy for the chemical industry
Henry Wiggin & Co. Ltd, Hereford (GB)

[22] Anonymous
At Pittsburgh Steel Company: NI-O-NEL alloy solves 15 year problem ... hot sulfuric acid mixtures
Chem. Eng. *67* (1960) 9, p. 40

[23] Rabald, E.
Corrosion guide, 2nd ed., p. 13
Elsevier Publishing Company, Amsterdam – London – New York, 1968

[24] Product Information
Inconel Alloy 601
Bulletin No. 3662 G, Sept. 1974
Henry Wiggin & Company Ltd., Hereford (GB)

[25] Product Information
Incoloy Alloy 825
Bulletin No. 3564, Dec. 1971, revised Nov. 1976
Henry Wiggin & Company Ltd., Hereford (GB)

[26] Product Information
Inconel Alloy 625
Bulletin 3700 G, Nov. 1977
Henry Wiggin & Company Ltd., Hereford (GB)

[27] von Kutepow, N.; Himmele, W.; Hohenschutz, H.
Die Synthese von Essigsäure aus Methanol und Kohlenoxyd
(The synthesis of acetic acid from methanol and carbon monoxide) (in German)
Chem.-Ing.-Tech. *37* (1965) 4, p. 383

[28] Kilchevskaya, T. E.; Anokhika, T. P.
Corrosion resistance of construction materials used in media for the production of acetic acid from hydrocarbons (in Russian)
Khim. Neft. Mashinostr. (1979) 11, p. 24

[29] Dillon, C. P.
Nickel alloy resistant to corrosion in distillation columns
Inco Nickel Topics *14* (1961) 9, p. 3

[30] Perin, Yu. I.; Valieva, R. A.; Burmistrova, L. K.; Kuznetsova, G. E.
Corrosion resistance of high-alloy steels and the alloy KhN65MV in certain media for the synthesis of terephthalic acid by the oxidation of p-xylene (in Russian)
Khim. Neft. Mashinostr. (1976) 10, p. 27

[31] Perin, Yu. I.; Valieva, R. A.; Burmistrova, L. K.; Kuznetsova, G. E.
The corrosion resistance of high-alloy steels and the (nickel) alloy KhN65MV in media for the synthesis of terephthalic acid by the oxidation of p-xylene (in Russian)
Khim. Neft. Mashinostr. (1976) 10, p. 27

[32] Perin, Y. I.; Valieva, R. A.; Burmistrova, L. K.; Kuznetsova, G. E.
Corrosion resistance of high-alloy steels and the alloy KhN65MV in certain media for the synthesis of terephthalic acid by the oxidation of p-xylene (in Russian)
Khim. Neft. Mashinostr. (1976) 10, p. 27
(C.A. *86* (1977) 94508x)

[33] Colombier, L. (Product Information)
Molybdän in rost- und säurebeständigen Stählen und Legierungen
(Molybdenum contained in stainless and acid-resistant steels and alloys) (in German)
Climax Molybdenum GmbH (Amax/American Metal Climax, Inc.), D-4000 Düsseldorf

[34] Freedman, A. J.; Tisinai, G. F.; Troscinski, E. S.
Selection of alloy for refinery processing equipment
Corrosion *16* (1960) p. 19t

[35] Anonymous
Nickelberichte *11* (1953) p. 106 (in German)

[36] Holmberg, E. G.
Valves for corrosive fluids
Corrosion *11* (1955) p. 406t

[37] Anonymous
Valve corrosion
Mater. Protection *4* (1965) 6, p. 28

[38] Swandby, R. K.
Corrosion charts: Guides to materials selection
Chem. Eng. *69* (1962) 23, p. 186

[39] Valieva, R. A.; Perin, Yu.; Klochkova, M. R.; Tsinman, A. I.
Effect of oxygen on the corrosion of stainless steels and alloy EP-567 in acetic acid at elevated temperatures (in Russian)
Zashch. Met. *11* (1975) 4, p. 475
(C.A. *83* (1975) 167668z)

[40] Tsinman, A. I.; Pishchick, L. M.
Corrosion-electrochemical behavior of metals in organic media. III. Anodic behavior and corrosion resistance of molybdenum-containing alloys and their components in an acetic acid solution of a chloride (in Russian)
Elektrokhimiya *11* (1975) 1, p. 127
(C.A. *83* (1975) 17551s)

[41] Valiyeva, R. A.; Perin, Yu. I.; Klochkova, M. R.; Tzinman, A. I.
Effect of oxygen on the corrosion of stainless steels and EP-567 alloy in acetic acid at high temperatures (in Russian)
Zashch. Met. *11* (1975) 4, p. 475

[42] Shvarts, G. L.; Makarova, L. S.; Gerasimenko, G. J.
Use of nickel-molybdenum and nickel-chromium-molybdenum alloys in the chemical industry (in Russian)
Khim. Neft. Mahinostr. (1969) 5, p. 20
(C.A. *71* (1969) z32952u)

[43] Eisenbrown, Ch. M.; Barbis, P. R.
Corrosion of metals by acetic acid
Chem. Eng. *70* (1963) 9, p. 148

[44] Product Information
Jacob & Korves GmbH, D-4730 Ahlen

[45] Weisert, E. D.
The corrosion forum: Hastelloy® B
Chem. Eng. *59* (1952) 7, p. 314

[46] Teeple, H. O.
Materials of construction data. Nickel and high-nickel alloys
Ind. Eng. Chem. *42* (1950) p. 2046

[47] Weisert, E. D.
The corrosion forum: Hastelloy® C
Chem. Eng. *59* (1952) 6, p. 297

[48] Luce, A. W.
The corrosion forum: Chlorimet® 3
Chem. Eng. *61* (1954) 3, p. 254

[49] Luce, A. W.
The corrosion forum: Chlorimet® 2
Chem. Eng. *61* (1954) 2, p. 246

[50] Koch, A.
Chem. Eng. *54* (1947) 6, p. 216

[51] Swales, G. L.
High-nickel alloys in the petroleum industry. Part 2
Corrosion Technology 6 (1959) 4, p. 119

[52] Anonymous
Corrosion by acetic acid and acetic acid mixtures under heat transfer conditions
Mater. Protection 6 (1967) p. 77

[53] Product Information
Druckschrift Nickel Korrosionstabellen (Bulletin nickel corrosion charts) (in German)
Vereinigte Deutsche Nickel-Werke AG, D-5840 Schwerte

[54] Togano, H.; Osato, K.
Corrosion test on the construction materials for acetic acid synthesis from methanol and carbon monoxide. (1) Corrosion tests by acetic acid solutions at elevated temperatures (in Japanese)
Corrosion Eng. (Boshoku Gijutsu) *10* (1961) 12, p. 529

[55] Katz, W.
Eigenschaften und Beständigkeit von Nickel-Werkstoffen mit Molybdän (Properties and corrosion resistance of nickel materials containing molybdenum) (in German)
Werkst. Korros. *13* (1962) 8, p. 469

[56] Hodge, F. G.; Kirchner, R. W.
An improved Ni Mo-alloy for hydrochloric acid service
Mater. Performance *15* (1976) 8, p. 40

[57] Anonymous
Improved version of Hastelloy® alloy B
Anticorros. Methods Mater. *22* (1975) 11, p. 19

[58] Anonymous
Materials
Br. Chemical Eng. Equipment Suppl. (1968) Nov., p. 105

[59] Weller, J.; Weissgerber, R.; Koehler, W.
Comparison of the corrosion resistance of titanium, zirconium, tantalum, and high molybdenum-nickel alloys in acetylation mixtures (in German)
Vortr. Diskuss. Kolloq. "Eigenschaften und Anwendungen hochschmelzender reaktiver Metalle" 1967, published 1968, p. 306

[60] Campbell, J. L.
Metal resist hot sulfuric acid
Ind. Eng. Chem. *16* (1961) 11, p. 79

[61] Musina, A. S.; Lange, A. A.; Bukhman, S. P.; Semashko, T. S.
Changes in the state of the surface of Permalloy and Kovar during chemical treatment (in Russian)
Izv. Akad. Nauk Kazakh. SSR Khim. (1984) 3, p. 12

[62] Weisert, E. D.
The corrosion forum: Hastelloy® D
Chem. Eng. 59 (1952) 8, p. 258

[63] Rabald, E.
Corrosion Guide, Second Edition
Elsevier Publishing Company, Amsterdam, London, New York, 1968

[64] Anonymous
NACE Publication 5A 180
Corrosion of metals by aliphatic organic acids
Mater. Perform. *19* (1980) 9, p. 65

[65] Rabald, E.
Dechema-Werkstoff-Tabelle, Caprylsäure
(Dechema Corrosion Data Sheets, Caprylic Acid) (in German)
DECHEMA, Frankfurt am Main, 1955

[66] Product Information
WIGGIN, Korrosionsbeständige Legierungen
(WIGGIN, corrosion resistant alloys) (in German)
Henry Wiggin & Company Limited, Hereford, England

[67] Heitz, E.
Korrosion von Metallen in organischen Lösungsmitteln
(Corrosion of metals in organic solvents) (in German)
Oberfläche-Surface *15* (1974) 11, p. 33

[68] Vehlow, J.
Ermittlung der Korrosionsgeschwindigkeiten von hochlegierten Stählen und Hastelloy® C in niederen Carbonsäuren mit Hilfe der Radionuklidtechnik
(Determination of the corrosion rates of high-alloy steels and Hastelloy® C in lower carboxylic acids by change of the radioactivity) (in German)
Werkstoffe u. Korrosion *31* (1980) 2, p. 120

[69] Vehlow, J.
Ermittlung der Korrosionsgeschwindigkeiten von hochlegierten Stählen und Hastelloy C in niederen Carbonsäuren mit Hilfe der Radionuklidtechnik
(Determination of the corrosion rates of high-alloy steels and Hastelloy C in lower carboxylic acids by change of the radioactivity) (in German) (unpublished manuscript)

[70] Product Information
Corrosion Resistant Charts
Du Pont de Nemours Internat. S.A., CH-Geneva, Information Service, April 1975

[71] Kosmenko, Yu. L.; Baidin, I. I.
Corrosion resistance of materials in synthetic acid production
(translated from Khimicheskoe i Neftyanoe Mashinostroenie (1985) No. 10, p. 19)

[72] Rabald, E.
Dechema-Werkstoff-Tabelle, Buttersäure
(Dechema Corrosion Data Sheets, Butyric Acid) (in German)
DECHEMA, Frankfurt am Main, 1955

[73] Rabald, E.
Dechema-Werkstoff-Tabelle, Capronsäure
(Dechema Corrosion Data Sheets, Caproic Acid) (in German)
DECHEMA, Frankfurt am Main, 1955

[74] Rabald, E.
Dechema-Werkstoff-Tabelle, Caprylsäure
(Dechema Corrosion Data Sheets, Caprylic Acid) (in German)
DECHEMA, Frankfurt am Main, 1955

[75] Rabald, E.
Dechema-Werkstoff-Tabelle, n-Caprinsäure
(Dechema Corrosion Data Sheets, n-Capric Acid) (in German)
DECHEMA, Frankfurt am Main, 1955

[76] Rabald, E.
Dechema-Werkstoff-Tabelle, Fettsäuren (höhere, ab C_6)
(Dechema Corrosion Data Sheets, Fatty Acids) (in German)
DECHEMA, Frankfurt am Main, 1957

[77] Product Information
Eigenschaften und Anwendungsgebiete korrosionsbeständiger Wiggin-Nickellegierungen
(Properties and applications of corrosion resistant Wiggin nickel alloys) (in German)
Henry Wiggin & Company Limited, Hereford, England

[78] Product Information
Hastelloy® B-2, A nickel-molybdenum alloy with outstanding resistance to hydrochloric and sulfuric acids in the as-welded condition
Cabot Wrought Products Division, Kokomo, Indiana 46901

[79] Product Information
Corrosion resistance of Hastelloy®-Alloys
Stellite Division, Cabot Corporation
International Department, Kokomo, Indiana 46901

[80] Rabald, E.
Werkstoffe und Korrosion
(Materials and Corrosion) (in German), Vol. 1, p. 667 ff.
Verlag Otto Spamer, Leipzig, 1931

[81] Swandby, R. K.
Acetic acid, formic acid
Chem. Eng. 69 (1962) 23, p. 187

[82] Newton, L. E.; Hausler, R. H.
CO_2-Corrosion in oil and gas production
Selected papers, abstracts and references
1984 National Association of Corrosion Engineers, 1940 South Creek Drive, Houston, Texas, 77084

[83] Newton, jr., L. E.; Hausler, R. H.
CO_2-corrosion in oil and gas production
National Association of Corrosion Engineers (Houston) (1984), p. 19

[84] Newton, jr., L. E.; Hausler, R. H.
CO_2-Corrosion in oil and gas production
National Association of Corrosion Engineers, Houston, Tex. (1984) p. 22

[85] Flowers, L. C.; Kolley, J. B.
Corrosion of electro-deposited nickel.
Resistance to water containing dissolved carbon dioxide and air Industrial
Eng. Chem. 42 (1950), p. 719

[86] Rabalt, E.
Corrosion Guide
Elsevier Publishing Company, Amsterdam-London-New York, 1968

[87] Anonymous
Nickel, Korrosionstabellen (Corrosion tables) (in German)
Bulletin of Vereinigte Deutsche Nickel-Werke AG, Schwerte, Germany

[88] Product Information
Eigenschaften und Anwendungsgebiete korrosionsbeständiger Wiggin-Nickellegierungen (Properties and applications of corrosion resistant Wiggin nickel alloys) (in German)
Bulletin of Henry Wiggin & Company, Ltd., Hereford, GB

[89] Brill, U.
Nickel, Cobalt und Nickel- und Cobalt-Basislegierungen
in: Egon Kunze
Korrosion und Korrosionsschutz, 1. Aufl., Bd. 2
Wiley-VCH, Berlin, 2001, pp. 1076–1168

[90] Hussain, N.; Schanz, G; Leistikow, S.; Shahid, K. A.
High-temperature oxidation and spalling behavior of Incoloy 825
Oxidation of Metals 32 (1989) 5–6, pp. 405–431

[91] Asphahani, A. I.
Evaluation of highly alloyed stainless materials for CO_2/H_2S environments
Corrosion 37 (1981) 6, pp. 327–335

[92] Franke, E.
Neue korrosionsbeständige Werkstoffe im Ausland VI (New corrosion resistant materials abroad. Pt. VI) (in German)
Werkst. Korros. 2 (1951) 2, p. 68

[93] Wesley, W. A.; Copson, H. R.
J. Electrochem. Soc. (1949), p. 226

[94] Marshall, T.; Hugill, A. J.
Corrosion by low-pressure geothermal steam
Corrosion (Houston) 13 (1957) 5, p. 329t

[95] Kane, R. D.; Boyd, W. K.
Materials technology for oil and gas production CO_2 Corrosion in Oil and Gas Production
National Association of Corrosion Engineers, Houston, Tex. (1984), p. 546

[96] Lang, F. S.; Mason jr., J. F.
Corrosion in amine gas treating solutions
CO_2 Corrosion in Oil and Gas Production
National Association of Corrosion Engineers, Houston Tex. (1984), p. 546

[97] Montrone, E. D.; Long, W. P.
Choosing materials for CO_2 absorption systems
Chem. Eng. 78 (1971) Jan., p. 94

[98] Asphahani, A. I.
Evaluation of highly alloyed stainless materials for CO_2/H_2S environments
Corrosion (Houston) 37 (1981) 6, p. 327

[99] Onoyama, M.; Shitani, K.; Hayashi, N.; Murata, M.
Resistance of some high alloys to CO_2-H_2S corrosion at elevated temperatures
Corrosion '84 (Proc. Conf.), New Orleans, La. (USA) 2–6 Apr. 1984, National Assoc. of Corrosion Engineers, Houston, Tex. (1984) p. 291

[100] Johnson, H. T.; McAlpin, A. W.; Schwenk, A. M.
Food Technologie (1955) p. 643

[101] Product Information
Silverin, Korrosionstabellen (Silverin, corrosion tables) (in German)
Vereinigte Deutsche Nickel-Werke AG, Schwerte, Germany

[102] Reuter, M.
Schäden und Vorschriften bei Anlagen zur Förderung, Reinigung und zum Transport von Sauergas (Damage of and rules for installations for the recovery, purification and transportation of sour gas) (in German)
Technische Überwachung 15 (1974) 3, p. 65

[103] Vijh, A. K.
The behavior of metals in anhydrous hydrogen fluoride
Surface Technol. *4* (1974) p. 401

[104] Jarry, R. L.; Fischer, J.; Gunther, W. H.
The mechanism and kinetics of the reaction between nickel and fluorine
ANL-6684, Argonne National Laboratory, Argonne, III., 1963

[105] Jarry, R. L.; Fischer, J.; Gunther, W. H.
The mechanism of the nickel fluorine reaction
J. Electrochem. Soc. *110* (1963) p. 346

[106] Lukyanychev, Ya. A.; Astakhov, I. I.; Nikolaev, N. S.
Mechanism of the formation of fluoride films on nickel and their properties (in Russian)
Bull. Akad. Sci. USSR, Div. Chem. Sci. (1965) p. 577
(Izv. Akad. Nauk. SSSR Ser. Khim. (1965) p. 588)

[107] Singleton, A. H.; Tompkins jr., J. F.; Kleinberg, S.; Sterner, C. J.
Corrosion of metals by liquid fluorine
Ind. Eng. Chem. *57* (1965) 3, p. 47

[108] Kleinberg, S.; Tompkins, J. F.
The compatibility of various metals with liquid fluorine
ASD-TDR-62-250 Air Force Systems Command, Wright-Patterson Air Force Base, Ohio, 1962

[109] Brown, P. E.; Crabtree, J. M.; Duncan, J. F.
The kinetics of the reactions of elementary fluorine with copper metals
J. Inorg. Nucl. Chem. *1* (1955) p. 202

[110] Myers, W. R.; DeLong, W. B.
Fluorine corrosion. High temperature attack on metals by fluorine and hydrogen fluoride. Behavior of insulated steel parts in fluorine cells
Chem. Eng. Progr. *44* (1948) p. 359

[111] Comninellis, Ch.; Plattner, E.
Dissolution and passivity of nickel in HF-H_2O-HCl system
Electrochim. Acta *24* (1979) p. 883

[112] Graver, D. L.
Nickel and high-nickel alloys
Process Ind. Corrosion (1975) p. 109

[113] Weisert, E. D.
Hastelloy® alloy C
Chem. Eng. *59* (1952) June, p. 297

[114] Mallaire, F. R.
Fluorine lubricated bearing technology Final Report
Contract NAS 3–11201, NASA-Cr-1220820, June 1973

[115] Heppekausen, J.
Die stromlose (chemische) Vernickelung in der Vakuum- und Feinwerktechnik (Electroless (chemical) nickel plating in the vacuum and precision industries) (in German)
Galvanotechnik *70* (1979) p. 401

[116] Movchan, B. A.; Kuzmin, G. S.; Mochalova, T. F.; Karatish, V. V.; Tihonovskij. A. L.; Yagupolskaya, L. N.
Corrosion of nickel of different purity in gaseous hydrogen fluoride (in Russian)
Zashch. Metallov *7* (1971) p. 32

[117] Arslambekov, V. A.; Guseva, N. S.; Mogilevtsev, P. K.
Reaction of oxidized nickel with hydrogen fluoride (in Russian)
Zh. Neorg. Khim. (1979) p. 1758; (J. Inorg. Chem. *24* (1979) p. 973)

[118] Watanabe, N.; Haruta, N.
An ESCA study of the anodic film on nickel in liquid hydrogen fluoride
Electrochim. Acta *25* (1980) p. 461

[119] Cid, M.; Jouanneau, A.; Nganga, D.; Petit, M. C.
Comparison between the dissolution and passivity of nickel in sulfuric and hydrofluoric acid
Electrochim. Acta *23* (1978) p. 945

[120] Osterwald, J.; Trunk, H.
Passivität und Transpassivität von Nickel in flußsauren Elektrolyten
(Passivity and transpassivity of nickel in electrolytes with hydrofluoric acid) (in German)
Werkstoffe und Korrosion *28* (1977) p. 321

[121] Nickel, O.
Eigenschaften und Anwendungen von Reinnickel, Nickel-Basislegierungen und nickelhaltigen NE-Werkstoffen im Rohrleitungsbau, Rohrleitungstransport
(Properties and application of pure nickel, nickel-base alloys and nickel-containing NF-materials for construction of pipings and transport of pipes) (in German)
3 R *14* (1975) p. 419

[122] Nickel, O.
Eigenschaften und neuere Anwendungen nickelhaltiger Werkstoffe
(Properties and new applications of nickel-containing materials) (in German)
VDI-Z. *121* (1979) p. 313

[123] Anonymous
Materials for receiving, handling and storing hydrofluoric acid
Mater. Protect. Performance *10* (1971) 3, p. 33

[124] Kuzmin, G. S.; Mochalova, T. F.
Hydrogen fluoride oxidation of titanium, carbon- and silicon-alloyed nickel welds (in Russian)
Svar. Proizvod. (1980) 2, p. 9

[125] Jackson, R. B.
Corrosion of metals and alloys by fluorine
NP-8845 Allied Chem. Corp. Morristown, N.J., 1960

[126] Steindler, M. J.; Vogel, R. C.
Corrosion of materials in the presence of fluorine at elevated temperatures
ANL-5662, Argonne National Laboratory, Argonne III., 1957

[127] Daniel, P. L.; Rapp, R. A.
Halogen corrosion of metals
in: M. G. Fontana, R. W. Staehle "Adv. Corrosion Sci. and Techn.", vol. 5, p. 152
Plenum Press, New York, 1976

[128] Jackson, J. D.
Corrosion in Cryogenic Liquids
Chem. Eng. Progr. *57* (1961) p. 61

[129] Gunther, W. H.; Steindler, M. J.
Laboratory investigations in support fluid-bed fluoride volatility processes, part XVI – The corrosion of nickel and nickel alloys by fluorine, uranium, hexafluoride and selected volatile fission product fluorides at 500°C
ANL-7241, Argonne National Laboratory, Argonne, III., 1966

[130] Reddy, S. N. S.; Rapp, R. A.
Internal Fluoridation of Ni5.6 Pct Cr Alloy
Metallurgical Trans. (B) *11* (1980) p. 533

[131] Anonymous
Nickel and high-nickel alloys
Mat. Eng. *80* (1974) 4; Materials Selector 1975, p. 164

[132] Ohmichi, T. et al.
Thermodynamical consideration for corrosion of metals in molten fluorides. Behavior of Hastelloy® N in molten lithium tetrafluoroberyllat
Report JAERI-M-5393 (C.A. *82* (1975) 117744c)

[133] Fishman, V. A.; Dobrolyubov, V. V.; Rogova, V. A.; Grabylnokova, V. A.; Munitsyna, R. G.; Fedorenko, G. V.
Corrosion resistance of metallic and non-metallic materials in technical media for the production of defluorinated phosphates (in Russian)
Khim Prom, Moscow (1979) 2

[134] Wirth, P.; Köhler, W.; Lehmann, H.
Korrosion metallischer Werkstoffe in Fluorwasserstoffsäure und wasserfreiem Fluorwasserstoff, Teil I: Eisen- und Nikkelwerkstoffe
(Corrosion of metallic materials in hydrofluoric acid and water-free hydrogen fluoride, part I: iron and nickel materials) (in German)
Chem. Techn. *22* (1970) p. 608

[135] Svistuna, T. V.; Petrovskaya, V. A.; Dobrolyubov, V. V. et al.
Development of a corrosion-resistant alloy based on nickel for usage in fluor-containing media (in Russian)
Tr. Nll Po Udobr. i. Insektofungitsidam (1978) 233, p. 74

[136] Paige, B. E.
Corrosion of nickel alloys in nuclear fuel reprocessing
Mater. Performance *15* (1976) 12, p. 22

[137] Anonymous
Nickel alloy for use in as-welded condition
Mater. Eng. *90* (1979) 1, p. 22

[138] Enke, Ch. G.
Schweißen in der Kerntechnik
(Welding in nuclear technology) (in German)
Chem.-Anl. Verf. (1979) 4, p. 115

[139] Anonymous
A new nickel alloy for gas turbines
Mat. Eng. *75* (1972) 5, p. 28

[140] Ross R. W.jr.; Scarberry, R. C.
Analysis of corrosion products as a tool in the investigations of nickel alloy failures
Corrosion *30* (1974) p. 211

[141] Danachie M. J.jr.
Phase extraction and analysis in superalloys — Second summary of investigations by ASTM Subcommittee E04.91
J. Testing & Evaluation (1978) p. 189

[142] Zarubin, P. I.; Zamyatina, E. A.
The selection of construction materials for the mixer in the reaction of a fluoride concentrate with sulfuric acid (in Russian)
Tr. Ural. Nauchno.-Issled. Khim. Inst. (1976) 35/36, p. 122

[143] Konstantinova, E. V.; Nazarova, N. P.
Corrosion of 47CrNiMo-2 alloy in nitric fluoride media (in Russian)
Zashch. Metallov *14* (1978) p. 187

[144] Konstantinova, E. V.; Chechetina, N. A.; Lomovsev, V. I.
Structural materials for hot nitrofluoride solutions (in Russian)
Zashch. Metallov *16* (1980) p. 66

[145] Solok, A. M.; Khodos, R. S.; Kusnetzov, V. M.
Corrosion resistance of alloys EI-435 and El-437 in mixture of nitric and hydrofluoric acid (in Russian)
Zashch. Metallov *9* (1973) p. 437

[146] Kramer, L. D.; Michael, S. T.
Service experience and stress corrosion of Inconel® 600 below expansion joints in turbine-steam environments
Mater. Performance *14* (1975) 9, p. 13

[147] Eiselstein, H. L.; Hosier, J. C.
Nickel-chromium alloys
Brit. Pat. 1336409 (7.11.1973)

[148] Kireeva, T. S.; Svistunova, T. V.; Runova, Z. K.
A corrosion-resistant alloy for operating in hydrofluoric-nitric acid etching solution (in Russian)
Stal (russ) (1979) p. 141

[149] Babchenko, A. M.; Khodos, R. S.; Kistanov, A. I.; Maksimo, A. S.; Nozdrachev, A. V.
The effect of the method of melting on the pickling of the alloy El-437 B in nitric-hydrofluoric acid mixture (in Russian)
Zashch. Metallov *14* (1978) p. 463

[150] Sharkova, S. et al.
Effect of heat treatment on the mechanical properties and corrosion resistance of Hastelloy-S (in Russian)
Mashinostroene *23* (1974)

[151] Svistunova, T. V.; Runova, Z. K.; Kireeva, T. S.; Tolmacheva, G. A.
Intercrystalline corrosion of the alloy KhN60V (in Russian)
Metallurgiya Moscow (1976) p. 68

[152] Crook, P., Meck, N.S., Rebak, R.B.
The influence of composition upon surface degradation and stress corrosion cracking of the Ni-Cr-Mo alloys in wet hydrofluoric acid
NACE CORROSION 2007, Mar 11–15 2007, Nashville, Tennessee, USA (2007) Paper No. 07481, p. 113

[153] Pawel, S.J.
Corrosion of high-alloy materials in aqueous hydrofluoric acid environments
Corrosion 50 (1994) 12, p. 963–971

[154] Rebak, R.B., Dillman, J.R., Crook, P., Shawber, C.V.V.
Corrosion behavior of nickel alloys in wet hydrofluoric acid
Materials and Corrosion 52 (2001), p. 289–297

[155] Jennings, H.S.
Corrosion by hydrogen fluoride and hydrochloric acid
ASM Handbook Corrosion: Environments and Industries 13C (2006), p. 690–703

[156] Anonymous
Nickel and its alloys—wrought, cast
Mat. Eng. *92* (1980) 4, Mater. Selector 1981, p. 84

[157] Godwin, T. W.; Lorenzo, C. L.
Ignition of several metals in fluorine
American Rocket Soc. Meeting, New York, November 1958, Paper-No. 740–58

[158] Tricoli, A.; Battarra, A.; Rebua, G.; Bostetti, L.
Cell for the electrolytic manufacture of fluorine (in Italian)
DOS 216820 (1971), (Montecatin Edison S.p.A.) (C.A. *76* (1972) 67490x)

[159] Noe, P.; Delafosse, D.; Deletre, G.
Electromagnetic circulation pumps for corrosive gases (in French)
Comm. Energie Atom. (France) Rapp. CEA-R-2744, 1965

[160] Hise, D. R.
Corrosion from halogens, process industries
Corrosion (1975) p. 240

[161] Anonymous
Nickel-copper alloys fight corrosion; useful over wide temperature range
Mater. Eng. *80* (1974) 3, p. 42

[162] Corella, J.
Kinetics of the non-catalytic gas-solid reactions in semicontinuous integral fluidized bed reactor. Application to the determination of the kinetics of the fluorination of uranium tetrafluoride
Chem. Eng. Sci. *35* (1980) p. 25

[163] Crawford, J. G.
Valve life stretched 3–4 tims at 1/10th the unit cost
Chem. Process *39* (1976) 10, p. 14

[164] Company publication
Corrosion Engineering Bulletin, CEB-5, 1975
INCO, New York (USA)

[165] Jennings, H.S., Monel 400 corrosion characteristics in hydrofluoric acid (HF), Materials Science and Technology 2010 Conference and Exhibition, Houston, TX, USA (2010) p. 340–349

[166] Degnan, T. F.; Burbridge, H. G.
Materials for receiving, handling and storing hydrofluoric acid
Mat. Prot. Perform. *10* (1971) 3, p. 33

[167] Anonymous
Anti-corrosion methods and materials, 4: Forms of corrosion, part 3: Corrosion cracking
Anti-Corrosion *19* (1972) 12, p. 9

[168] Zotikov, V. S.; Bakhmutova, G. B.; Bocharova, N. A.; Bogdanova, N. O.; Kotchetkov, N. I.; Petukhov, V. V.; Semenyuk, E. Ya.
Corrosion of some non-ferrous metals in hydrogen fluoride media (in Russian)
Khim. Prom, Moscow (1975) p. 773

[169] Neroslavskaja, L. L.; Artemev, V. I.; Nazvic, M. G.
Selection of materials for gas filters in the production of magnesium and aluminium (in Russian)
Zashch. Metallov *12* (1974) p. 633

[170] Crum, J.R., Smith, G.D., McNallan, M.J., Hirnyj, S.
Characterization of corrosion resistant materials in low and high temperature HF environments
CORROSION 99 (1999) Paper No. 382, p. 1–11

[171] Jennings, H.
Materials for storing and handling commercial grades of aqueous hydrofluoric acid and anhydrous hydrogen fluoride
NACE International Publication 5A171 (2007 Edition) Item No. 24057, p. 1–23

[172] Crook, P.
Corrosion of Nickel and Nickel-Base Alloys
ASM Handbook Corrosion: Materials 13B (2005), p. 228–251

[173] Matusevich, V. S.; Lebedev, A. N.; Fopkin, M. N.; Konstantinova, E. V.
Investigation of the corrosion of the nickel-molybdenum alloy EP-496 in hydrofluoric acid containing fluorine ions (in Russian)
Zashch. Metallov *8* (1972) p. 317

[174] Kirchner, R. W.
Superalloys
Proc. Ind. Corr. (1975) p. 114

[175] Svistunova, T. V.
Corrosion resistant nickel-base alloys (in Russian)
Metalloved. Term. Obrab. Met. (1980) 7, p. 24

[176] Alenchikova, I. F.; Neporezov, V. S.; Sukhoverkhov, V. F.
Kinetics of interaction of fluorine with iron-nickel alloys (in Russian)
Zashch. Metallov *9* (1973) p. 190

[177] Levinskii, N. N.; Krutikokva, I. M.; Tyapkina, V. V.; Mogilvtsev, P. K.
Corrosion stability of NA-1 nickel alloy and NP-2 in fluorine (in Russian)
Zashch. Metallov *13* (1977) p. 494

[178] Chilenskas, A. A.; Gunderson, G. E.
Engineering development of fluid-bed fluoride volatility processes, part VII—The corrosion of nickel in process environments
ANL-6979, Argonne National Laboratory, Argonne, Ill., 1965

[179] Swisher, J. H.; Schrey, F.; Gallagher, P. K.; Fuchs, E. O.
Insulating films on Ni-Fe-Mo alloy powders by fluorination
J. Electrochem. Soc. *117* (1970) p. 537

[180] Kelly, J. J.; Koel, G. J.
Electrochemical aspects of the beveling of sputtered Permalloy® films
J. Electrochem. Soc. *125* (1978) p. 860

[181] Lackey, J. Q.; Degnan, T. F.
Formic acid corrosion of common materials of construction
Mater. Performance *13* (1974) No. 7; p. 13

[182] Miller, R. F.; Treseder, R. S.; Wachter, A.
Corrosion by acids at high temperatures
Corrosion (Houston) *10* (1954) p. 7

[183] Teeple, H. O.
Corrosion by some organic acids and related compounds
Corrosion (Houston) *8* (1952) No. 1, p. 14

[184] Swandby, R. K.
Corrosion charts: Guides to materials selection
Chem. Eng. *69* (1962) Nov., p. 186

[185] Asphahani, A. I.
Corrosion resistance of high performance alloys
Mater. Performance *19* (1980) No. 12, p. 33

[186] Schütze, K. G.
Korrosionsverhalten metallischer Werkstoffe in Ameisensäure
(Corrosion behavior of metallic materials in formic acid) (in German)
Werkstoffe und Korrosion *38* (1987) No. 1, p. 41

[187] Product Information
Hastelloy® C276 is the best answer to broad spectrum corrosion problems
Cabot Corporation, High Technology Materials Division,
Kokomo, IN 46901
Mater. Eng. *92* (1980) No. 6, Circle *55*

[188] Sachsze, W.
Ullmanns Encyklopädie der technischen Chemie
(Ullmann's Encyclopedia of Industrial Chemistry) (in German)
4th Ed., Vol. 7 (1974) p. 362
Verlag Chemie, 6940 Weinheim

[189] DECHEMA-WERKSTOFF-TABELLE
Salzsäure
(Hydrochloric Acid) (in German)
DECHEMA e.V., Frankfurt a.M. (1976)

[190] Materials Selector for Hazardous Chemicals, vol. 3
Hydrochloric Acid, Hydrogen Chloride and Chlorine
MTI Publication MS-3, Elsevier Science, 1999, p. 15

[191] Heubner, U. et al.
Nickelwerkstoffe und hochlegierte Sonderedelstähle, 4. ed., Vol. 153
(Nickel materials and high-alloyed special stainless steels) (in German)
expert verlag, Renningen (2009)

[192] Company publication
Corrosion Resistant Alloy Data, Bulletin 151, 2003
Rolled Alloys Inc., Temperance, Michigan (USA)

[193] Company publication
Nicrofer® 3127 hMo – alloy 31 Material Data Sheet, No. 4031, August 1997

[194] Company publication
HASTELLOY® alloy C-276 Bulletin H-2002B
Haynes International, Kokomo, Indiana (USA)

[195] Company publication
Technical data BLUE SHEET
Allegheny Ludlum Corporation, Pittsburgh, Pennsylvania (USA)

[196] Köhler, M., Heubner, U., Eichenhofer, K.W., Renner, M., Progress with alloy 33 (UNS R20033), A new corrosion resistant chromium-based austenitic material, CORROSION 96, Paper No. 428, NACE International, Houston, Texas, 1996

[197] Company publication
INCONEL alloy 625, Bulletin T-42
Huntington Alloys Inc., Huntington, West Virginia (USA)

[198] Manning, P. E.; Smith, J. D.; Nickerson, J. L.
New Versatile Alloys for the Chemical Process Industry
Mater. Performance 27 (1988) 6, p. 67–73

[199] Class, I.; Graefen, H.; Schell, E.
Highly Corrosion-Resistant Ni-Cr-Mo Alloy with Improved Resistance to Intergranular Corrosion
U. S. Patent No. 3, 203, 792 (1965)

[200] Badran, B. M.; Abdel Fattah, A. A.; Abdul Azim, A. A.
New corrosion inhibitors based on fatty materials – II. Epoxidized fatty materials modified with aromatic amines
Corros. Sci. 22 (1982) 6, p. 525–536

[201] Manning, P. E.; Schoebel, J. D.
Hastelloy alloy C-22 – A New and Versatile Material for the Chemical Process Industries
Werkst. Korros. 37 (1986), p 137

[202] Sridhar, N.; Wu, J. B. C.; Corey, S. M.
The effect of acid mixtures on corrosion of nickel-base alloys
Mater. Performance 26 (1987) 10, p. 17–23

[203] Axsom, J. F.
On-stream cleaning of refineries pays off
Oil and Gas Journal 71 (1973), p. 106

[204] Asphahani, A. I.
Corrosion Resistance of High Performance Alloys
Mater. Performance 19 (1980) 12, p. 33–43

[205] Asphahani, A. I.
Localized Corrosion of High Performance Alloys
Mater. Performance 19 (1980) 8, p. 9–21

[206] Corbett, R. A.; Morrison, W. S.
Comparative Corrosion Resistance of Some High-Nickel, Chromium-Molybdenum Alloys
Mater. Performance 28 (1989) 2, p. 56–59

[207] Konys, J.; Fodi, S.; Hausselt, J.; Schmidt, H.; Casal, V.
Corrosion of high-temperature alloys in chloride-containing supercritical water oxidation systems
Corrosion 55 (1999) 1, p. 45–51

[208] Company publication
MONEL alloys, Bulletin T-5
Huntington Alloys Inc., Huntington, West Virginia (USA)

[209] Company publication
HASTELLOY Alloy B-2, Bulletin H2006C (1997)
Haynes International, Kokomo, Indiana (USA)

[210] Agarwal, D. C.
UNS N10629: A New Ni-28 % Mo Alloy
Mater. Performance 33 (1994) 10, p. 64–68

[211] Brooks, C. R.; Spruiell, J. W.; Stansbury, E. E.
Int. Metals Rev. 29 (1984) 3, p. 210

[212] Lei, T. S.
The Effect of 0 – 8 Wt % Fe on Transformation of Phases in Ni-Mo-Fe Alloys
University of Tennessee, Knoxville, Tennessee, 1979, PhD diss.

[213] Zucchi, F.; Trabanelli, G.; Brunoro, G.
Corrosion inhibition of duplex steel in hot HCl solutions
Eur. Congr. Corros. 9 (1989), p. 1–6

[214] Anonymous
Improved version of Hastelloy Alloy B
Anticorros. Methods Mater. 22 (1975) 11, p. 19–20

[215] Ogino, K.; Hida, A.; Kishima, S.; Kumanomido, S.
Susceptibility of type 431 Stainless Steel to erosion-corrosion by vibration cavitation in corrosive media
Wear 116 (1987) 3, p. 299–307

[216] Golovanenko, S. A.; Svistunova, T. V.
Relaxation between intercrystallite corrosion resistance of Nickel-Molybdenum alloys and thermodynamic activity of carbon
Zashch. Met. 20 (1984) 5, p. 690–697

[217] Moller, G. E.; Franson, I. A.; Nichol, T. J.
Experience With Ferritic Stainless Steel In Petroleum Refinery Heat Exchangers
Mater. Performance 20 (1981) 4, p. 41–50

[218] Rebak, R. B.; Srivastava, S. K.
Corrosion performance of a nickel-molybdenum-chromium alloy: effects of aging, alloying elements and electrolyte composition
Corrosion 55 (1999) 4, p. 412–421

[219] Groysman, A.; Kaufman, A.; Feldman, B.; Man, Y.
Cavitation corrosion in a continuous catalytic reformer unit
Mater. Performance39 (2000) 6, p. 62–65

[220] Corrosion in Acid Mixtures
in: ASM International
Metals Handbook, 9th Edition, Vol. 13
ASM International, Metals Park, Ohio, USA, 1987, pp. 646–656

[221] Sridhar, N.; Wu, J. B. C.; Corey, S. M.
The effect of acid mixtures on corrosion of nickel-base alloys
Materials Performance; NACE (1987) 10, pp. 17–23

[222] Franke, E.
Hastelloy-Legierungen
(Hastelloy alloys) (in German)
Werkst. Korros. 2 (1951), pp. 140

[223] DECHEMA-WERKSTOFF-TABELLE
45. Ergänzungslieferung “Mischsäuren”
(45th supplement “Mixed acids”)
(in German)
DECHEMA e.V., Frankfurt a.M. (1961)

[224] Company publication
Rockel, M. B.
Nicrofer 5923 hMo – Alloy 59 ein Hochleistungswerkstoff für die chemische Verfahrenstechnik und andere Prozeßindustrien, VDM Case History 5, April 1999
(Nicrofer 5923 hMo – Alloy 59 a high performance material for chemical process engineering and other process industries)
(in German)
KRUPP VDM GmbH, Werdohl

[225] Heubner, U. et al.
Nickelwerkstoffe und hochlegierte Sonderedelstähle, 4. ed., Vol. 153
(Nickel materials and high-alloyed special stainless steels) (in German)
expert verlag, Renningen (2009)

[226] Company publication
Langley Corrosion Resistance, GA8692/3M7481
Langley Alloys Ltd., Slough (UK)

[227] Manning, P. E.; Smith, J. D.; Nickerson, J. L.
New Versatile Alloys for the Chemical Process Industry
Mater. Performance 27 (1988) 6, pp. 67–73

[228] Company publication
High Performance Alloys for Resistance to Aqueous Corrosion, Publication No. SMc-026/9M/2000. 2000
Special Metals, Huntington (WV/USA)

[229] Corbett, R. A.; Morrison, W. S.
Comparative Corrosion Resistance of Some high-Nickel, Chromium-Molybdenum Alloys
Mater. Performance 28 (1989) 2, pp. 56–59

[230] Company publication
Hastelloy Alloy C-22 – A New NiCrMo Alloy with Superior Corrosion Resistance, 35, (5), Pp 4, Oct. 1984
Cabot Wrought Products Div.Dig., Kokomo (Indiana/USA)

[231] Company publication
Rockel, M. B.; White, F. E.
A Critical Comparison of some NiCrMo Alloys for Chemical Industry Applications, Internal Report, 1992
KRUPP VDM, Werdohl

[232] Agarwal, D. C.; Herda, W. R.
The ‘C’- family of NiCrMo Alloys’ partnership with the Chemical Process Industry: the last 70 years
Materials and Corrosion 48 (1997), pp. 542–548

[233] Company publication
Hastelloy C-22 Alloy, H-2019E – 091099, 1999
Haynes International, Kokomo (Indiana, USA)

[234] Company publication
Rockel, M. B.
Nicrofer 3127 hMo – Alloy 31 ein neuer, hoch legierter Sonderedelstahl für die chemische Verfahrenstechnik, VDM Case History 6, March 2001
(Nicrofer 3127 hMo – Alloy 31 a new, high alloyed special stainless steels for chemial engineering) (in German)
KRUPP VDM GmbH, Werdohl

[235] Heubner, U. et al.
Nickelwerkstoffe und hochlegierte Sonderedelstähle, 2. ed., Vol. 153
(Nickel materials and high-alloyed special stainless steels) (in German)
expert verlag, TAE Esslingen (1993)

[236] Agarwal, D. C.; Heubner, U.; Kirchheiner, R.; Koehler, M.
Cost-effective Solutions to CPI Corrosion Problems with a New Ni-Cr-Mo Alloy
Corrosion 91 NACE (1991) 179, pp. 1–26

[237] Company publication
Inconel Alloy 686, Publication No. SMC-024, 2000
Special Metals, Huntington (WV/USA)

[238] Company publication
Inconel Alloy C-276, Publication No. SMC-019, 1999
Special Metals, Huntington (WV/USA)

[239] Company publication
Agarwal, D. C.; Herda, W. R.
Alloying effects and innovations in nickel base alloys for combating aqueous corrosion, VDM Report No. 23, 1996
KRUPP VDM GmbH, Werdohl

[240] Horn, E.-M.; Manning, P. E.; Renner, M.
Korrosion nichtrostender Stähle und Nickelbasislegierungen in Salpetersäure-Flußsäure-Gemischen
(Corrosion of stainless steels and nickel-base alloys in solutions of nitric acid and hydrofluoric acid) (in German)
Werkst. Korros. 43 (1992) 5, pp. 191–200

[241] Company publication
Hastelloy G-30 Alloy, H-2028D 061097, 1997
Haynes International, Kokomo (Indiana/USA)

[242] Asphahani, A. I.
Corrosion Resistance of High Performance Alloys
Mater. Performance 19 (1980) 12, pp. 33–43

[243] Company publication
Incoloy Alloy 825, Publication No. SMS-0.30. 2000
Special Metals, Huntington (WV/USA)

[244] Singh, V. B.; Arvind, U.
Active, passive and transpassive dissolution of a nickel base super alloy in concentrated acid mixture solution
Materials and Corrosion 46 (1995), pp. 590–595

[245] Murdock, D.; Gaines, A.
Phos-acid pilot plant selects pumps of Hastelloy C for corrosive/erosive service
Chem. Processing (Chicago) 45 (1982) 7, p. 151

[246] Chlorimet corrosion-resistant cast nickel-base Alloy
ALLOY DIGEST Pp 2 Ni-12 (1996) 5 (USA)

[247] Nimofer 6629 (nickel-molybdenum Alloy) (2/1996)
ALLOY DIGEST Pp 2 (USA)

[248] Company publication
ULTIMET Alloy, H-2082B – 091099, 1999
Haynes International, Kokomo (Indiana, USA)

[249] Schulz, W.-D.
(Some questions concerning corrosion protection with sprayed metal layers against predominantly medial corrosion) (in German)
Schweisstechnik (Berlin) *30* (1980) 12, p. 539

[250] Feldstein, N.
Composite coatings: plating plus particles. Tiny particles in a nickel-alloy matrix stand up to tough service environments
Mater. Eng. *94* (1981) 1, p. 38

[251] Razygraev, V. P.; Mirolyubov, E. N.; Pisarenko, T. A.
Mechanism of nickel corrosion in nitric acid solutions (in Russian)
Zashch. Met. 9 (1973) 1, p. 48

[252] Khedr, M. G. A.; Mabrouk, H. M.; Abd El Haleem, S. M.
Autocatalytic dissolution of Ni in HNO_3
Corros. Prov. Control *30* (1983) 2, p. 17

[253] Stupniek-Lisac, E.; Karšulin, M.
Electrochemical behaviour of nickel in nitric acid
Electrochim. Acta *29* (1984) 10, p. 1339

[254] Shams El Din, A. M.; Al-Kharafi, F. M.; Al-Fahd, Z.; El-Tantawy, Y. A.
Corrosion behaviour of nickel in nitric acid solutions
Corros. Prev. Control *32* (1985) 5, p. 92

[255] Tousek, J.
Die aktivierende Wirkung der Nitrationen auf die Geschwindigkeit der Metallauflösung im passiven Zustand
(The conditions of activation of pit corrosion of metals by nitrate ions were investigated) (in German)
Corros. Sci. *12* (1972) p. 799

[256] Anonymous
Erfolgreich im Kampf gegen Korrosion und Verschleiß: Dicknickelschichten
(Successful in the fight against corrosion and wear: thick nickel layers) (in German)
Metalloberfläche *33* (1979) 11, p. 521

[257] Metzger, W.; Pappe, G.; Schmitz, M.
Electrolytic metal stripping in solutions containing carboxylic acids
Trans. Inst. Met. Finish. *53* (1975) p. 184

[258] Landolt, D.
Mechanistische Gesichtspunkte der elektrochemischen Metallbearbeitung
(Views on the mechanism of electrochemical treatment of metals) (in German)
Chem.-Ing. Tech. *45* (1973) 4, p. 188

[259] Takei, T.
The effect of various acids on the behavior of nickel electrodeposited from Ni $(C.F_3COO)_2$-halide-MeOH bath (in Japanese)
J. Met. Finish Soc. Jpn *36* (1985) 4, p. 166

[260] Anonymous
Stripping nickel layers
USA Pat. 3 856 694 (Oxy Metal Finishing Corp., Warren (Mich.))

[261] Sriveeraraghavan, S.; Krishnan, R. M.; Natarajan, S. R.; Parthasaradhy, N. V.; Udupa, H. V. K.
Immersion stripping of nickel deposits
Met. Finish. *77* (1979) 12, p. 67

[262] Datta, M.; Landolts, D.
Surface brightening during high rate nickel dissolution in nitrate electrolytes
J. Electrochem. Soc. *122* (1975) p. 1466

[263] Vdovenko, I. D.; Vakulenko, L. I.; Shedenko, L. I.; Vakulenko, L. V.
Effect of oxalic acid on the pickling of Chromel (in Russian)
Zh. Prikl. Khim. *57* (1984) p. 2624

[264] Svistunova, T. V.; Kireyeva, T. S.; Runova, Z. K.
Structure and corrosion behavior of ChN 60 type in a nitrofluoride solution (in Russian)
Zashch. Met. *19* (1983) 2, p. 212

[265] Knotek, O.; Lugscheider, E.; Wichert, W.
On the structure and properties of wear- and corrosion-resistant nickel-chromium-tungsten-carbon-(silicon) alloys
Thin Solid Films *53* (1978) p. 303

[266] Page, R. A.
Stress corrosion cracking of Alloys 600 and 690 and Nos. 82 and 182 weld metals in high temperature water
Corrosion *39* (1983) 10, p. 409

[267] Solok, A. M.; Babchenko, V. A.; Vydra, E. I.; Kuznetsov, V. M.
Use of molten sodium hydroxide for removing high-temperature scale from certain heat-resistant alloys
Sov. Appl. Chem. *46* (1973) p. 2044

[268] Konstantinova, E. V.; Chechetina, N.A.; Lomovtsev, V. I.
Construction materials for hot nitrofluoride solutions (in Russian)
Zashch. Met. *16* (1980) 1, p. 66

[269] Svistunova, T. V.
Nickel alloy
UdSSR Pat. 450 844 (Nov. 25, 1974)

[270] Kireeva, T. S.; Svistunova, T. V.; Runova, Z. K.
A corrosion-resistant alloy used for work with hydrogen fluoride-containing nitric acid etching solutions
Stal' (1979) 2, p. 141

[271] Ivanov, E. S.; Komolova, L. F.; Makovskii, R. D.
Investigation of the surface of 36 alloys after etching in nitric acid/fluoride solutions (in Russian)
Zashch. Met. *19* (1983) 6, p. 935

[272] Kravchinskii, A. P.; Isaev, N. I.; Baluev, V. N.; Zakharov, A. I.; Shumilov, V. N.; Revyakin, A. V.
The anodic behavior of various metals in a magnetic field
Russ. Metall. *25* (1984) 6, p. 30

[273] Svistunova, T. V.
Corrosion-resistant alloys based on nickel
Met. Sci. Heat Trat. (USSR) (1980) p. 483

[274] Babchenko, A. M.; Khodos, R. S.; Kistanov, A. I.; Maximenko, A. S.; Nozdrachev, A. V.
Effect of the smelting method on the etching of alloy EI437B in a mixture of nitric and hydrofluoric acids (in Russian)
Zashch. Met. *14* (1978) 4, p. 463

[275] Maness, R. F.
Korrosionsbeständige Nickellegierung
(Corrosion-resistant nickel alloy) (in German)
DOS 2 002 652 (United States Atomic Energy Commission; Nov. 19, 1970)

[276] Page, R. A.; McMinn, A.
Relative stress corrosion susceptibilities of Alloys 690 and 600 in simulated boiling water reactor environments
Metall. Trans. A *17A* (1986) May, p. 877

[277] Crum, J. R.
Effect of composition and heat treatment on stress corrosion cracking of Alloy 600 steam generator tubes in sodium hydroxide
Corrosion *38* (1982) 1, p. 40

[278] Briant, C. L.; O'Toole, C. S.; Hall, E. L.
The effect of microstructure on the corrosion cracking of Alloy 600 in acidic and neutral environments
Corrosion *42* (1986) 1, p. 15

[279] Spyra, W.; Klupsch, H.
Über die Entwicklung einer Nickelhartlegierung zum Auftragsschweißen
(The development of a hard nickel alloy for deposition welding) (in German)
Thyssen Edelst. Tech. Ber. *8* (1982) 2, p. 162

[280] Okada, Y.; Yoshikawa, K.; Yukitoshi, T.
Measurement of sensitization behavior in Ni-based Alloy 600 by means of magnetic property (in Japanese)
J. Jpn. Inst. Met. *45* (1981) 5, p. 496

[281] Copson, H. R.; van Rooyen, D.; McIlree, A. R.
Stress corrosion behavior of Ni-Cr-Fe alloys in high temperature aqueous solutions
5th Intern. Congress Metallic Corrosion, Tokyo 1972, publ.
NACE, Houston (1974) p. 376

[282] Sato, A.; Kon, K.; Tsujikawa, S.; Hisamatsu, Y.
Effects of aging and misorientation on intergranular corrosion of Inconel® Alloy 600 (in Japanese)
J. Jpn. Inst. Met. *43* (1979) 7. p. 664

[283] McIlree, A. R.; Michels, H. T.; Morris, P. E.
Effects of variations of carbon, sulfur and phosphorus on the corrosion behavior of Alloy 600
Corrosion *31* (1975) 12, p. 441

[284] Scarberry, R. C.; Pearman, S. C.; Crum, J. R.
Precipitation reactions in Inconel® alloy 600 and their effect on corrosion behavior
Corrosion *32* (1976) 10, p. 401

[285] Lee, III, T. S.; Hodge, F. G.
Resistance of Hastelloy® alloys to corrosion by inorganic acids
Mater. Performance *15* (1976) 9, p. 29

[286] Paige, B. E.
Corrosion of nickel alloys in nuclear fuel reprocessing
Mater. Performance *15* (1975) 12, p. 22

[287] Kramer, L. D.; Michael, S. T.; Pement, F. W.
Service experience and stress corrosion of Inconel® 600 bellows expansion joints in turbine steam environments
Mater. Performance *14* (1975) 9, p. 15

[288] Mah, R.; Terada, K.; Cash, D. L.
Corrosion of distillation equipment by HNO_3-HF solutions during HNO_3 recovery
Mater. Performance *14* (1975) 11, p. 28

[289] rum, J. R.; Scarberry, R. C.
Corrosion testing of Inconel® alloy 690 for PWR steam generators
J. Mater. Energy Systems *4* (1982) 3, p. 125

[290] Vermilyea, D. A.; Tedmon jr., C. S.; Broecker, D. E.
Effect of phosphorus and silicon on the intergranular corrosion of a nickel-base alloy
Corrosion *31* (1975) 6, p. 222

[291] Kurosawa, F.; Taguchi, I.; Matsumoto, R.
Determination of gamma prime in Ni-base by non-aqueous electrolyte-potentiostatic method (in Japanese)
J. Jpn. Inst. Met. *44* (1980) 10, p. 1187

[292] Hodge, F. G.; Kirchner, R. W.
An improved Ni-Cr-Mo alloy for corrosion service
Corrosion *32* (1976) 8, p. 332

[293] Yoshida, T.; Takizawa, Y.; Sekine, I.
The characteristics and applications of recently improved Hastelloy® alloys (in Japanese)
Corros. Eng. (Boshoku Gijutsu) *31* (1982) 2, p. 81

[294] Anonymous
Nickel alloy for use in as-welded condition
Mater. Eng. *90* (1979) 1, p. 22

[295] Kirchheiner, R. R.; Heubner, U.; Hofmann, F.
Increasing the life-time of nitric acid equipment using improved stainless steels and nickel alloys
Corrosion 88, St. Louis-Missouri, 1988, Nr. 318, 22 p.

[296] Dilthey, U.; Wanke, R.
Hochleistungs-Schweißplattierverfahren für den Chemie-Apparatebau
(Highly efficient plating in chemical plant construction) (in German)
Chem.-Ing.-Tech. *46* (1974) 11, p. 467

[297] Shvarts, G. L.; Moroz, V. A.; Gerasimenko, G. I.; Akshentseva, A. P.; Afanasenko, E. A.
Fields of use of the nickel-chromium-molybdenum alloy ChN60MB (in Russian)
Khim. Neft. Mashinostr. (1975) 8, p. 20

[298] Manning, P. E.; Schöbel, J.-D.
Hastelloy® Alloy C 22 – ein neuer und vielseitig anwendbarer Werkstoff für die chemische Industrie
(Hastelloy® alloy C 22 – a new and versatile material for the chemical process industries) (in German)
Werkst. Korros. *37* (1986) p. 137

[299] Konstantinova, E. V.; Zazarova, N. P.
Corrosion of the 47ChNM-2 alloy in nitric fluoride media (in Russian)
Zashch. Met. *14* (1978) 2, p. 187

[300] Solonin, S. M.; Chernyshev, L. I.; Fedorchenko, I. M.
Investigation of the properties of porous materials made of nickel-molybdenum and nickel-chromium-molybdenum alloys (in Russian)
Poroshk. Metall. *12* (1972) 6, p. 36

[301] Abelev, M. M.; Shvarts, G. L.; Gerasimenko, G. I.; Belinkij, A. L.; Besednyi, V. A.; Kostyuchenko, V. A.
Fabrication of fractionating columns of nickel-chromium-molybdenum alloy EP567
Chem. & Petroleum Eng., New York (1971) p. 699

[302] Jesso Steel Comp., Washington, Pa
When you need strength plus corrosion resistance, specify Jesso JS 625
Mater. Eng. 99 (1984) 5, p. 17

[303] Nickel, O.
Eigenschaften und Anwendungen von Reinnickel, Nickel-Basislegierungen und nickelhaltigen NE-Werkstoffen im Rohrleitungsbau
(Properties and uses of pure nickel, nickel-base alloys and nickel-bearing non-ferrous metals in piping) (in German)
3R international *14* (1975) 8, p. 419

[304] Nickel, O.
Eigenschaften und neuere Anwendungen nickelhaltiger Werkstoffe
(Properties and new applications of nickel-containing materials) (in German)
VDI Z *121* (1979) 7, p. 313

[305] Anonymous
Le leghe Wiggin di nichel negli impianti chimici ed industriali (The Wiggin nickel alloy in chemical and industrial plants) (in Italian)
Ingegneria Chim. *27* (1978) 2, p. 12

[306] Zall, D. M.; Bolander, E. H.
In situ identification of alloys. Spot testing alloys for submarine construction
Mater. Protection *6* (1967) 7, p. 37

[307] Nickel, O.
Eigenschaften und neuere Anwendungen nickelhaltiger Werkstoffe
(Properties and new applications of nickel-containing materials) (in German)
VDI Z *121* (1979) 7, p. 313

[308] Asphahani, A. I.
Corrosion resistance of high performance alloys
Mater. Performance *19* (1980) 12, p. 33

[309] Mallory, G. O.
Properties of electroless nickel-molybdenum alloys
Plat. Surf. Finish. *63* (1976) 6, p. 34

[310] Steigerwald, R. F.
New molybdenum stainless steels for corrosion resistance: A review of recent developments
Mater. Performance *13* (1974) 9, p. 9

[311] Taylor, E.
Corrosion resistance of multiphase alloys
Mater. Protection 9 (1970) 3, p. 29

[312] Anonymous
A new nickel alloy for gas turbines
Mater. Eng. *75* (1972) 5, p. 28

[313] Williams, K. J.; Evans, T. E.
Neue Legierung auf NiSi-Basis für Schwefelsäure
(New NiSi-based alloy for sulfuric acid) (in German)
Chem. Process Eng. *51* (1970) 9, p. 57

[314] Masui, K.; Yamada, T.; Hisamatsu, Y.
Effects of phosphorus content and heat-induced structural change of electro-deposited Ni-P alloys on acid corrosion resistance (in Japanese)
J. Met. Finish. Soc. Jpn. *32* (1981) 8, p. 410

[315] Schneider, W.; Weinlich, W.
Nickel-Beryllium aushärtbare Federlegierung mit besonderen Eigenschaften
(Nickel-beryllium age-hardenable spring alloy with special properties) (in German)
Metall *36* (1982) 11, p. 21

[316] Volokhova, V. I.; Vakhidov, R. S.; Lukyanitsa, A. I.
Corrosion stability of nickel-phosphorus galvanic coatings (in Russian)
Zashch. Met. *11* (1975) 3, p. 370

[317] Musina, A. S.; Lange, A. A.; Bukhman, S. P.; Semashko, T. S.
Change in the surface state of the alloys Permalloy® and Kovar® after chemical treatment (in Russian)
Izv. Akad. Nauk Kaz. SSR. Ser. Khim. (1984) 3, p. 12

[318] Bartosiewicz, L.
Preferential electrolytic etching technique to reveal the phase in nickel base alloys
Prakt. Metallogr. *10* (1973) p. 450

[319] Zhu Hun-De; Wang Jing-Yun
The morphology of carbides and α-phase in cast-nickel-base superalloys
Prakt. Metallogr. *17* (1980) p. 680

[320] Meisel, H.; Johner, G.; Scholz, A.
Metallographic development of the structure of nickel based-alloys
Prakt. Metallogr. *17* (1980) p. 261

[321] Kohlhaas, E.; Fischer, A.
The use of the carbon-extraction technique in investigations of superalloys
Prakt. Metallogr. *6* (1969) p. 291

[322] Rickes, M.
The contrasting of high strength, high temperature nickel alloys with protective coatings by etching
Prakt. Metallogr. *22* (1985) p. 430

[323] Fabian, R. J.; Vaccari, J. A.
How materials stand up to corrosion and chemical attack
Mat. Eng. 73 (1971) 2, p. 36

[324] Product Information
Eigenschaften von Nickel
(Properties of nickel) (in German), November 1971
International Nickel Deutschland GmbH, Düsseldorf

[325] Kny, E.; Bader, T.; Hohenrainer, C.; Schmid, L.; Glätzle, R.
Korrosionsresistente, hochverschleißfeste Hartmetalle
(Corrosion and highly wear resistant hard metals) (in German)
Werkst. Korros. 37 (1986) 5, p. 230

[326] Hirschfeld, D.; Ostermann, G.
Interessante Eigenschaften von Hartmetallen für die Chemische Industrie
(Interesting properties of hard metals for the chemical industry) (in German)
Z. Werkstofftech. 4 (1973) 7, p. 367

[327] Anonymous
Corrosion Data Survey, Metals Section, 6th ed., p. 95
NACE, Houston (Texas/USA), 1985

[328] Hiermaier, M.; Simon, H.
Chemisch Abtragen, chemisch und elektrolytisch Entgraten
(Chemical machining, chemical and electrolytic deburring) (in German)
Galvanotechnik 74 (1983) 4, p. 401

[329] Itou, K.; Suzuki, H.; Iida, T.; Yamada, T.
Corrosion behavior of metals-Copper group and platinum group-in condensed phosphoric acid at higher temperature (in Japanese)
Nagoya Kogyo Daigaku Gakohu 25 (1973) 20, p. 377

[330] Schorr, M.; Weintraub, E.; Andrasi, D.; Finkelstein, N. P.
Erosion-corrosion in wet process phosphoric acid (WPA) production
International Congress on Metallic Corrosion (Proc. Conf.), Mainz, September 1981, vol. 2, p. 1384
DECHEMA e.V., Frankfurt am Main

[331] Khvostov, V. P.; Anoshchenko, I. P.
Stainless steel corrosion in phosphoric acid (in Russian)
Zashch. Met. 16 (1980) 3, p. 310

[332] Product Information
Eigenschaften und Anwendungsgebiete korrosionsbeständiger Wiggin-Nickellegierungen
(Properties and fields of application of corrosion-resistant Wiggin nickel alloys) (in German), January 1964, p. 24
Henry Wiggin & Company Limited, Hereford (GB)

[333] DECHEMA-WERKSTOFF-TABELLE "Phosphorsäure" (DECHEMA-WERKSTOFF-TABELLE "Phosphoric Acid") (in German) DECHEMA e. V., Frankfurt am Main, December 1966

[334] Rabald, E. Corrosion Guide, 2nd revised edition, p. 556 Elsevier Publishing Company, Amsterdam-London-New York, 1968

[335] Anonymous Fertigungs- und Standardlieferprogramm M 3-M 56. Schrauben, Muttern, Zubehör nach DIN, ISO und EURONORM (Production- and standard delivery programs M 3-M 56. Screws, nuts, fittings according to DIN, ISO and EURONORM) (in German) Werkstoffe und ihre Veredlung 2 (1980) 5, p. 250

[336] Product Information Korrosionbeständigkeit nickelhaltiger Werkstoffe gegenüber Phosphorsäure und Phosphaten (Corrosion resistance of nickel containing materials against phosphoric acid and phosphates) (in German), 1st ed., No. 61, March 1970, p. 1 International Nickel Deutschland GmbH, Düsseldorf

[337] Product Information Corrosion resistance of nickel-containing alloys in phosphoric acid, p. 1 The International Nickel Company, Inc., New York (USA)

[338] Marcus L.; Ahrens, R. R. Chemical resistance of solid materials to concentrated phosphoric acid Am. Ceram. Soc. Bull. 60 (1981) 4, p. 490

[339] Johnson, C. E.; Mullen, J. L.; Lashmore, D. S. Corrosion tests of electrodeposited coatings in boiling 100 percent phosphoric acid Mechanical Properties. Performance and Failure Modes of Coatings (Proc. Conf.), Gaithersburg (USA), May 1983, p. 124 Cambridge University Press, Cambridge (GB)

[340] Nassif, N. Influence of impurities in phosphoric acid on the corrosion resistance of some commercially produced alloys Surf. Tech. 26 (1985) 3, p. 189

[341] Product Information Korrosionsbeständigkeit nichtrostender Stähle und Nickellegierungen in Phosphorsäure (Corrosion resistance of stainless steels and nickel alloys in phosphoric acid) (in German), 1988 Mannesmann Edelstahlrohr GmbH, Langenfeld; Vereinigte Deutsche Metallwerke AG, Duisburg

[342] Effertz, P.-H.; Forchhammer, P. Aus der Schadenforschung. Korrosionsverhalten von Rohrwerkstoffen für DWR-Dampferzeuger in konditioniertem Sekundärwasser bei 350 °C-Kapselprüfungen (Corrosion patterns in tube materials for DWR-steam generating plants in conditioned secondary water at 350 °C-capsule-tests) (in German) Der Maschinenschaden 50 (1977) 5, p. 180

[343] Manning, P. E.; Sridhar, N.; Asphahani, A. I. New developmental NiCrMo alloys Corrosion '83 (Proc. Conf.), Anaheim (Calif./USA), April 1983, p. 21/1 NACE, Houston (Texas/USA)

[344] Alves, H., Behrens, R., Paul, L., Evolution of nickel base alloys – modifications to traditional alloys for specific applications, NACE Corrosion 2014, San Antonio, Texas, USA. Paper No. 4317 (2014)

[345] Alves, H., Agarwal, D.C., Stenner, F., Hoxha, A Alloys suitable for phosphoric acid applications CORROSION 2006 (2006) Paper No. 06221, p. 1–17

[346] Alves, H., Baessler, R. New experiences with nickel alloys and special stainless steels for an optimized material selection in acids at high temperatures CORROSION 2008 (2008) Paper No. 08189, p. 1–15

[347] Alves, H., Baessler. R. New results on the selection of nickel alloys and a 6 % Mo special stainless steel for high temperature acid service CORROSION 2009 (2009) Paper No. 09369, p. 1–18

[348] Alves, H., Behrens, R., Paul. L.
Review of corrosion issues and material solutions in the CPI
CORROSION 2010 (2010) Paper No. 10338, p. 1–26

[349] Crook, P., Caruso, M.L.
A new nickel alloy resistant to "wet process" phosphoric acid
CORROSION 2004, Paper No. 04221, (2004) p. 1–9

[350] Product Information
Hastelloy® alloy C-4, 1983
Cabot Wrought Products Division, Cabot Corporation, Kokomo (Ind./USA)

[351] McDowell, D. W.
Handling phosphoric acids and phosphate fertilizers-II
Chem. Eng. 82 (1975) 18, p. 121

[352] Anonymous
Bisphenol-Polyester-Harze für die Herstellung von korrosionsbeständigen GfK-Konstruktionen
(Polyester bisphenol resins for the production of corrosion-resistant GRP-constructions) (in German)
Kunststoffberater 24 (1979) 1/2, p. 20

[353] Burbridge, J. F.
Corrosion-resistant reinforced polyester for process plant
Anticorros. Methods Mater. 23 (1976) 2, p. 7

[354] Product Information
Properties of Delore Stellite® alloys, March 1974
Delore Stellite (UK) Limited, Swindon (GB)

[355] Product Information
Corrosion Guide. Langley corrosion resistant alloys
Langley Alloys Ltd., Slough (GB)

[356] Umemura, F.; Kawamoto, T.
Corrosion of stainless steel in crude phosphoric acid (in Japanese)
Boshoku Gijutsu 31 (1982) 4, p. 275

[357] Product Information
VDM Hochleistungswerkstoffe
(VDM high-power materials) (in German)
Vereinigte Deutsche Metallwerke AG, Werdohl

[358] Mitsuhashi, A.; Asami, K.; Kawashima, A.; Hashimoto, K.
The corrosion behavior of amorphous nickel base alloys in a hot concentrated phosphoric acid
Corros. Sci. 27 (1987) 9, p. 957

[359] Lee III, T. S.; Hodge, F. G.
Resistance of "Hastelloy®" alloys to corrosion by inorganic acids
Mater. Performance 15 (1976) 9, p. 29

[360] Product Information
Hastelloy® alloy C-276, 1983
Cabot Wrought Products Division, Cabot Corporation, Kokomo (Ind./USA)

[361] Product Information
Sonderlegierungen für die Rauchgasentschwefelung
(Special alloys for flue gas desulfurization) (in German), ed. 1988
Edelstahlwerk Schmees, Langenfeld

[362] Manning, P. E.; Smith, J. D.; Nickerson, J. L.
New versatile alloys for the chemical process industry
Mater. Performance 27 (1988) 6, p. 67
(Materials Selection & Design)

[363] Berg, F. F.
Korrosionsschaubilder
(Corrosion Diagrams) (in German), 2nd ed., p. 2
VDI-Verlag GmbH, Düsseldorf, 1969

[364] Product Information
Corrosion resistance of Hastelloy® alloys, 1984, p. 3
Haynes International, Inc., Kokomo (Ind./USA)

[365] Bishop, C. R.
Corrosion tests at elevated temperatures and pressures
Corrosion 19 (1963) 9, p. 308

[366] Schillmoller, C. M.
Alloy selection in wet process phosphoric acid plants
in: Process Industries Corrosion – The theory and practice, p. 161
NACE, Houston (Texas/USA), 1986

[367] Uhlig, H. H.; Asphahani, A. I.
Corrosion behavior of cobalt base alloys in aqueous media
Mater. Performance 18 (1979) 11, p. 9

[368] Anonymous
Verschleißfest und abriebbeständig: LP-Legierungen Tribaloy® für Gußteile und zum Beschichten
(Wear and abrasion resisting: LP alloys Tribaloy® for cast parts and for coating) (in German)
Werkstoffe und ihre Veredlung I (1979) 2, p. 24

[369] Product Information
Incoloy® alloy 825, Nov. 1976
Henry Wiggin & Company Limited, Hereford (GB)

[370] Anonymous
Hastelloy® alloy G-3
Alloy Dig. Ni-277, June 1982

[371] Ward, E.; Maylor, J. B.
Ni-O-Nel® 825 – Eine neue korrosionsfeste Legierung für die chemische Industrie (Ni-O-Nel® 825 – a new corrosion-resistant alloy for the chemical industry) (in German)
Werkst. Korros. 16 (1965) 5, p. 382

[372] Product Information
Cabot alloy No. 625, 1982
Cabot Wrought Products Division, Cabot Corporation, Kokomo (Ind./USA)

[373] Product Information
Inco alloys international. Serving the chemical processing and related industries, 1986
Inco Alloys International, Inc. Huntington (West Virginia/USA)

[374] Product Information
Wiggin korrosionsbeständige Legierungen (Wiggin corrosion-resistant alloys) (in German), March 1977, p. 14
Henry Wiggin & Company Limited, Hereford (GB)

[375] Product Information
Wiggin Nickellegierungen in der chemischen Verfahrenstechnik
(Wiggin nickel alloys in chemical process engineering) (in German), November 1976, p. 27
Henry Wiggin & Company Limited, Hereford (GB)

[376] Product Information
Hastelloy® alloy G
Cabot Wrought Products Division, Cabot Corporation, Kokomo (Ind./USA)

[377] Product Information
Inconel® alloy 625, November 1977, p. 12
Henry Wiggin & Company Limited, Hereford (GB)

[378] Singleton, D. J.; Blickensderfer, R.
Wear and corrosion of 12 alloys during laboratory milling of phosphate rock in phosphoric acid waste water, p. 16
U.S. Bureau of Mines, Pittsburgh (PA./USA), 1984

[379] Jallouli, E. M.; Fikrat, M.
Investigation on abrasive corrosion of some stainless steel alloys in industrial phosphoric media
Bull. Cercle Etud. Métaux 15 (1988), p. 18.7.

[380] Jeyaprabha, C., Muralidharan, S., Venkatachari, G.
Corrosion behaviour of Monel in phosphoric acid
Anti-Corrosion Methods and Materials 47 (5) 2000, p. 285–287

[381] Swandby, R. K.
Corrosion charts: Guides to materials selection
Chem. Eng. (New York) 69 (1962) November, p. 186

[382] Company publication
HASTELLOY Alloy B-2, Bulletin H2006C (1997)
Haynes International, Kokomo, Indiana (USA)

[383] Crook, P.
Development of a new Ni-Cr-Mo alloy
CORROSION 1996 (1996) paper No. 412, p. 412/1–412/17

[384] Product Information
Hastelloy® alloy B-2
Cabot Wrought Products Division, Cabot Corporation, Kokomo (Ind./USA)

[385] Crook, P., Corrosion resistant nickel alloys – Part 4, Advanced Materials and Processes, (2007) Vol. 165, no. 9, p. 45–47

[386] Katz, W.
Gießen für die Chemie. Phosphor und Phosphorsäure
(Casting in chemistry. Phosphorus and phosphoric acid) (in German), p. 58
VDI–Verlag GmbH, Düsseldorf, 1976

[387] Strassburg, F. W.
Schweißen und Schweißeignung von Nickel-Chrom-(Molybdän)-Legierungen
(Welding and weldability of nickel-chromium-(molybdenum)-alloys) (in German)
Werkst. Korros. 39 (1988) 5, p. 251

[388] Hönig, A.; Schlenk, K. W.
Aushärtbare Federlegierungen und ihre Anwendungen
(Curable spring alloys and their use) (in German)
Chemie-Technik 6 (1977) 10, p. 409

[389] Schneider, W.; Weinlich, W.
Nickel-Beryllium – aushärtbare Federlegierung mit besonderen Eigenschaften
(Nickel-beryllium – age-hardenable spring alloys with special properties) (in German)
Metall 36 (1982) 1, p. 2

[390] Zürn, H.; Morach, E.
Thermische Spritzen im Chemie-Apparatebau
(Thermal spraying in chemical apparatus engineering) (in German)
Chemie-Ing.-Techn. 46 (1974) 7, p. 281

[391] Mitsuhashi, A.; Kawashima, A.; Asami, K.; Hashimoto, K.
The corrosion behavior of amorphous Ni-Cr-Ta-P and Ni-Cr-Mo-P alloys in a hot concentrated phosphoric acid
Boshoku Gijutsu 37 (1988) 1, p. 3

[392] Zhivotovsky, E. A.; Zhivotovskaya, G. P.; Kudreyko, I. I.
Corrosion of titanium-nickel alloy in mixtures of sulfuric and phosphoric acids (in Russian)
Zashch. Met. 16 (1980) 6, p. 732

[393] Shoji, Sh.; Tochigi, N.
Acid aqueous pickling solution for metal treatment (in German)
German DOS 2364162 (21 Dec. 1973)

[394] Product Information
Eigenschaften von Nickel
(Properties of nickel) (in German), 1st ed., Nov. 1971
International Nickel Deutschland GmbH, D-4000 Düsseldorf

[395] Product Information
Eigenschaften und Anwendungsgebiete korrosionsbeständiger Wiggin-Nickellegierungen (Properties and fields of application of corrosion-resistant Wiggin nickel alloys) (in German), Information No. 3046, 1.1964
Henry Wiggin u. Comp. Ltd., Hereford (GB)

[396] Product Information
Beständigkeit von nickelhaltigen Legierungen und Stählen gegenüber Schwefelsäure
(Resistance of nickel-containing alloys and steels to sulfuric acid) (in German), October 1961
Nickel-Informationsbüro GmbH, D-4000 Düsseldorf

[397] Corrosion Data Survey, Metals Section, 6th ed., 1985
National Association of Corrosion Engineers, Houston (USA)

[398] Rabald, E.
Corrosion Guide, 2nd revised edition
Elsevier Publishing Company, Amsterdam-London-New York, 1968

[399] Anonymous
Corrosion resistance of materials
Mater. Eng. *74* (1971) 4; (Mater. Selector), p. 7

[400] Fabian, R. J.; Vaccari, J. A.
How materials stand up to corrosion and chemical attack
Mater. Eng. *73* (1971) 2, p. 36

[401] Rabald, E.
Einiges über das Verhalten von Werkstoffen gegenüber Schwefelsäure
(Some information about the behavior of materials towards sulfuric acid) (in German)
Werkst. Korros. *7* (1956) 11, p. 652

[402] Swandby, R. K.
Corrosion charts: Guides to materials selection
Chem. Eng. *69* (1962) Nov. 12, p. 186

[403] Abd El Rehim, S. S.; Abd El Wahaab, S. M.; Abdel Maguid, E. A.
Electrochemical behaviour of nickel anode in H_2SO_4 solutions and the effect of halide ions
Werkst. Korros. *37* (1986) 10, p. 550

[404] Berg. F. F.
Korrosionsschaubilder
(Corrosion diagrams) (in German)
VDI-Verlag GmbH, D-4000 Düsseldorf, 1969

[405] Nickel, O
Eigenschaften und Anwendungen von Reinnickel, Nickel-Basislegierungen und nickelhaltigen NE-Werkstoffen im Rohrleitungsbau
(Properties and uses of pure nickel, nickel-based alloys and nickel-bearing non-ferrous metals in piping) (in German)
3 R international *14* (1975) 8, p. 419

[406] Product Information
Wiggin korrosionsbeständige Legierungen
(Wiggin corrosion-resistant alloys) (in German), edition Dec. 1977
Henry Wiggin & Company Ltd., Hereford (GB)

[407] Product Information
Korrosionsbeständigkeit nickelhaltiger Werkstoffe gegenüber Fluor, Fluorwasserstoff, Flußsäure und anderen Fluorverbindungen
(Corrosion resistance of nickel-containing materials to fluoride, hydrogen fluoride, hydrofluoric acid and other fluorine compounds) (in German), Information No. 59, 1st ed., March 1970
International Nickel Deutschland GmbH, D-4000 Düsseldorf 1

[408] Heitz, E.; v. Meysenbug, C. M.
Die Korrosion von Eisen, Nickel, Kupfer und Aluminium in organischen Lösungsmitteln mit geringem Mineralsäuregehalt
(The corrosion of iron, nickel, copper and aluminium in organic solvents containing a small amount of mineral acid) (in German)
Werkst. Korros. *16* (1965) 7, p. 578

[409] Erdmann-Jesnitzer, F.; Wessel, A.
Untersuchungen zur Wasserstoffversprödung einiger kubisch-flächenzentrierter Werkstoffe
(Investigation into the hydrogen embrittlement of some cubic face-centered materials) (in German)
Arch. Eisenhüttenwesen *52* (1981) 2, p. 77

[410] Klyuchnikov, N. G.; Verizhskaya, E. V.
Corrosion of certain metals in sulfuric acid solutions passed through a magnetic field (in Russian)
Zashch. Met. *8* (1972) 6, p. 700

[411] Uhlig, H. H.
Korrosion und Korrosionsschutz
(Corrosion and corrosion protection) (in German)
Akademie-Verlag, D-1000 Berlin, 1970

[412] McClain, G. E.; Mueller, W. A.
Corrosion problems in acid flow control
Chem. Eng. Progr. (CEP) *78* (1982) 2, p. 48

[413] Clayton, jr. C. H.; Johnson, T. E.
Engineering materials for pumps and valves
Chem. Eng. Progr. *74* (1978) 9, p. 54

[414] Groth, V. J.; Hafsten, R. J.
Corrosion of refinery equipment by sulfuric acid and sulfuric acid sludges
Corrosion *10* (1954) 11, p. 368

[415] Mayer, K. H.; König. H.
Spannungsrißkorrosionsuntersuchungen an warmfesten Schraubenwerkstoffen für Dampfturbinen in turbinenspezifischen Lösungen
(Investigations into stress corrosion cracking of heat-resistant screw stock for steam turbines in turbine-specific solutions)(in German)
Final report FE-Project-No. O-B 2.3/5, Dec. 1984
DECHEMA, D-6000 Frankfurt

[416] Mayer, K. H.; König. H.
Spannungsrißkorrosionsuntersuchungen an warmfesten Schraubenwerkstoffen für Dampfturbinen in turbinenspezifischen Lösungen
(Investigations into stress corrosion cracking of heat-resistant screw stock for steam turbines in turbine-specific solutions) (in German)
Werkst. Korros. *36* (1985) 11, p. 524

[417] McIlree, A. R.; Michels, H. T.; Morris, P. E.
Effects of variations of carbon, sulfur and phosphorus on the corrosion behavior of alloy 600
Corrosion *31* (1975) 12, p. 441

[418] Anonymous
Fertigungs- und Standardlieferprogramm M3-M56 Schrauben, Muttern, Zubehör nach DIN, ISO und EURONORM
(Manufacturing and standard delivery program M3-M56 screws, nuts and equipment according to DIN, ISO and EURONORM) (in German)
Werkstoffe und ihre Veredlung *2* (1980) 5, p. 250

[419] Anonymous
Verschleißfest und abriebbeständig: LP-Legierungen Tribaloy® für Gußteile und zum Beschichten
(Wear-resistant and abrasion-resistant: LP-alloys Tribaloy® for cast parts and coating) (in German)
Werkstoffe und ihre Veredlung *1* (1979) 2, p. 24

[420] Vermilyea, D. A.
Susceptibility of iron- and nickel-base alloys to SCC in pH 2.5 H_2SO_4 at 289 °C
Corrosion *31* (1975) 12, p. 421

[421] Ward, E.; Maylor, J. B.
Ni-o-nel® 825 – Eine neue korrosionsfeste Legierung für die chemische Industrie (Ni-o-nel® 825 – a new corrosion-resistant alloy for the chemical industry) (in German)
Werkst. Korros. *16* (1965) 5, p. 382

[422] Product Information
Wiggin Nickellegierungen in der chemischen Verfahrenstechnik
(Wiggin nickel alloys in chemical process engineering) (in German), Information No. 3681G, Nov. 1976
Henry Wiggin & Co., Ltd., Hereford (GB)

[423] Lamb, B. R.
Sulphuric acid – technology moves platewards
Processing *27* (1981) 10, p. 56

[424] Hopkins, D. J.
Korrosionsbeständige Nickellegierungen und deren Anwendungsmöglichkeiten (Corrosion-resistant nickel-alloys and their possible applications) (in German)
Chem. Anl. + Verfahren (CAV) (1979) 9, p. 110

[425] Product Information
Inco® Alloys International – Serving the chemical processing and related industries, Information IAI-10/3M/1986
Inco Alloys International, Inc., Huntington (West Virginia/USA)

[426] Crum, J. R.; Adkins, M. E.
Comparison of electrochemical behavior and corrosion rate of alloy 625 in sulfuric acid
Mater. Performance *25* (1986) 2, p. 27

[427] NACE Publication
Corrosion in sulfuric acid
Proceedings of the Corrosion /85, Symposium on Corrosion in Sulfuric Acid, Catalog Card Number: 85-61573
National Association of Corrosion Engineers, Houston (Texas/USA), 1985

[428] Product Information
Incoloy® alloy 825, Publication No. 3564, Nov. 1976
Henry Wiggin & Co, Ltd., Hereford (GB)

[429] Henson, R., Chang, W., Thayer, M., Mackenzie, T.
Material selection for sulfuric acid service
ALTA 2001 Nickel/Cobalt-7; Perth, Australia, 15–18 May 2001 (2001), p. 1–6

[430] Hibner, E.L., Shoemaker, L.E.
A practical guide to the use of nickel alloys for sulfuric acid service
CORROSION NACExpo 2006, 61st Annual Conference & Exhibition, San Diego, CA, USA, 12–16 Mar. 2006 (2006) Paper No. 06225, p. 1–17

[431] Rodda, J.R., Ives, M.B., Kish, J.R.
Corrosion of cathodes in anodic protection systems in hot concentrated sulfuric acid. Part 1: conventional cathodes
Materials Performance 44 (2005) 7, p. 20–23

[432] Product Information
Hastelloy® Alloy G – Excellent resistance to hot sulfuric and phosphoric acids in the as-welded condition, edition 1983
Cabot Corporation, Kokomo (Indiana/USA)

[433] Rockel, M. B.; Weidemann, R.
Untersuchung über den Einfluß der Vorbehandlung auf das Korrosionsverhalten hochlegierter Sonderstähle in reiner und chloridhaltiger Schwefelsäure
(Study of the effect of pretreatment on the corrosion behavior of highly alloyed special steels in pure sulfuric acid and sulfuric acid containing chloride) (in German)
Werkst. Korros. *33* (1982) 1, p. 42

[434] Altpeter, E.; Rockel, M. B.
Untersuchungen über den Einfluß des Gefügezustandes auf das Korrosionsverhalten hochlegierter, rost- und säurebeständiger Stähle in reiner sowie in chloridhaltiger Schwefelsäure
(Study of the effect of pretreatment on the corrosion behavior of highly alloyed special steels in pure sulfuric acid and sulfuric acid containing chloride) (in German)
Werkst. Korros. *33* (1982) 1, p. 42

[435] Renner, M.
Hochlegierte Austenite für die Anwendung in ruhender und bewegter hochkonzentrierter Schwefelsäure (+ 95 %)
(High-alloy austenites for use in static and agitated sulfuric acid of high concentration (+ 95 %)) (in German), Final report FE-project No. 0-d 3.9/1, Sept. 1986
DECHEMA, D-6000 Frankfurt

[436] Manning. P. E.; Tuff, W. F.; Zordan, R. D.; Schuur, P. D.
Evaluation of high performance alloys used in the pulp and paper industry
Mater. Performance *23* (1984) 1, p. 19

[437] Wallen, B.; Liljas, M.; Olsson, J. (Product Information)
Avesta® 254 SMO – Performance of a high molybdenum stainless steel in gas cleaning systems with particular reference to the pulp and paper industry, January 1982
Avesta Jernverks AB, Avesta (Sweden)

[438] Product Information
Langley corrosion resistance, GA 8692/3M/481
Langley Alloys Ltd., Slough, Berks (GB)

[439] Alves, H., Kurumlu, D., Behrens, R.,
A newly developed Ni-Cr-Mo alloy with improved corrosion resistance in flue gas desulfurization and chemical process applications
NACE Corrosion 2013, Orlando, Florida, USA, Paper 2325

[440] Crook, P., Meck, N.S., Koon, N.E.
The corrosion characteristics of a uniquely versatile nickel alloy
CORROSION 2008 Conference & Expo, New Orleans, LA, USA 16-20 Mar. 2008 (2008) Paper No. 08190, p. 1–10

[441] Alves, H., Werner, H., Agarwal, D.C.
Corrosion performance and applications of alloy 31 and alloy 59 in sulfuric acid media
CORROSION NACExpo 2006 61st Annual Conference & Exposition, San Diego, CA, USA (2006) Paper No. 06222, p. 1–16

[442] Grubb, J.F., Botti, C., Ozturk, B., Dunn, J. J.
Corrosion properties of UNS N06110 for the CPI
CORROSION NACExpo 2006 61st Annual Conference and Exposition, San Diego, CA, USA (2006) Paper No. 06214, p. 1–10

[443] Renner, M. H.W.
Metallic materials forconcentrated sulfuric acid service
CORROSION 2001, Houston, texas, USA 11–16 Mar. 2001 (2001) Paper No. 01342, p. 1–22

[444] Katz, W.
Gießen für die Chemie – Das Verhalten metallischer Gußwerkstoffe unter korrosiver Beanspruchung
(Casting in chemistry – the behavior of metallic cast materials under corrosive attack) (in German)
VDI-Verlag GmbH, D-4000 Düsseldorf

[445] Manning, P. E.; Schöbel, J.-D.
Hastelloy® alloy C-22 – ein neuer und vielseitig anwendbarer Werkstoff für die chemische Industrie
(Hastelloy® alloy C-22 – a new and versatile material for the chemical process industries) (in German)
Werkst. Korros. *37* (1986) 3, p. 137

[446] Süry, P.
Schwefelsäurekorrosion metallischer Werkstoffe unter extremen Bedingungen
(Corrosion of metallic materials by sulfuric acid under extreme conditions) (in German)
Chemie-Technik 6 (1977) 10, p. 415

[447] Steigerwald, R. F.
New molybdenum stainless steels for corrosion resistance: a review of recent developments
Mater. Performance *13* (1974) 9, p. 9

[448] Product Information
Sonderlegierungen für die Rauchgasentschwefelung. Information über die ES-Sonderlegierungen K 0446.1; ESS-C und ESS-C4
(Special alloys for flue gas desulfuration. Information on the special ES alloy K 0446.1, ESS-C and ESS-C4) (in German)
Edelstahlwerk Schmees, D-4018 Langenfeld

[449] Product Information
Hastelloy® alloy C-276, August 1983
Cabot Corporation, Kokomo (Indiana/USA)

[450] Lee III, T. S.; Hodge, F. G.
Resistance of Hastelloy® alloys to corrosion by inorganic acids
Mater. Performance *15* (1976) 9, p. 29

[451] Product Information
Inco® alloy C-276, Information 2M, 9-87 IAI-23
Inco Alloys International Ltd., Hereford (GB)

[452] Product Information
Corrosion resistance of Hastelloy® alloys, Information F-30, 131F, 1978
Stellite Division, Cabot Corporation, Kokomo (Indiana/USA)

[453] Product Information
Sonderwerkstoff Hastelloy® Alloy C-22
(Special material Hastelloy® alloy C-22) (in German), edition 1985
Robert Zapp Werkstofftechnik GmbH u. Co, D-4000 Düsseldorf 1

[454] Bünger, J.
Die Korrosion durch Schwefelsäure (The corrosion caused by sulfuric acid) (in German)
Werkst. Korros. *7* (1956) 6, p. 322

[455] Mills, S. H.; Munson, W. H.
Graphite tube vs. plate exchangers in quench acid service
Chem. Eng. Progr. *78* (1982) 7, p. 33

[456] Hohenhinnebusch, W.
Email im Apparatebau – Fortschritte und Möglichkeiten
(Enamels in apparatus construction – Progress and applications) (in German)
Chemie-Technik *11* (1982) 6, p. 645

[457] Miller, R. F.; Treseder, R. S.; Wachter, A.
Corrosion by acids at high temperatures
Corrosion *10* (1954) 1, p. 7

[458] Product Information
Hastelloy® alloy G-3, Information H-2009A, 1983
Haynes International, Inc., Kokomo (Indiana/USA)

[459] Uhlig. H. H.; Asphahani, A. I.
Corrosion behavior of cobalt base alloys in aqueous media
Mater. Performance *18* (1979) 11, p. 9

[460] Wallen, B.; Liljas, M.; Olsson, J.
Performance of a high molybdenum stainless steel in gas cleaning systems with particular reference to the pulp and paper industry
Mater. Performance *21* (1982) 6, p. 40

[461] Rabald, E.
in: Ullmanns Encyklopädie der technischen Chemie
(Ullmann's encyclopedia of industrial chemistry) (in German), 3rd ed., vol. *1*, p. 920
Urban & Schwarzenberg, München-Berlin, 1951

[462] Ellison, B. T.; Schmeal, W. R.
Corrosion of steel in concentrated sulfuric acid
J. Electrochem. Soc. *125* (1978) 4, p. 524

[463] Tödt, F.
Korrosion und Korrosionsschutz
(Corrosion and corrosion protection) (in German), 2nd ed.
Verlag Walter de Gruyter & Co., D-1000 Berlin, 1961

[464] Gräfen, H. et al.
Die Praxis des Korrosionsschutzes, Kontakt und Studium,
(Corrosion protection in practice; dealings with and study of) (in German), vol. *64*
expert verlag, D-7031 Grafenau, 1981

[465] Wendler-Kalsch, E.
Nichteisenwerkstoffe im Apparatebau
(Non-ferrous materials in apparatus construction) (in German)
Lecture given on the Congress "Werkstoffe im Chemieanlagenbau", D-4300 Essen, Haus der Technik, Juni 1989

[466] Erdmann-Jesnitzer, F.
Untersuchungen zur Wasserstoffversprödung und Spannungsrißkorrosion austenitischer Stähle unter besonderer Berücksichtigung des Einflusses einer thermomechanischen Behandlung (HERF)
(Investigation of the hydrogen embrittlement and stress corrosion cracking of austenitic steels with special consideration of the influence of thermomechanical treatment) (in German)
Werkst. Korros. *33* (1982) 4, p. 221

[467] Volk, K. E.
Nickel und Nickellegierungen, Eigenschaften und Verhalten
(Nickel and nickel alloys, properties and behavior) (in German), p. 356
Springer-Verlag, Berlin-Heidelberg-New York, 1970

[468] Barker, W.; Evans, T. E.; Williams, K. J.
Effect of alloying additions on the microstructure, corrosion resistance and mechanical properties of nickel-silicon alloys
Br. Corros. J. *5* (1970) March, p. 78

[469] McDowell, D. W.
Handling sulfuric acid
Chem. Eng. *81* (1974) 24, p. 118

[470] Dilthey, U.; Wanke, R.
Hochleistungs-Schweißplattierverfahren für den Chemie-Apparatebau
(The high-performance cladding process in chemical apparatus engineering) (in German)
Chem.-Ing.-Tech. *46* (1974) 11, p. 467

[471] Product Information
Beständigkeit des Gußstahls 9.4462 (G-X 3 CrNiMoCuN 26 6 3) in Schwefelsäure
(Resistance of the cast steel 9.4462 (G-X 3 CrNiMoCuN 26 6 3) in sulfuric acid)
(in German), Investigation report, Dec. 1986
Eisen- und Stahlwerk Pleissner GmbH, D-3420 Herzberg

[472] Colombier, L. (Product Information)
Molybdän in rost- und säurebeständigen Stählen und Legierungen
(Molybdenum in stainless and acid-resistant steels and alloys) (in German)
Climax Molybdenum GmbH, D-4000 Düsseldorf-Oberkassel

[473] Pierce, R. R.; Semler, C. E.
Ceramic and refractory linings for acid condensation – part I
Chem. Eng. *90* (1983) 25, p. 81

[474] Shuker, F. S.
When to use refractory metals and alloys in the plant
Chem. Eng. *90* (1983) 9, p. 81

[475] Product Information
Hastelloy® alloy B-2, Information H-2006, 1983
Cabot Corporation, Kokomo (Indiana/USA)

[476] Nazarov, V. N.; Boiko, A. Z.; Demyanenko, N. S.; Elizov, A. M.; Zatulin, V. D.; Tishchenko, G. A.
Testing on the use of apparatus made of the nickel-molybdenum alloy N70M27F in the Chemical Combine Cherkassk (in Russian)
Khim. Prom. (1975) 8, p. 593

[477] Kumar, P.
Influence of microstructure on the corrosion behavior of a
Ni-Si alloy high-temperature ordered intermetallic alloys. Materials Research Society Symposium, Boston (USA), Nov. 1984
(Proceedings) *39* (1985) p. 537

[478] Nikandrov, E. A.; Maslenkov, S. B.
Corrosion resistance of the nickel-molybdenum-tungsten alloy system
(in Russian)
Zashch. Met. 9 (1973) 1, p. 40

[479] Trarbach, K. O.
Korrosion von Plattierungen bei Chemieanlagen und in der Meerestechnik
(Corrosion of platings in chemical plants and in marine technology) (in German)
Maschinenmarkt *90* (1984) 38, p. 936

[480] Ramchandran, T.; Namboodhiri, T. K. G.
A potentiostatic evaluation of the corrosion of a Ni-Co base amorphous alloy
Corrosion *40* (1984) 2, p. 73

[481] Hönig. A.; Schlenk, K. W.
Aushärtbare Federlegierungen und ihre Anwendungen
(Heat treatable spring alloys and their uses)
(in German)
Chemie-Technik 6 (1977) 10, p. 409

[482] Hart, A. C.
The corrosion behaviour of binary nickel-germanium alloys
Corros. Sci. *12* (1972) 2, p. 137

[483] Uhlig, H.
Korrosion und Korrosionsschutz
(Corrosion and corrosion protection)
(in German) p. 318 and 323
Akademie-Verlag, Berlin, 1970

[484] Anonymous
Die Nickel-Resist-Gußeisenwerkstoffe
(Nickel-resist cast iron materials)
(in German)
International Nickel, 3rd ed. (1968) p. 92

[485] Anonymous
Copper and its alloys – wrought
Materials Engineering (Materials Selector) 74 (1971) 4, pp. 17, 181

[486] Tsejtlin, K. L. et al.
High temperature corrosion of metals in gaseous ammonia
Werkst. Korros. 22 (1971), p. 258

[487] Fabian, R. J.; Vaccari, J. A.
How materials stand up corrosion and chemical attack
Materials Engineering 73 (1971) 2, p. 36

[488] Nelson, G. A.
Corrosion Data Survey, 2nd edition
NACE, Houston, 1971, p. Table A-6

[489] Anonymous
Nickel copper alloys fight corrosion; useful over wide temperature range
Materials Engineering 80 (1974) 3, p. 42

[490] Firmenschrift
Silverin-Korrosionstabellen
(Silverin corrosion tables) (in German)
Vereinigte Deutsche Nickelwerke, Schwerte

[491] TRD 451 (12/1996)
Anlagen zur Lagerung von druckverflüssigtem Ammoniak für Dampfkesselanlagen-Druckbehälter – (Plants for the storage of pressure-liquefied ammonia for boiler plant pressure vessels) (in German)
Beuth Verlag GmbH, Berlin

[492] Firmenschrift
Corrosion resistance of HASTELLOY alloys, 1978
Cabot Corporation, Kokomo (Indiana/USA)

[493] Andreaus, A.; Cicciarelli, R.; Schönhofer, J.
Evaluation of duplex steel for a chemical production plant
Duplex Stainless Steels´91Vol.2, Beuane, Bourgogne, Frankreich, 28–30 Oct. 1991
Publ.: Les Editions, de Physique, Avenue du Hoggar, Zone Industrielle de Courtaboeuf, B.P. 112, F-91944 Les Ulis Cedex A, France 1992, 1133–1139

[494] Kudryavseva, Z. I.; Shumilova, N. A.; Openkin, V. A.; Zhuchkova, N. A.; Krucheva, E. I.
Ellipsometric examination of oxygen adsorption and oxide layer formation on nickel, cobalt and silver in alkaline solution (in Russian)
Elektrokhimiya *13* (1977) p. 710

[495] de Vries, G.; Grantham, L. F.
Corrosion of metals in the molten LiOH-LiF-H_2O system
Electrochem. Technol. *5* (1967) p. 335

[496] Gurovich, E. I.
Reaktionen von geschmolzenem Lithium-, Natrium und Kaliumhydroxid mit Nickel, Kupfer, Eisen und Stahl
(Reactions of molten lithium, sodium and potassium hydroxide with nickel, copper, iron and steel) (in Russian)
Zhur. Priklad. Khim. *32* (1968) p. 817

[497] Pfeiffer, H.; Hauffe, K.
Über die Beeinflussung der Oxidationsgeschwindigkeit von Nickel und Titan durch Legierungszusätze und durch Behandlung von Metalloxiddampf
(Influencing the rate of oxidation of nickel and titanium by alloying additions and by treatment with metal oxide vapours) (in German)
Z. Metallkunde *43* (1952) p. 364

[498] Berry, W. E.
The corrosion behavior of FeCrNi alloys in high-temperature water
Edit.: High temperature, high pressure electrochemistry in aqueous solutions
Stähle, R. W.; Jones, D.; Slater, L.; Houston (1977) p. 48

[499] van Royen, D.; Copson, H. R.; Berry, W. E.
Corrosion behavior of nickel-chromium-iron Alloy-600 in borated pressurized water reactor environments
Corrosion *25* (1969) p. 194

[500] Weeks, J. R.
Materials performance in operating pressurized water reactor steam generators
Nuclear Technol. *28* (1976) p. 348

[501] Reinhardt, K. H.; Smith, I. J.
Gase und Korrosionsprodukte im Primärkreislauf der NS "Otto Hahn"
(Gases and corrosion products in the primary circuit of the Otto Hahn) (in German)
VGB Kraftwerkstechnik *57* (1977) p. 165

[502] Smith, G. P.; Hoffman, E. E.
Corrosion and metal transport in fused sodium hydroxide II.
Corrosion of nickel-molybdenum-iron alloy
Corrosion *13* (1957) p. 627t

[503] DIN 17740 (05/1998) draft
Nickel in Halbzeug – Zusammensetzung
(Nickel in semi-finished material – composition) (in German)
Beuth Verlag GmbH, Berlin

[505] Hopkins, D. J.
Korrosionsbeständige Nickellegierungen und deren Anwendungsmöglichkeiten
(Corrosion resistant nickel alloys and their applications) (in German)
Chem.-Anl. Verfahren (1979) 9, p. 110

[505] Firmenschrift
Resistance of Nickel and its Alloys to Corrosion by Caustic Alkalis, Corrosion Engineering Bulletin CEB-2
The International Nickel Company, New York (NY/ZSA)

[506] De, P. K.; Ghosal, S. K.; Elayaperumal, K.
Corrosion behavior of wrought and electrodeposited nickel in alkaline environments – Effects of ammonia, sulphide and chloride
Trans. Indian Inst. Metals 31 (1978), p. 444

[507] Combrade, P.
Study of the corrosion resistance of nickel and nickel base alloys in water electrolyzers working at 200°C
Synthesis of the results of the second R & D programme
Report No. EUR 9782
Commission of the European Communities, Luxembourg, 1985, pp. 134–137
and
Etude de la Resistance a la Corrosion du Nickel et d`Alliages a Base Nickel dans les Electrolyseurs d`Eau travaillant a 200°C
Rapport Final EUR 10033 FR
Commission of the European Communities, Luxembourg, 1985

[508] Abdel Aal, M. S.; Osman, A. H.
Anodic polarisation of tin, nickel and a 65/35 tin-nickel alloy in alkaline media containing halide ions
Corrosion NACE 36 (1980), p. 591

[509] Zhang, D.; Dong, R.; Li, Z. X.; Zhang, X.
An anticorrosion study on the application of electroless nickel to valves
Confer. Progress in the Understanding and Prevention of Corrosion
Barcelona, Spain, July 1993; Vol. 1, pp. 369–373

[510] Tracy, R. P.; Kuzyk, P.
Maintenance program for tank car valves achieves large cost savings with low-phosphorus electroless nickel
Plating and Surface Finishing 76 (1989) 1, pp. 30–31

[511] Drodten, P.
Sodium Hydroxide
in: Kreysa, G.; Schütze, M.
Corrosion Handbook, 2nd Edition, Vol. 1
DECHEMA e.V., Frankfurt, 2004

[512] Donat, H.
Anwendung von Nickel und Nickelbasislegierungen im Chemieapparatebau
(Use of nickel and nickel basis alloys in the chemistry apparatus engineering)
(in German)
Maschinenmarkt 86 (1980) 97, pp. 1931–1934

[513] Sussek, G.; Kesten, M.
(Eine Charakterisierung der Lochfraß-Korrosion des Nickels – I. Die Bestimmung der Lochfraß-Potentiale in neutralen und alkalischen Lösungen)
A characterisation of the pitting corrosion of nickel – I. the determination of the pitting corrosion potentials in neutral and alkaline solutions
Corrosion Science 15 (1975), p. 225

[514] Pement, F. W.; Wilson, I. L. W.; Aspden, R. G.
Stress corrosion cracking studies of high nickel austenitic alloys in several high-temperature aqueous solutions
Materials performance 19 (1980) 4, p. 43

[515] Weeks, J. R.
Materials performance in operating pressurized water reactor steam generators
Nucl. Technol. 28 (1976), p. 348

[516] Pement, F. W.; Wilson, I. L. W.; Aspden, R. G.
Stress corrosion cracking studies of high nickel austenitic alloys in several high-temperature aqueous solutions
Materials performance 19 (1980) 4, p. 43

[517] Kramer, L. D.; Michael, S. T.; Pement, F. W.
Service experience and stress corrosion of Inconel 600 bellows expansion joints in turbine steam environments
Materials Performance 14 (1975) 9, p. 15

[518] Gräfen H.; Gerischer K.; Gramberg, U.
Schadensverhütung in Chemiebetrieben durch systematische Schadensuntersuchung
(Damage prevention in chemistry enterprises by systematic damage investigation) (in German)
Z. Werkstofftechn. 8 (1977), p. 357

[519] Crum, J. R.
Effect of composition and heat treatment on stress corrosion cracking of alloy 600 steam generator tubes in sodium hydroxide
Corrosion 38 (1982) 1, pp. 40–45

[520] Rajan, V. B.; Sung, J. K.; Was, G. S.
Stress corrosion cracking of controlled purity Inconel 600 type alloys in high purity water and caustic solutions
Conference: Environmental degradation of materials in nuclear power systems – Water reactors, Traverse City, Michigan, USA, 30. Aug.–3. Sept. 1987
Publ.: The Metallurgical Society/AIME, Pennsylvania, USA, 1988

[521] Sung, J. K.; Koch, J.; Angeliu, Z.; Was, G. S.
The effect of grain-boundary chemistry on intergranular stress corrosion cracking of Ni-Cr-Fe alloys in 50 % NaOH at 140°C
Metallurgical Transactions A, 23A, (1992) 10, pp. 2887–2904

[522] Mignone, A.; Borello, A.; Masci, D.; Camona, C.
SCC behavior of Inconel 600 tubes. Sodium hydroxide effect of heat treatment, NaOH concentration and temperature
Eurocorr`87, European Corrosion Meeting, Karlsruhe, FRG, 6–10 April, 1987
DECHEMA, Frankfurt, Germany, 1987, pp. 159–165

[523] Borello, A.; Masci, D.; Mignone, A.
Spannungsrißkorrosion von Rohren aus Inconel 600: Einfluß von Wärmebehandlung, NaOH-Konzentration und Temperatur
(Stress corrosion cracking of pipes from Inconel 600: Influence of thermal treatment, NaOH concentration and temperature) (in German)
Werkst Korros. 39 (1988) 9, pp. 416–423

[524] Payne, S. M.; McIntyre, P.
Influence of grain boundary microstructure on the susceptibility of Alloy 600 to intergranular attack and stress corrosion cracking
Corrosion 44 (1988) 5, pp. 314–319

[525] Yamanaka, K.
The role of grain-boundary chromium-carbide precipitations in the caustic IGSCC resistance of nickel-base alloy 600
Corros. Eng. (JPN) 41 (1992) 7, 533–545
also in: Conference: Sixth International Symposium on Environmental Degradation of Materials in Nuclear Power Systems – Water Reactors, San Diego, California, USA, 1–5 Aug 1993
Publ.: The Minerals, Metals & Materials Society, Warrendale, Pennsylvania, US, 1993

[526] Shei, S. A.; Yang, W.J.
Stress corrosion crack tip microstructure in nickel-base-alloys
NACE Corrosion '94, Houston (1994) Paper No. 154

[527] Le Coze, J.; Bonningue, M.; Hocquellet, D.; Fraczkiewcz, J.
Stress corrosion cracking of Fe-Ni-Cr alloys in NaOH solutions at 350°C
Conf. Mechanical Behaviour of Materials, V. Beijing, China, 3.–6. June 1987, Vol. 2, pp. 833–839

[528] Forchhammer, P.
Das Korrosionsverhalten (Ik-SpRK und IK) der Rohrwerkstoffe Inconel® 600 und Incoloy® 800 für nukleare Dampferzeuger in konditioniertem Sekundärwasser bei Kapselprüfungen im Temperaturbereich von 270 bis 350°C
(The corrosion behavior (IC-SCC and IC) of the tubing materials Inconel® 600 and Incoloy® 800 for nuclear steam generators in conditioned secondary water with cap examinations in the temperature range from 270 to 350°C) (in German)
Der Maschinenschaden 58 (1985) 6, pp. 217–225

[529] Hickling, J.; Forchhammer, P.; Effertz, P.-H.
Experimental work on the SCC of steam generator tubing material in modified PWR environments
in: Conference: Environmental Degradation of Materials in Nuclear Power Systems – Water Reactors, Myrtle Beach, S.C.; USA, 22–25 Aug. 1983
NACE, Houston, Texas, USA, 1984, pp. 658–674

[530] Liu, X.; Shao, J.; Cragnolino, G.; MacDonald, D. D.
Selective grain boundary dissolution of Inconel® alloy 600 in a concentrated caustic solution at 140°C
Conference: Corrosion of Nickel-Base Alloys, Cincinnati, Ohio, USA, 23.–25. Oct. 1984
Publ.: American Society for Metals, Metals Park, Ohio, USA, 1985, pp. 211–225
also in: Journal of Materials for Energy Systems 7 (1985) 3, pp. 223–236

[531] Airey, G. P.; Pement, F. W.
A comparison of intergranular attack in Inconel® alloy 600 observed in the laboratory and in operating steam generators
Corrosion, 39 (1983) 2, pp. 46–55

[532] Donati, J. R.; Guttmann, M.; Rouillon, Y.; Saint-Paul, P.; van Duysen, J. C.
Stess corrosion cracking behaviour of nickel-base alloy with 19 % Chromium in high temperature water
Conference: Environmental degradation of materials in nuclear power systems – water reactors, Traverse City, Michigan, USA, 30. Aug.–3. Sept. 1987
Publ.: The Metallurgical Society/AIME, Pennsylvania, USA, 1988

[533] Kawamura, H.; Hirano, H.
Intergranular attack and stress corrosion cracking propagation behavior of alloy 600 in high-temperature caustic solution
Corrosion 55 (1999) 6, pp. 566–575

[534] Yamanaka, K.
Effects of chromium on electrode strain behaviour for high-nickel alloys in hot caustic solutions
Corros. Eng. (JPN) 37 (1988) 11, pp. 605–611

[535] Crum, J. R.
Stress corrosion cracking testing of Inconel® alloys 600 and 690 under high-temperature caustic conditions
Corrosion 42 (1986) 6, pp. 368–372

[536] Sarver, J. M.; Crum, J. R.; Mankins, W. L.
Carbide precipitation and SCC behavior of Inconel alloy 690
CORROSION 44 (1988) 5, pp. 288–289

[537] Wilson, I. L. W.; Aspden, R. G.
The influence of specimen type and heat treatment on the caustic stress corrosion cracking of some stainless alloys
CORROSION-NACE, Vol. 32, (1976) 5, pp. 193–201

[538] Scarberry, R. C.; Graver, D. L., Stephens, C. D.
Alloying for Corrosion Control, Properties and Benefits of Alloy Materials
Mat. Protection 6 (1967) 6, p. 54

[539] Firmenschrift
Corrosion Resistance of Hastelloy® Alloys
Haynes International, Inc., Kokomo (IN/USA) 1984

[540] Nickel, O.
Eigenschaften und neuere Anwendungen nickelhaltiger Werkstoffe
(Properties and newer applications of nickel containing materials) (in German)
VDI-Ztg. 121 (1979), p. 313

[541] Pathania, R. S.; Chitty, J. A.
Stress Corrosion Cracking of Steam Generator Tube Materials in Sodium Hydroxide Solutions
Corrosion NACE 34 (1978), p. 369

[542] Sugimoto, K.; Belanger, G.; Piron, D. L.
The corrosion of NbNi alloys in a KOH solution
Corrosion NACE 36 (1980), p. 437

[543] Lu, P. W. T.; Srinivasan, S.
Nickel-based alloys as electrocatalysts for oxygen evolution from alkaline solutions
J. Electrochem. Soc. 125 (1978), p. 26

[544] Gras, J. M.; Pernot, M.
Oxygen overvoltage on 50:50 Ni-Fe in KOH 40 W/O
in: McIntyre, J. D. E.
Proc. Symp. Electrode Mat. & Proc. for Energy Conversion and Storage
Princeton N.J. (1977), p. 425

[545] Gmelin Handbuch, Nickel, vol. 2, p. 57, 1969

[546] Corrosion Engineering Bulletin CEB-2, The International Nickel Company, Inc, One New York Plaza, New York, N.Y., 10004 (1973)

[547] Carlson, G. A.; Estep, E. E.; Jacqueau D.
Eine poröse Kathodenzelle für die Laugenreinigung
(Porous Cathode Cell for the Purification of Caustic Alkali Solutions) (in German)
Chem.-lng.-Techn. 45 (1973) p. 217

[548] Hopkins, D. J.
Korrosionsbeständige Nickellegierungen und deren Anwendungsmöglichkeiten (Corrosion resistant nickel alloys and possibilities of their application)
Chem.-Anl. Verfahren (1979) 9, p. 110

[549] Palmer, J. D.
Alkalis Can Cause Unexpected Problems
Canad. Chem. Processing 53 (1969) 9, p. 7

[550] Zhang, D.; Dong, R.; Li, Z.X.; Zhang, X.
An anticorrosion study on the application of electroless nickel to valves
Confer. Progress in the Understanding and Prevention of Corrosion
Barcelona, Spain, July 1993; Vol. 1, p. 369–373

[551] Tracy, R.P.; Kuzyk, P.
Maintenance program for tank car valves achieves large cost savings with low-phosphorus electroless nickel
Plating and Surface Finishing 76 (1989) 1, p. 30–31

[552] Donat, H.
Anwendung von Nickel und Nickelbasislegierungen im Chemieapparatebau
(Application of nickel and nickel base alloys in chemical apparatus construction)
(in German)
Maschinenmarkt 86 (1980) 97, p. 1931–1934

[553] Rebak, R., Chambers, G., ASM Materials Solutions Conference 2001, Indianapolis, IN, November 6, 2001.

[554] Crum, J.R., Shoemaker, L.E., Corrosion resistance of nickel alloys in caustic solutions, CORROSION 2006, Paper 06219, (2006) p. 1–13

[555] Kramer, L. D.; Michael, S. T.; Pement, F. W.
Service Experience and Stress Corrosion of Inconel® 600 Bellows Expansion Joints in Turbine Steam Environments
Mat. Performance 14 (1975) 9, p.15

[556] Gräfen, H.; Gerischer, K.; Gramberg, U.
Schadensverhütung in Chemiebetrieben durch systematische Schadensuntersuchung
(Damage prevention in chemical plants by systematic damage analyses) (in German)
Z. Werkstofftechn. 8 (1977) p. 357

[557] Crum, J.R.
Effect of composition and heat treatment on stress corrosion cracking of alloy 600 steam generator tubes in sodium hydroxide
Corrosion 38 (1982) 1, p. 40–45

[558] Rajan, V.B.; Sung, J.K.; Was, G.S.
Stress corrosion cracking of controlled purity Inconel 600 type alloys in high purity water and caustic solutions
Conference: Environmental degradation of materials in nuclear power systems – Water reactors, Traverse City, Michigan, USA, 30. Aug.–3. Sept. 1987
Publ.: The Metallurgical Society/AIME, 420 Commonwealth Dr. Warrendale, Pennsyslvania 15086, USA (1988)

[559] Sung, J.K.; Koch, J.; Angeliu, Z.; Was, G.S.
The effect of grain-boundary chemistry on intergranular stress corrosion cracking of Ni-Cr-Fe alloys in 50% NaOH at 140°C
Metallurgical Transactions A, 23A (1992) 10, p. 2887–2904

[560] Mignone, A.; Borello, A.; Masci, D.; Camona, C.
SCC Behaviour of Inconel 600 tubes. Sodium hydroxide effect of heat treatment, NaOH concentration and temperature
Eurocorr'87, European Corrosion Meeting, Karlsruhe, FRG, 6–10 April, 1987
DECHEMA, Frankfurt, Germany (1987) p. 159–165

[561] Borello, A.; Masci, D.; Mignone, A.
Spannungsrißkorrosion von Rohren aus Inconel 600: Einfluß von Wärmebehandlung, NaOH-Konzentration und Temperatur
(Stress corrosion cracking of Inconel 600 tubing: Influence of heat treatment, NaOH concentration and temperature)
(in German)
Werkst. Korros. 39 (1988) 9, p. 416–423

[562] Lee, K.H.; Cragnolino, G.; Macdonald, D.D.
Effect of heat treatment applied potential on the caustic stress corrosion cracking of Inconel 600
Corrosion 41 (1985) 9, p. 540–553

[563] Payne, S.M.; McIntyre, P.
Influence of grain boundary microstructure on the susceptibility of Alloy 600 to intergranular attack and stress corrosion cracking
Corrosion 44 (1988) 5, p. 314–319

[564] Yamanaka, K.
The role of grain-boundary chromium-carbide precipitations in the caustic IGSCC resistance of nickel-base alloy 600
Corros. Eng. (JPN) 41 (1992) 7, p. 533–545
auch in Conference: Sixth International Symposium on Environmental Degradation of Materials in Nuclear Power Systems – Water Reactors, San Diego, California, US. 1–5 Aug 1993
Publ.: The Minerals, Metals & Materials Society, 420 Commonwealth Dr., Warrendale, Pennsylvania 15086, USA (1993)

[565] Shei, S.A.; Yang, W.J.
Stress corrosion crack tip microstructure in nickelbase-alloys
NACE Corrosion '94, Houston (1994) Paper No. 154

[566] Le Coze, J.; Bonningue, M.; Hocquellet, D.; Fraczkiewcz, J.
Stress corrosion cracking of Fe-Ni-Cr alloys in NaOH solutions at 350°C
Conf. Mechanical Behaviour of Materials, V. Beijing, China, 3.–6. June 1987, Vol. 2, p. 833–839

[567] Younes, C.M.; Morrissey, F.H.; Allen, G.C.; McIntyre, P.
Effect of heat treatment on the grain boundary chemistry and the resistance to intergranular corrosion of alloy 600 and 690
in: Second International Conference on Corrosion-Deformation Interactions, CDI 96, Nice, France, 24–26 Sept. 1996
Institute of Materials,, 1 Carlton House Terrace, London, SW1Y 5DB, UK, 1997, p. 421–434

[568] Forchhammer, P.
Das Korrosionsverhalten (Ik-SpRK und IK) der Rohrwerkstoffe Inconel® 600 und Incoloy® 800 für nukleare Dampferzeuger in konditioniertem Sekundärwasser bei Kapselprüfungen im Temperaturbereich von 270 bis 350°C
(The corrosion behaviour (stress corrosion and intergranular corrosion) of the tube materials Inconel 600 and Incoloy 800 for nuclear steam generators in treated secondary cycle waste during tube burst tests in the temperature range from 270 to 350°C) (in German)
Der Maschinenschaden 58 (1985) 6, p. 217–225

[569] Hickling, J.; Forchhammer, P.; Effertz, P.-H.
Experimental work on the SCC of steam generator tubing material in modified PWR environments
in: Conference: Environmental Degradation of Materials in Nuclear Power Systems – Water Reactors, Myrtle Beach, S.C.; USA, 22–25 Aug. 1983
NACE, Houston, Texas, USA (1984) p. 658–674

[570] Navas, M.; Gómez-Briceno, D.; Garcia-Mazario, M.; McIlree, A.R.
Effect of silicon compounds on stress corrosion cracking of alloy 600 in caustic solutions
Corrosion 55 (1999) 7, p. 674–685

[571] Konda, N.; Toyama, K.; Yamanaka, K.; Tokimasa, K.
Environmental effects on the crack growth properties of alloy 600
International Journal of Pressure Vessels and Piping
52 (1992) 2, p. 217–226

[572] Liu, X.; Shao, J.; Cragnolino, G.; MacDonald, D.D.
Selective grain boundary dissolution of Inconel® alloy 600 in a concentrated caustic solution at 140 0C
Conference: Corrosion of Nickel-Base Alloys, Cincinnati, Ohio, USA, 23.–25. Oct. 1984
Publ.: American Society for Metals, Metals Park, Ohio, USA, 1985, 211–225
auch in: Journal of Materials for Energy Systems 7 (1985) 3, p. 223–236

[573] Airey, G.P.; Pement, F.W.
A comparison of intergranular attack in Inconel® alloy 600 observed in the laboratory and in operating steam generators
Corrosion 39 (1983) 2, p. 46–55

[574] Donati, J.R.; Guttmann, M.; Rouillon, Y.; Saint-Pual, P.; van Duysen, J.C.
Stress corrosion cracking behaviour of nickel-base alloy with 19 % Chromium in high temperature water
Conference: Environmental degradation of materials in nuclear power systems – water reactors, Traverse City, Michigan, USA, 30. Aug.–3. Sept. 1987
Publ.: The Metallurgical Society/AIME, 420 Commonwealth Dr. Warrendale, Pennsylvania 15086, USA (1988)

[575] Kawamura, H.; Hirano, H.
Intergranular attack and stress corrosion cracking propagation behavior of alloy 600 in high-temperature caustic solution
Corrosion 55 (1999) 6, p. 566–575

[576] Yamanaka, K.
Effects of chromium on electrode strain behaviour for high-nickel alloys in hot caustic solutions
Corros. Eng. (JPN) 37 (1988) 11, p. 605–611

[577] Crum, J.R.
Stress corrosion cracking testing of Inconel® alloys 600 and 690 under high-temperature caustic conditions
Corrosion 42 (1986) 6, p. 368–372

[578] Sarver, J.M.; Crum, J.R.; Mankins, W.L.
Carbide precipitation and SCC behavior of Inconel alloy 690
Corrosion 44 (1988) 5, p. 288–289

[579] Wilson, I.L.W.; Aspden, R.G.
The influence of specimen type and heat treatment on the caustic stress corrosion cracking of some stainless alloys
Corrosion-NACE 32 (1976) 5, p. 193–201

[580] Ju, S., Zhang, Y., Shang, D., Chen, F., Wang, Y., Zhang, Y
Corrosion of several typical alloys in concentrated sodium hydroxide solution with chloride between 180°C and 240°C
CORROSION 67 (2011) 3, p. 036001-1-0366001-5

[581] Scarberry, R. C.; Graver, D. L., Stephens, C. D.
Alloying for Corrosion Control, Properties and Benefits of Alloy Materials
Mat. Protection 6 (1967) 6, p. 54

[582] Nickel, O.
Eigenschaften und neuere Anwendungen nickelhaltiger Werkstoffe
(Properties and newer applications of nickel containing materials) (in German)
VDI-Ztg. 121 (1979), p. 313

[583] Pathania, R. S.; Chitty, J. A.
Stress Corrosion Cracking of Steam Generator Tube Materials in Sodium Hydroxide Solutions
Corrosion NACE 34 (1978) p. 369

[584] Borom, M. P.; Will, F. G.
Stress Corrosion Cracking of Turbine Blades During Autoclave Caustic Leaching
Metallurgical Trans. (A) 12 (1981) p. 913

[585] Yasuda, M.; Fukumoto, K.; Ogata, Y.; Hine, F.
Corrosion behaviour of nickel and nickel-based alloys in concentrated NaOH-solutions at high temperatures
J. Electrochem. Soc. 135 (1988) 12, p. 2982–2987

[586] Smith, G. P.; Hoffman, E. E.
Corrosion and Metal Transport in Fused Sodium Hydroxide.
Part 2 – Corrosion of Nickel-Molybdenum-Iron Alloys
Corrosion NACE 13 (1957) 627 t.

[587] Smith, G. P.; Steidlitz, M. E.; Hoffman, E. E.
Corrosion and Metal Transport in Fused Sodium Hydroxide,
Part 1 – Experimental Procedures
Corrosion NACE 13(1957) 561 t.

[588] Manning, P. E.
Corrosion '82, Preprint No. 176, National Association of Corrosion Engineers
Houston, Texas (1982)

[589] Rebak, R. B.
Corrosion of Non-Ferrous Alloys. I. Nickel-, Cobalt-, Copper-, Zirconium- and Titanium-Based Alloys
Materials Science and Technology (2013), p. 69–111

[590] Galen Hodge, F.
The history of solid-solution-strengthened Ni alloys for aqueous corrosion service
JOM 58 (2006) 9, p. 28–31

[591] Company publication
Hastelloy® D-205 alloy, preliminary data sheet, H-2103, 1993
Haynes International, Inc., Kokomo, IN (USA)

Index of materials

Subject index

h

i

k

l

m

n

o

p